Natural Polymers
Volume 1: Composites

RSC Green Chemistry

Series Editors:
James H Clark, *Department of Chemistry, University of York, UK*
George A Kraus, *Department of Chemistry, Iowa State University, Ames, Iowa, USA*
Andrzej Stankiewicz, *Delft University of Technology, The Netherlands*
Peter Siedl, *Federal University of Rio de Janeiro, Brazil*
Yuan Kou, *Peking University, People's Republic of China*

Titles in the Series:
1: The Future of Glycerol: New Uses of a Versatile Raw Material
2: Alternative Solvents for Green Chemistry
3: Eco-Friendly Synthesis of Fine Chemicals
4: Sustainable Solutions for Modern Economies
5: Chemical Reactions and Processes under Flow Conditions
6: Radical Reactions in Aqueous Media
7: Aqueous Microwave Chemistry
8: The Future of Glycerol: 2nd Edition
9: Transportation Biofuels: Novel Pathways for the Production of Ethanol, Biogas and Biodiesel
10: Alternatives to Conventional Food Processing
11: Green Trends in Insect Control
12: A Handbook of Applied Biopolymer Technology: Synthesis, Degradation and Applications
13: Challenges in Green Analytical Chemistry
14: Advanced Oil Crop Biorefineries
15: Enantioselective Homogeneous Supported Catalysis
16: Natural Polymers Volume 1: Composites
17: Natural Polymers Volume 2: Nanocomposites

How to obtain future titles on publication:
A standing order plan is available for this series. A standing order will bring delivery of each new volume immediately on publication.

For further information please contact:
Book Sales Department, Royal Society of Chemistry, Thomas Graham House, Science Park, Milton Road, Cambridge, CB4 0WF, UK
Telephone: +44 (0)1223 420066, Fax: +44 (0)1223 420247
Email: booksales@rsc.org
Visit our website at http://www.rsc.org/Shop/Books/

Natural Polymers
Volume 1: Composites

Edited by

Maya J John
CSIR Materials Science and Manufacturing, Port Elizabeth, South Africa and Department of Textile Science, Faculty of Science, Nelson Mandela Metropolitan University, Port Elizabeth, South Africa
Email: mjohn@csir.co.za

Thomas Sabu
School of Chemical Sciences, Mahatma Gandhi University, Kottayam, India

RSC Publishing

RSC Green Chemistry No. 16

ISBN: 978-1-84973-402-8
ISSN: 1757-7039

A catalogue record for this book is available from the British Library

Published by The Royal Society of Chemistry,
Thomas Graham House, Science Park, Milton Road,
Cambridge CB4 0WF, UK

Registered Charity Number 207890

For further information see our web site at www.rsc.org

Printed and bound in Great Britain by CPI Group (UK) Ltd, Croydon, CR0 4YY, UK

Preface

Natural polymers offer an alternative solution to the growing environmental threat and looming petroleum crisis. The use of natural polymers for many applications would therefore contribute to creating a sustainable economy. In contrast, the feedstocks for polymers derived from petrochemicals will in the long run eventually perish. Polymer chemists, physicists and engineers show great interest in the development of eco-friendly micro- and nano-structured functional materials based on natural polymers. In recent years, natural polymers have generated much interest due to their unique morphology and physical properties. The growing interest among academics and industrial researchers in the field of natural polymers is the driving force behind the present book.

The book is divided into two volumes: the first covers natural polymer composites and the second deals with natural polymer nanocomposites. Volume 1 comprises two sections reviewing (1) natural fibres and composites and (2) protein fibres and composites. Under natural fibres and composites, the characterization and new sources of natural fibres are discussed. It also looks into whether natural fibres can indeed be a replacement for synthetic fibres in industrial applications. Under protein fibres and composites, important advancements in the field of silk, spider silk and mussel fibres are discussed. Volume 2 deals with the properties and characterization of cellulose, chitosan, furanic, starch and silk nanocomposites. A final chapter touches upon the industrial applications of natural polymer nanocomposites.

This book is unique in the sense that it deals exclusively with some of the important polymers found in nature, modifications of natural polymers and tailoring them into composites and nanocomposites. In addition, it covers novel topics related to the properties and characterization of mussel fibres and sea shells.

RSC Green Chemistry No. 16
Natural Polymers, Volume 1: Composites
Edited by Maya J John and Thomas Sabu
© The Royal Society of Chemistry 2012
Published by the Royal Society of Chemistry, www.rsc.org

As the editors of this book, we have enjoyed working with the individual contributors from different parts of the world and appreciate their diligence and patience. We would also like to thank all the publishers who generously gave their permission to reprint material.

Maya Jacob John and Sabu Thomas

Contents

Volume 1: Composites

RSC Green Chemistry No. 16
Natural Polymers, Volume 1: Composites
Edited by Maya J John and Thomas Sabu
© The Royal Society of Chemistry 2012
Published by the Royal Society of Chemistry, www.rsc.org

Volume 2: Nanocomposites

CHAPTER 1
Natural Polymers: An Overview

MAYA JACOB JOHN*[a,b] AND SABU THOMAS[c]

[a] CSIR Materials Science and Manufacturing, Polymers and Composites Competence Area, P.O. Box 1124, Port Elizabeth 6000, South Africa; [b] Department of Textile Science, Faculty of Science, Nelson Mandela Metropolitan University, P.O. Box 1600, Port Elizabeth 6000, South Africa; [c] School of Chemical Sciences, Mahatma Gandhi University, Priyadarshini Hills P.O., Kerala, India
*Email: MJohn@csir.co.za

1.1 Introduction

The scarcity of natural polymers during the world war years led to the development of synthetic polymers like nylon, acrylic, neoprene, styrene–butadiene rubber (SBR) and polyethylene. The increasing popularity of synthetic polymers is partly due to the fact that there are unlimited and economic avenues for modification of chemical structures to obtain a product with specific properties. However, this rampant use of petroleum products has created a twin dilemma: depletion of petroleum resources (Figure 1.1) and entrapment of plastics in the food chain and environment.[1] The exhaustive use of petroleum-based resources has initiated efforts to develop biodegradable plastics. This is based on renewable bio-based plant and agricultural products that can compete in the markets currently dominated by petroleum-based products. Table 1.1 presents a selected list of the common synthetic polymers.

Another issue is that the disposal of plastics in landfills creates a serious aesthetic problem in large urbanized areas of the world. The chemical stability of plastic prevents plastic waste from decomposing into the environment at a

RSC Green Chemistry No. 16
Natural Polymers, Volume 1: Composites
Edited by Maya J John and Thomas Sabu
© The Royal Society of Chemistry 2012
Published by the Royal Society of Chemistry, www.rsc.org

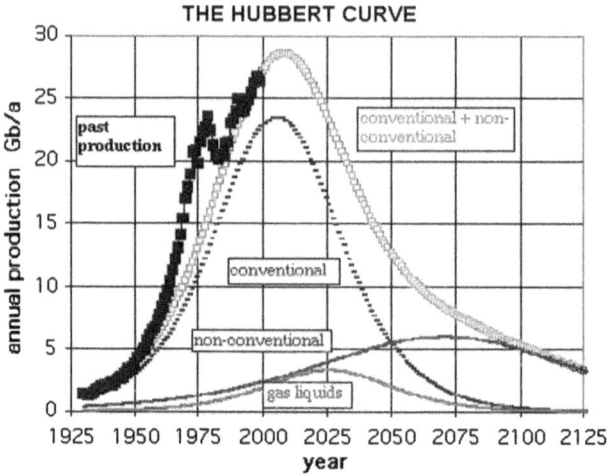

Figure 1.1   Chart of world oil reserves.

Table 1.1   List of selected synthetic polymers.

Synthetic polymer
Poly(ethylene terephthalate)
Polyethylene
Poly(vinyl chloride)
Polypropylene
Polystyrene
Poly(tetrafluoroethylene)
Polyurethane
Polyamide
Polyacrylamide

rate comparable to the rate of waste generation. In the long run, the incentive to preserve the local environment is reduced and the costs of cleaning and recovery of contaminated sites rise. Large streams can also transport excess plastic waste to other areas, creating a mobile contamination problem. Plastic waste comprises 60–80% of the marine debris litter accumulated in ocean shores. The problem of marine waste is aggravated by the low reliability of removal mechanisms aimed at reducing marine plastic residual concentration in the oceans. The effects of plastic waste on marine life include the entanglement and ingestion of harmful plastics by marine vertebrates and the bioaccumulation of toxicants along the food chain.

Natural polymers are those which are present in, or created by, living organisms. These include polymers from renewable resources that can be polymerized to create bio-plastics. There are two main types of natural polymers: those that come from living organisms (these include carbohydrates and proteins) and those which need to be polymerized but come from renewable resources (*e.g.* lactic acid and triglycerides). Both types are used in the production of bio-plastics.

Among the different types of natural polymers, the best known resources capable of making biodegradable plastics are starch and cellulose. Cellulose is the most abundant carbohydrate in the world (40% of all organic matter is cellulose). It is the main constituent of plants, serving to maintain their structure, and is also present in bacteria, fungi, algae and even in animals. Cellulose from trees and cotton plants is a substitute for petroleum feedstocks to make cellulose plastics.

Starch is a condensation polymer made up of hundreds of glucose monomers, which release water molecules as they chemically combine. Starch is a member of the basic food group of carbohydrates and is found in cereal grains and potatoes. It is also referred to as a polysaccharide, because it is a polymer of the monosaccharide glucose. Starch molecules include two types of glucose polymers, *i.e.* amylose and amylopectin, the latter being the major starch component in most plants, making up about three-quarters of the total starch in wheat flour. Amylose is a straight-chain polymer with an average of about 200 glucose units per molecule. Starch is one of the least expensive biodegradable materials available in the world market today. It is a versatile polymer with immense potential for use in non-food industries. The annual world production of starch is well over 70 billion pounds weight, with much of it being used for non-food purposes, like making paper, cardboard, textile sizing and adhesives.

Chitin, a polysaccharide similar to cellulose, is Earth's second most abundant polysaccharide. It is present in the cell walls of fungi and is the fundamental substance in the exoskeletons of crustaceans, insects and spiders. The structure of chitin is identical to that of cellulose, except for the replacement of the OH group on the C-2 carbon of each of the glucose units with an $-NHCOCH_3$ group. The principal source of chitin is shellfish waste. Commercial uses of chitin waste include the making of edible plastic food wraps and the cleaning up of industrial wastewater.

Chitin is the main source of production of chitosan, which is used in a number of applications, such as a flocculating agent, a wound healing agent, a sizing and strengthening agent for paper, and a delivery vehicle for pharmaceuticals and genes. Chitin deacetylation leads to the formation of chitosan. The process involves the use of strong alkali solutions for the removal of *N*-acetyl groups, both at room and elevated temperatures. The amount of chitin obtained annually from harvested shellfish is estimated to be over 39 000 tonnes. At least 10 billion tonnes of chitin are produced in the biosphere each year, chiefly in marine environments.[2]

Collagen is one of the most plentiful proteins present in the bodies of mammals, including humans. In fact, it makes up about 25% of the total amount of proteins in the body. It has found increasing applications in tissue engineering and repair.[3] The ability of collagen to polymerize into a three-dimensional fibrous matrix makes it an appealing material for extensive therapeutic applications, including medical implants.[4]

Table 1.2   List of common natural polymers.

Natural polymer
Polysaccharides
Starch
Cellulose
Chitin
Proteins
Collagen/gelatin
Casein, albumin, fibrogen, silks
Polyesters
Poly(hydroxyalkanoates)
Other polymers
Lignin
Lipids
Shellac
Natural rubber

Some of the other important natural polymers that are under scrutiny by the research community, but beyond the scope of this book, include lignin, shellac and natural rubber. In the category of natural polymers which need to be polymerized is the interesting development of biodegradable plastics from edible and non-edible vegetable oils like soybean oil, peanut oil, walnut oil, sesame oil, sunflower oil, tung oil and castor oil.

Table 1.2 presents a selected list of the common natural polymers.[5]

The production of 100% bio-based materials as substitutes for petroleum-based products is not an economical solution. Some of the possible solutions are blending biopolymers with synthetic polymers and reinforcing natural fibres with synthetic polymers (termed bio-composites), which are a viable alternative to glass fibre composites.

1.2   Natural Polymer Research

The aim of this book is to examine the research conducted worldwide on the use of different types of natural polymers. The book looks at the different processing techniques of natural polymers as well as applications in advanced industrial sectors. The structure, mechanical and thermal characteristics of selected natural polymers are highlighted.

1.2.1   Natural Fibres

The history of fibre-reinforced plastics began in 1908 with cellulose fibre in phenolics, later extending to urea and melamine and reaching commodity status with glass fibre-reinforced plastics. Natural fibres are subdivided based on their origins, coming from plants, animals or minerals. All plant fibres are composed of cellulose, while animal fibres consist of proteins (hair, silk and wool). Plant fibres include bast (or stem or soft sclerenchyma) fibres, leaf or hard fibres, seed, fruit, wood, cereal straw and other grass fibres. Knowledge of the structure of natural fibres is crucial in understanding the structural parameters (number, size and shape of cells, chemical constituents) and fracture mechanisms in fibres.[6]

Some of the important natural fibres used as reinforcement in composites are listed in Table 1.3.

Over the last few years, a number of researchers have been involved in investigating the exploitation of natural fibres as load-bearing constituents in

Table 1.3 List of important natural fibres.

Fibre source	Species	Origin
Abaca	*Musa textilis*	Leaf
Agave	*Agave americana*	Leaf
Alfa	*Stippa tenacissima*	Grass
Bagasse	–	Grass
Bamboo	(>1250 species)	Grass
Banana	*Musa indica*	Leaf
Broom root	*Muhlenbergia macroura*	Root
Cantala	*Agave cantala*	Leaf
Caroa	*Neoglaziovia variegata*	Leaf
China jute	*Abutilon theophrasti*	Stem
Coir	*Cocos nucifera*	Fruit
Cotton	*Gossypium* spp.	Seed
Curaua	*Ananas erectifolius*	Leaf
Date palm	*Phoenix dactylifera*	Leaf
Flax	*Linum usitatissimum*	Stem
Hemp	*Cannabis sativa*	Stem
Henequen	*Agave fourcroydes*	Leaf
Isora	*Helicteres isora*	Stem
Istle	*Samuela carnerosana*	Leaf
Jute	*Corchorus capsularis*	Stem
Kapok	*Ceiba pentranda*	Fruit
Kenaf	*Hibiscus cannabinus*	Stem
Kudzu	*Pueraria thunbergiana*	Stem
Mauritius hemp	*Furcraea gigantea*	Leaf
Nettle	*Urtica dioica*	Stem
Oil palm	*Elaeis guineensis*	Fruit
Piassava	*Attalea funifera*	Leaf
Pineapple	*Ananas comosus*	Leaf
Phormium	*Phormium tenas*	Leaf
Roselle	*Hibiscus sabdariffa*	Stem
Ramie	*Boehmeria nivea*	Stem
Sansevieria (bowstring hemp)	*Sansevieria*	Leaf
Sisal	*Agave sisalana*	Leaf
Sponge gourd	*Luffa cylindrica*	Fruit
Straw (cereal)	–	Stalk
Sun hemp	*Crorolaria juncea*	Stem
Cadillo/urena	*Urena lobata*	Stem
Wood	(>10 000 species)	Stem

composite materials. The use of such materials in composites has increased due to their relative cheapness, their ability to recycle and the fact that they can compete well in terms of strength per weight of material.

Volume 1 focuses on different sources and applications of natural fibres. One chapter deals with novel renewable sources from which natural fibres can be extracted. Another chapter looks at relating the structural anisotropy of natural fibres to mechanical properties. One of the challenges of using natural fibres in aerospace applications is the airworthiness requirements. Currently, natural fibres are being explored for use in secondary structures in aircraft for which flame, smoke and toxicity (FST) requirements are very stringent. This has led to a lot of developmental research being undertaken in this field. A further chapter therefore explores the flammability properties of natural fibre reinforced composites. A crucial problem associated with the use of natural fibres in composites is their hydrophilic properties. This aspect is dealt with in a chapter on probing the water sorption characteristics of natural fibres. The chemical modification of natural fibres has been well documented in the literature, but ideally it would also be desirable that the chemicals used for modification should also be from renewable resources as it would preserve the biodegradable nature of natural fibres. A chapter therefore focuses on environmentally friendly coupling agents for natural fibre-reinforced composites. Other chapters include examining the characterization techniques of the interfacial properties of natural fibre-reinforced composites and the increasing applications of natural fibre composites in the automotive sector.

1.2.2 Protein Fibres

The book also deals with the properties of selected protein fibres. Protein fibres are formed by natural animal sources through condensation of α-amino acids to form repeating polyamide units with various substituents on the α-carbon atom. The sequence and type of amino acids making up the individual protein chains contribute to the overall properties of the resultant fibre.[7] In general, protein fibres possess moderate strength, resiliency and elasticity. They have excellent moisture absorbency and transport characteristics and do not build up static charge. Some of the common protein fibres include wool, spider silk, cashmere, *etc.* Among natural fibres, silk exhibits exceptional properties, especially in toughness and biocompatibility properties. A chapter therefore focuses on the studies and properties of silk fibre-reinforced composites. Other chapters include studies on collagenous waste-based composites and exploring the properties and applications of mussel byssus fibres. Important advancements in the field of zein fibres are also discussed in another chapter.

Volume 2 deals with the properties and characterization of selected natural polymer nanocomposites. Cellulose nanowhiskers (CNWs) have emerged as one of the most interesting bio-based nano-reinforcements in the last decade.[8,9] Cellulose nanowhiskers can be generated from various plant sources with transverse dimensions as small as 3–30 nm, giving a high surface-to-volume ratio.

It has also been shown that since the nanowhiskers are rod-like, they can be self-assembled into chiral nematic liquid crystalline structures, not only in solution but also in the dry state. The volume begins by exploring nanocellulose as a potential reinforcement in composites. Chitosan (a natural polymer) is a good candidate for the development of conventional and novel drug delivery systems. Chitosan has been found to be used as a support material for gene delivery, cell culture, and tissue engineering. However, practical use of chitosan has been mainly confined to the unmodified forms. For a breakthrough in utilization, especially in the field of controlled drug delivery, graft copolymerization onto chitosan will be a key point, which will introduce desired properties and enlarge the field of the potential applications of chitosan by choosing various types of side chains. The properties and applications of chitosan and soy protein-based nanocomposites are discussed in subsequent chapters. Other chapters include studies on furanic-based nanocomposites, the characterization of molecular interactions in amylose/starch nanocomposites, and unique properties of nacre from mollusc shells.[10,11] The last two chapters touch upon the industrial and biomedical applications of natural polymer nanocomposites.

References

1. M. J. John and S. Thomas, *Carbohydr. Polym.*, 2008, **71**, 343–364.
2. K. D. Sturm and K. J. Hesse, *Ocean Challenge*, 2000, **10**, 20.
3. D. A. Wahl and J. T. Czernuszka, *Eur. Cells Mater.*, 2006, **11**, 43–56.
4. K. Madhavan, D. Belchenko, A. Motta and W. Tan, *Acta Biomater.*, 2010, **6**, 1413–1422.
5. E. S. Stevens, *Green Plastics*, Princeton University Press, Princeton, 2002.
6. K. G. Satyanarayana and F. Wypych, in *Handbook of Engineering Biopolymers: Homopolymers, Blends and Composites*, ed. S. Fakirov and D. Bhattacharyya, Hanser, Munich, 2007, pp. 3–47.
7. http://www.textileschool.com/School/Fiber/NaturalProteinFibers.aspx
8. K. Oksman, A. P. Mathew, D. Bondeson and I. Kvien, *Compos. Sci. Technol.*, 2006, **66**, 2776–2784.
9. A. P. Mathew, A. Chakraborty, K. Oksman and M. Sain, in *Cellulose Nanocomposites: Processing, Characterization and Properties*, ed. K. Oksman and M. Sain, ACS Symposium Series 938, Oxford University Press, Oxford, 2006, pp. 114–131.
10. R. K. Pai, L. Zhang, D. Nykpanchuk, M. Cotlet and C. S. Korach, *Adv. Eng. Mater.*, 2011, **13**, 415–422.
11. A. Gandini, *Macromolecules*, 2008, **41**, 24.

CHAPTER 2

Biomimetics: Inspiration from the Structural Organization of Biological Systems

KALPANA S. KATTI,* CHUNJU GU AND
DINESH R. KATTI

Department of Civil Engineering, North Dakota State University, Fargo,
ND 58105, USA
*Email: kalpana.katti@ndsu.edu

2.1 Introduction: Hierarchy and Structural Order

Structural materials found in nature exhibit hierarchical structures that span a structural order over length scales from molecular to macroscopic.[1–5] Optimization of properties and structural redundancy as well as growth are all ramifications of the hierarchical structures that result from eons of evolution. The following sections elucidate the hierarchical details of structure in bone, teeth, seashells and spider silk and fabrication attempts at duplication of these structures.

2.2 Biological Materials Systems

2.2.1 Bone

Bone is a remarkable biological material, providing skeletal stability, support and protection of vital organs. The structure of bone has been extensively

RSC Green Chemistry No. 16
Natural Polymers, Volume 1: Composites
Edited by Maya J John and Thomas Sabu
© The Royal Society of Chemistry 2012
Published by the Royal Society of Chemistry, www.rsc.org

studied historically as well as recently, owing to its perfect adaption of mechanical properties to metabolic functions. In addition, as a protein/mineral nanocomposite, bone combines the optimal properties of both components: stiffness and toughness. This rather unusual combination of material properties is a combination of rigidity and resistance against fracture.[6] The unique mechanical properties of bone and the structure–property relationship have attracted significant attention.

Research on the structure of bone dates back to the early 17th century, when the compound microscope was invented. Clopton Havers is generally credited with providing the first description of the porous nature of bone in 1691, but due to the poor quality of the magnifying lenses, the initial descriptions dealt primarily with the canal system and the "laminar" structure of bone without the presence of osteonal bone. In the 18th and 19th century, some observations were described and defined in detail, such as the Haversian system of lamellae, and the orientation and disposition of lacunae and canaliculi.[7] By utilizing polarized light microscopy, Schmidt found that the crystallographic "*c*" axis of mineral in bone is well aligned with the collagen fibrils.[8] Further detailed studies came after the invention of scanning electron microscopy (SEM) and transmission electron microscopy (TEM) in 1930, which enabled examination of structures on the nanometer scale. With the help of these high-resolution instruments and other techniques such as X-ray diffraction, polarized optical microscopy, sonic velocity, as well as mechanical tests, the hierarchical structure of bone was discovered and depicted. Atomic force microscopy (AFM), which appeared in 1980, made it possible to investigate the structure of bone in the ambient environment on the nanometer scale.[9–11] Although the overall structure from nano to macro scales of bone has been extensively studied, more detailed research is also needed.

Bone refers to a family of materials having in common a basic building block, the mineralized collagen fibril; however, the structural organization of the fibrils is different in different bone types. The family of bone also contains dentin, cementum and mineralized tendon, which have various proportions of bone components.[12] Disregarding the different shapes, bone is generally mechanically divided into two types, compact bone and cancellous bone,[13] which will be introduced later.

Primarily, bone tissue is composed of collagen fibres, crystals of a calcium phosphate complex also known as carbonated apatite $[Ca_{10}(PO_4,CO_3)_6(OH)_2]$ (namely, mineral) and a cement containing mucopolysaccharides, among other biopolymers. The chemical analysis of bone shows that there are three primary components in bone: collagen, mineral and water. The collagen part accounts for nearly 1/3 and the mineral part accounts for nearly 2/3 of the dry weight of bone matrix.[14] The water component is about 10–12 wt% of cortical bone and 20% of the bone matrix.[15] The mineral crystals and grow in the triple helical collagen fibres and replace some of the water while mineralization takes place. The bone structure has been described in terms of up to six or seven hierarchical levels of organization from nanoscale collagen and mineral to macroscale femur bone (Figure 2.1).[12,16,17] Recently, new studies indicate hierarchical

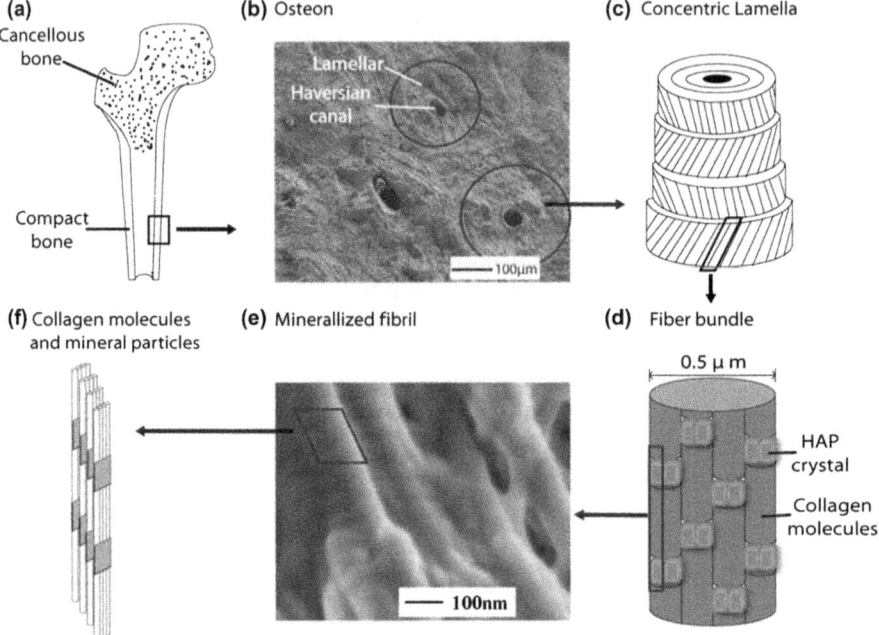

(a) Cancellous bone

Compact bone

(b) Osteon

Lamellar

Haversian canal

100μm

(c) Concentric Lamella

(f) Collagen molecules and mineral particles

(e) Minerallized fibril

100nm

(d) Fiber bundle

0.5 μm

HAP crystal

Collagen molecules

Figure 2.1 Hierarchical organization of a human femur bone from macro- to nanoscale. (a) Macroscale organ level: human femur bone. (b) Macroscale tissue level: osteon. (c) Microscopic level: bone lamellae (adapted from Giraudguille[17]). (d) Mesoscopic level: fibre bundle. (e) Nanoscale level: mineralized fibril. (f) Molecular level: collagen molecule and mineral particle. (Adapted from Fratzl.[16])

structures in the collagen molecule in bone with three levels of hierarchy within the molecule.[18]

2.2.1.1 Level 1: Collagen Fibrils and Minerals

2.2.1.1.1 Collagen. Type I collagen molecules, also called triple helices, are supercoiled assemblies of three polypeptide chains, two identical α1-chains and one α2-chain, each with over 1000 amino acid residues. Collagen type I accounts for nearly 90% of its total organic content. The main part of a collagen chain consists of Gly-X-Y repeats, in which X and Y can be any amino acid but are frequently proline and hydroxyproline. A triple-helical molecule is cylindrically shaped, with an average diameter of about 1.5 nm and length of 300 nm.[19] Besides the main helical part, collagen triple helices also comprise short nonhelical end sequences called telopeptides with both N- and C-terminal ends. Telopeptides account for 2% of the molecule and are critical for fibril formation in a self-assembly process.[20] Figure 2.2 shows a schematic of the general structure and triple helical motifs of the collagen molecules.

(a)

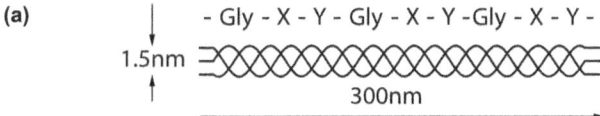

(b)

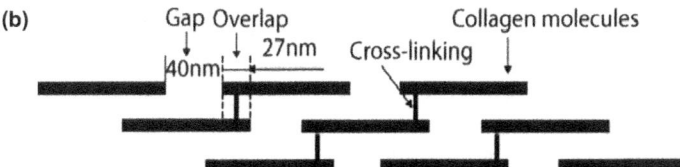

Figure 2.2 Collagen molecules and intermolecular cross-linking. (a) Triple-helical structural motif of collagen molecules (adapted from Giraud-Guille[19]). (b) Lysyl oxidase cross-linking. (Adapted from Kadler *et al.*[20])

During the formation of a fibril the collagen spontaneously self-assembles into cross-striated fibrils that occur in the extracellular matrix of connective tissues. The fibrils are stabilized by covalent cross-linking which is initiated by oxidative de-amination of specific lysine and hydroxylysine residues in collagen by lysyl oxidase.[20] The intermolecular cross-linking provides the fibrillar matrices with various mechanical properties, such as tensile strength and viscoelasticity. Both high-performance liquid chromatography (HPLC) and Fourier transform infrared spectroscopy (FTIR) have been frequently adopted to analyze the cross-links.[21]

2.2.1.1.2 Mineral. The mineral in bone primarily consists of poorly crystalline non-stoichiometric carbonated hydroxyapatite (dahllite), which has a plate-shaped hexagonal crystal structure. Bone crystallites are probably the smallest biogenic crystals. They are only 1.5–4 nm thick, 25 nm wide and 50 nm long on average,[12] but very little is understood about the atomic structure. The *c*-axis of the unit cells of these crystallites in bone is usually aligned parallel with the long axis of the adjacent collagen fibres.[22] Studies using AFM indicate that many of these mineral plates appear to be aligned, forming larger aggregates (475–600 nm long × 75–90 nm thick) that also retain collagen periodicity along their exposed edges.[11]

2.2.1.2 Level 2: Mineralized Fibril

The triple-helical collagen molecules assemble into the fibril in a staggered arrangement. Within the fibril there is a linear shift of ∼67 nm (D-period) between neighboring molecules. The D-period is divided by overlaps and gaps, where an overlap is around 27 nm and a gap is 40 nm (shown in Figure 2.2b). Robinson and Watson[22,23] pioneered a TEM study and reported the 68-nm banding pattern in collagen fibrils. The assembly of collagen molecules into fibrils is an entropy-driven process, driven by the loss of solvent molecules from the surface of protein molecules, resulting in assemblies with a circular cross-section, which minimizes the surface area/volume ratio of the final assembly.[20]

Mineralized fibrils are the basic building blocks of bone. Minerals are intimately associated with the collagen framework in which they form, resulting in a highly complex but ordered mineral/organic composite material. Studies of crystal growth show that crystals are first formed in the gap, and then they continue to grow and penetrate into the overlap zone, thus pushing aside the triple-helical collagen molecules and even breaking cross-linking and other bonds.[24] Because the density of crystals is higher in the gap region, a periodic mineral density profile with around 67 nm spacing can be observed by electron microscopy. The width of a mineralized fibril is about 100 nm.

It is also important to mention that the spaces between mineralized collagen fibrils (extrafibrillar) are filled with non-collageneous macromolecules and minerals. The extrafibrillar volume is at least 60% of the total, while the fibrils are no more than 40%. More of the mineral appears to be extrafibrillar than within the fibrils, and is cemented together by non-collagenous organic matter.[25]

2.2.1.3   Level 3: Fibrillar Arrays

The mineralized fibrils are self-organized into fibre bundles along their length and the bundles may fuse with neighboring bundles. How the mineralized fibrils are aligned is of great interest, which leads to a great difference in mechanical properties. Two arrangements have been suggested.[12] One is an arrangement of mineralized collagen fibrils aligned both with respect to the crystal layers and the fibril axes. The other arrangement of mineralized collagen fibrils is with only the fibril axes aligned. Sonic velocity measurements in three orthogonal directions of macroscopic specimens show significant differences,[26] implying that orthotropic order at the fibril level may well extend to millimeter distances.

2.2.1.4   Level 4: Fibrillar Array Patterns

The fibril bundles continue to be organized into layers or lamellae with a few microns thickness, and they in turn are arranged in a variety of ways into higher-order structures, depending on the bone type. Four of the most common patterns have been presented. In a parallel array pattern, all the fibrils are parallel to each other. In a woven array pattern, individual fibrils or fibril bundles are randomly organized. A radical fibril array is characteristic of the bulk of dentine, which also belongs to the bone family.[12] Further, plywood-like structures are common in nature, which is believed to have a structure–function relationship.[19] SEM and TEM studies showed that lamellar bone is made up of alternating collagen-rich and collagen-poor layers, all having an interwoven arrangement of fibres.[27] The successive layers in a thin/thick lamellar unit proceed by an angle of roughly 30° from one layer to the next.[28]

2.2.1.5 Level 5: Osteon

The osteon, also called the Harversian system, is the fundamental structural unit of compact bone. Each osteon consists of concentric layers or lamellae that surround a central canal, the Haversian canal. The Haversian canal, parallel to the long axis of the bone, contains the bone's nerve and blood supplies. Between adjoining osteons there are angular intervals that are occupied by interstitial lamellae. These lamellae are remnants of osteons, the greater parts of which have been destroyed.[14] The osteon takes part in the remodeling process, whereby tunnels are eroded and then filled in again with cylinders of bone.

2.2.1.6 Level 6: Compact and Cancellous Bone

Compact bone is solid, with the only porosities existing for canaliculi, osteocyte lacunae, blood channels and erosion cavities, while cancellous bone has porosity that is easily visible to the naked eye. Overall, bone serves a protective function by the construction of two compact plates separated by intervening cancellous (spongy or trabecular) bone. This construction is beneficial for the maximum absorption of energy with the minimum trauma to the bone itself.

2.2.1.7 Mechanical Advantages of the Hierarchical Structure

As pointed out by Gao *et al.*,[29] bone is much less sensitive to flaws because of its hierarchical structure. The hierarchical design distributes stresses throughout the levels of a structure, thereby minimizing dangerous stress concentrations that could precipitate failure and fracture. The hierarchical feature of bone controls the fracture properties, particularly the toughness.[13] Because the collagen fibres in neighboring lamellae are oriented at some angle to each other, fracture surfaces show considerable roughness. The work of driving a crack across the interfaces consisting of the plates, sheets and Haversian systems of bone is much greater than it would be if the material were homogeneous.[30] Recently, many efforts have been made to model the structure of bone and its constituents through continuum and molecular modeling methods. The mechanical properties of bone and the relationship of the properties to the hierarchal structural organization have been extensively investigated.[29,31–41] Continuum methods, using finite element modeling (FEM), are used to calculate the fundamental mechanical properties of bone.[42–45] Attempts at multiscale modeling techniques for collagen have been made.[18,36,46–50] Recent experimental studies using fluorescence resonance energy transfer has indicated molecular interactions between apatite and collagen. Simulations using steered molecular dynamics have indicated the nature of the interactions (primarily non-bonded), as well as their role on mechanics of the collagen molecule.[50,51] In addition, it has been shown that the load-deformation of tropocollagen interacting with HAP[50] is significantly influenced by water. Molecular and nanomechanics of bone have also been described in a number of studies[5,18,34,52–55] and different nanoscale deformation mechanisms have been proposed in the literature. It has also been shown recently that the collagen molecule, when observed in its full length through simulations, exhibits a new level of hierarchy of structure which has significant contributions to the mechanics. The three levels are: level 1, the helicity of the individual polypeptide chain; level 2, the helical organization of the three peptide chains; and a third new hierarchy, the helicity of the overall triple helix that can only be observed in a full length collagen molecule.[18]

2.2.2   Teeth

Teeth are small, calcified, structures found in the jaws (or mouths) of many vertebrates that have the primary function of breaking down food. Teeth are anchored within alveolus bone sockets and are held in place by a thin cementum interlayer adjoining the periodontal ligament. The part of the tooth that projects into the mouth is called the tooth crown, while the part that is set into the jaw is called its root.[56] The tooth interior generally consists of three layers. The outer layer of enamel (96 wt% mineral), which is the hardest tissue in the body, covers part or all of the crown of the tooth. The middle layer of the tooth is made up of dentine (or dentin), with an enamel/dentine junction (EDJ) several micrometers thick separating it from enamel. Dentine is less hard than enamel and similar in composition to bone (70 wt% mineral, 20% organic, 10% water). The third and the innermost layer is pulp, containing blood vessels and nerves. The hard, brittle enamel coat protects the soft, tough dentine and pulp interior.[57]

Enamel is not only hard (resistant to permanent surface deformation), but also very tough (resistant to crack propagation and brittle fracture).[58] Unlike bone, which is a collagen-based ceramic composite, enamel is a tissue mineralized with calcium phosphate, containing no collagen or cells, but rather having long, thin strands of hydroxyapatite (HAP) that are woven into a fabric-like ceramic. Enamel retains less than 0.5% protein and a very small amount of water and holds the mineral fibres together.[59] Additionally, in contrast to bone, enamel is not remodeled during its lifetime since the ectodermally derived cells that create enamel are lost once the tooth erupts into the oral cavity.[60]

2.2.2.1   Hierarchical Structure of Enamel

The enamel mineral is a form of non-stoichiometric carbonated calcium hydroxyapatite $[Ca_{10}(PO_4)_6(OH)_2]$. The Ca/P ratio of enamel apatite is slightly lower than that of hydroxyapatite.[61] Enamel exhibits a hierarchical organization that spans nanoscale to macroscale levels. Some literature regards the nanospheres as the first level, virtually as the first step of biomineralization. At the nanoscale, amelogenin molecules undergo self-assembly to form 15–20 nm spheres.[60] In mature teeth, apatite crystallites are the least complex and smallest structural unit.[58] The crystals resemble long bars with a hexagonal cross-section at a final size of about 30–40 nm across.[56] Here we list the five specific levels of structure based on structural complexity of mature teeth.[62] These orders of scale are interdependent and coalesce with one another, thereby creating a structural continuum in the tooth organization.

2.2.2.1.1 Level 1: HAP Crystallites, a Few Ångstroms Wide.

Unlike the minute, short crystallites of bone, the crystallites of dahllite in enamel are needle-like, narrow and extremely long. Reports of crystallite length vary from a fraction of a micron to 100 μm. Each crystallite is surrounded by nanospheres favoring growth on their c-axis and proper spacing among the crystallites.[60]

In the early mineralized enamel crystals, ribbon-like crystals appear near the ameloblast at the DEJ. As maturation occurs, the enamel crystals become more densely packed and more highly oriented.[63] Cross sections of fully mineralized crystallites are usually hexagonal[64] or rhomboidal.[65] Many other growth habits, including rectangular or irregular crystals (30–40 nm in width) which contain "notches", have also been observed.[64] Simmer proposed a three-stage enamel crystal growth mechanism which involved a crystal precursor, octacalcium phosphate (OCP). The three-stage mechanism includes: (1) formation of the incipient seed of the crystals; (2) two-dimensional growth of the seed (a- and b-axes); (3) three-dimensional growth, which involves growth along the c-axis and then one-unit-cell thickness of OCP hydrolyzes to a two-unit-cell thickness of HAP, thereby leading to a contraction of the lattice in one direction and growth in the c-axis produces a regular hexagonal prism. Further, enamel incorporates some ions in its apatite lattice, such as HPO_4^{2-}, CO_3^{2-}, Na^+, F^-, *etc.*, different from ideal hydroxyapatite. Therefore, some ions fit in the interstices and cause distortions in the close-packed lattice.[65]

2.2.2.1.2 Level 2: Prisms.

Concerning the arrangement and orientation of crystallites, two major classes of mammalian enamel are defined: prismless (also named as aprismatic, nonprismatic or preprism in various publications) and prismatic enamel.[58] Prismless enamel refers to a discontinuous structure separated by distinct boundary planes, and prismatic enamel has the cross-sections of prisms that are demarked by sheaths that usually have curved outlines.[62] In primates, as in most mammals, prismatic enamel is the dominant component of enamel patterns and the ameloblast cells weave the hydroxyapatite crystallites into bundles called "rods" that decussate one another.[60] The diameter of a prism ranges from 2 to 10 μm.[58] The sheath, where protein and water accumulate, is approximately 100 nm thick.[57] The cross-sectional morphology of the prism sheath was determined by the shape of the Tomes process on the tip of the ameloblast,[62] and various morphologies were shown, such as circular prisms with complete sheaths, highly derived prisms with key-hole cross-sections, *etc.*, but it is still not known if the differences in morphology of the prism cross-sections affect functions.

2.2.2.1.3 Level 3: Enamel Types.

Units of enamel in which the prisms have similar orientations are defined as enamel types. The orientations of prisms are independent of their cross-section morphology. In order to study the orientation of prisms, the enamel/dentine junction (EDJ) is regarded as a reference plane.[62] The two prismatic enamel types that occur in primate enamel are radial enamel and decussating enamel.

In radial enamel, all prisms are roughly parallel to one another as they rise radially from the enamel/dentine junction and occlusally toward the enamel surface.[58] Radial enamel can be distinguished into different subtypes by differences in orientation of the interprismatic matrix (IPM) crystallites relative to the prisms, which can vary from almost parallel to intersections at angles of approximately 90°.[62]

In decussating enamel, prisms are arranged in regularly organized, alternating layers or groups that rise from the enamel/dentine junction to the surface at different orientations.[58] Decussation is manifested optically in tooth sections as so-called Hunter–Schreger bands (HSBs), due to variations in light reflection from differently oriented prism bundles. Hunter–Schreger bands, like radial enamel, are commonly observed in mammalian enamel but vary in degree from species to species.[57] Complete crossing of prisms at 90° to those in adjacent groups ("true decussation") does not occur in all species. More often, the angle between prisms in adjacent bands is less than 90°, and changes as prisms pursue a slightly sinuous course from the enamel/dentine junction to the outer tooth surface. In primates and most other mammals the layers are usually several prisms thick, but in some rodents each layer is only one prism thick. The most common type of decussating enamel is horizontal decussation, in which HSBs are stacked on top of one another from crown to root, with long axes of prisms in adjacent horizontal layers of prisms extending toward the outer enamel surface at different angles. This is the type of decussating enamel found in primate teeth.[58] Tangential enamel and irregular decussation of prism bundles were also recognized in some molars of rodents.[62]

2.2.2.1.4 Level 4: Enamel Patterns. Enamel patterns refer to the arrangement of the enamel types within a tooth. Normally, mammalian teeth are capped with enamel that is composed of two or more enamel types and have a characteristic distribution through the enamel of the crown. The enamel pattern was designated "schmeltzmuster" by Koenigswald in 1977.[62] In rodents, the enamel pattern regularly includes an inner layer of HSBs and the outer layer of radial enamel. In molars of small-bodied primates the enamel pattern usually consists of an inner layer of radial enamel and a much thinner outer layer of prismless parallel crystal. In larger primates the enamel pattern is usually composed of an inner layer of horizontal HSBs, a middle radial layer and an outer prismless layer.[58]

2.2.2.1.5 Level 5: Dentition. Dentition is the most complex and largest-scale hierarchical level that describes the variation of enamel patterns from tooth to tooth. Animals whose teeth are all of the same type, such as most non-mammalian vertebrates, are said to have homodont dentition, whereas those whose teeth differ morphologically are said to have heterodont dentition. At the dentition level, primates appear to have relatively little variation in enamel microstructure in comparison with rodents.[58] In human teeth, enamel is the most dense (96% mineral) close to the outer surface and less dense (84% mineral) near the enamel/dentine junction.[66]

Further, each level of complexity only provides limited information on one aspect of the total structure. The thorough analysis should also take the study of systematic interrelationships and biomechanical functions into account.

Although protein and water occur only a very small amount in enamel, they are crucial to the development and toughness of bone. Amelogenin and enamelin are the main proteins present in enamel. Initially, amelogenins make up about 90% of the protein and are most important. They are hydrophobic and probably control crystal size and orientation.[56] As the mineral crystals grow in size during maturation, amelogenins disappear and then suck out vast amounts of protein (degraded with proteases) and water. Enamelin is the largest known enamel protein and its expression is highly restricted to developing teeth. There are also some other proteins that have been observed, such as tuftelin, which may play a role in the nucleation of enamel crystallites.[67] Enamelins and tuftelins in the spaces between mineral crystals also serve as a "glue",[68] and some other proteins such as ameloblastin and enamelysin are believed to be essential for proper enamel mineral formation.[69] The water in enamel was found to be bound in two different ways. A small part is very loosely bound, whereas the greater part is firmly bound to the mineral phase.[70]

2.2.2.1.6 Dentine. Dentine forms most of the volume of the tooth (as shown in Figure 2.3). Dentine contains tubules that intersect the EDJ

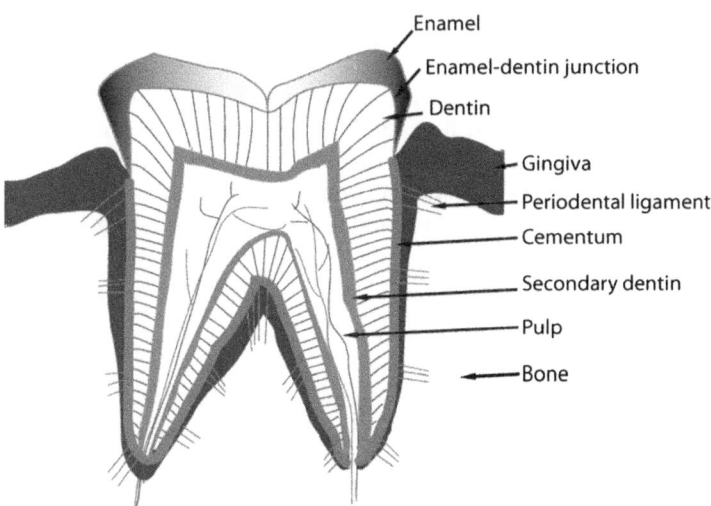

Figure 2.3   Schematic representation of the basic structure of the human tooth. (Adapted from Lawn *et al.*[57])

approximately perpendicularly, enabling transport of nutrients from the pulp through the tooth interior.[57] Dentine is rather like bone, with about 48% of its volume composed of mineral and has an organic matrix based on type I collagen fibres.

The microstructure of dentine was also suggested to be hierarchical. The mineral crystallites are needle-like near the pulp and the shape continuously progresses to be plate-like near the DEJ.[71] The edge view of plate-like mineral crystallites in dentine is as long as in enamel (up to 100 nm),[72] and with thickness in the range 2–3.5 nm. The collagen fibrils, approximately 30% by volume, are roughly 50–100 nm in diameter; they are randomly oriented in a plane perpendicular to the direction of dentine formation. The whole orientation of collagen seems to determine that of the crystals which initiate within and around them. Like bone structure, distinct bands from collagen fibrils in longitudinal sections of dentine were also seen by TEM[72,73] and AFM.[74]

At a higher level of organization, dentine is regarded as a composite, with the intertubular dentine as the matrix and the tubule lumens with their associated cuffs of peritubular dentine forming the cylindrical fibre reinforcement. Peritubular dentine exhibits hardness close to that of enamel.[56] At the greatest length scale are the effective, or continuum, properties of dentine.[75]

2.2.2.1.7 Mechanical Advantages of the Hierarchical Structure. From a micro-structural point of view, enamel and dentine are both very hetero-geneous tissues. There are obvious mechanical advantages of their intricate make-up. At the cost of a low stiffness and the loss of some compressive "strength", dental tissues have much greater toughness than plain ceramics and survive much repeated loading.[56] Cracks spreading between adjacent prisms may be stopped if the rods change direction periodically (decussation).[76] The complex fabric that dissipates forces traveling through teeth also protects them from fracture.[58]

Many studies have been done on the hierarchical complexity of enamel and dentine. However, there is still a lot that remains to be done if we are to understand the adaptive significance of mammalian dental form and the relations between the hierarchical structure and function; a better understanding of their structure could definitely contribute to improving the processing of artificial composite materials.

2.2.3 Seashells

Seashells are natural mineralized nanocomposites with hierarchical structures. There are a large variety of seashells found in nature and they possess many different morphological types of shell structures,[77,78] but most of them are built of two $CaCO_3$ polymorphs, *viz.* an outer prismatic layer of rhombohedral calcite and an inner nacreous layer of orthorhombic aragonite, which are sandwiched by an organic matrix containing glycine- and alanine-rich proteins and polysaccharides, thus forming complicated multilayered microstructures.[79]

The outer calcite layer is hard and deployed by seashells to prevent penetration from the outside, but it is prone to brittle fracture. The thick inner layer of some mollusk shells is composed of nacre, also known as mother of pearl. Nacre is well known for its superior mechanical properties, particularly fracture toughness: a 3000-fold enhanced fracture resistance compared to a single crystal of aragonite, which is its major constituent.[80] Nacre consists of more than 95 wt% of aragonitic calcium carbonate ($CaCO_3$), which is a ceramic, and less than 5 wt% of organic material, primarily composed of proteins and polysaccharides. The literature shows that the main strengthening and toughening mechanisms of nacre are because of its unique microarchitecture.[80–83]

2.2.3.1   Hierarchical Structure of Seashell

Like bone and enamel, as well as many other biological materials, seashells exhibit a hierarchical structure with at least six levels.[84] The nacre structure is described by interlocked "brick/mortar" architecture, in which mineral is the brick and organic materials are the mortar.[78] Hierarchical levels starting from the molecular level are discussed here.

2.2.3.1.1   Level 1: Nanograins and Intracrystalline Matrix in an Aragonite Tablet.   AFM imaging has been used to indicate the presence of a nanograin structure in the nacre platelets.[84,85] The organic material network is called an intracrystalline matrix. As seen in Figure 2.4, an aragonitic tablet in nacre is composed of many individual nanograins. The mean size of the nanograins as reported in literature varies from about 32 nm[84,86] to 38 nm[87]

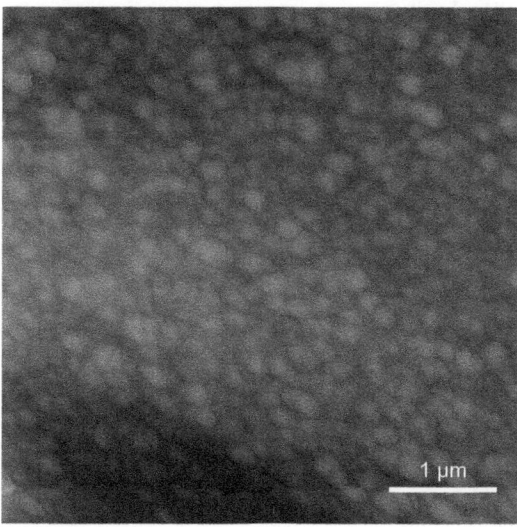

Figure 2.4   AFM image of a single nacre platelet showing the nanograin structure.

to about 45 nm with extensions.[88] Synchrotron spectromicroscopy and X-ray and TEM experiments indicate that at least part of the intracrystalline matrix is crystallized and responds like a "single crystal"[88] and that the thickness of the "intracrystalline" matrix is about 4 nm.[87]

2.2.3.1.2   Level 2: Polygonal Platelet and Intercrystalline Organic Matrix. At the micrometer level, the polygonal tablets are made of aragonite crystals (orthorhombic; $CaCO_3$), with a thickness of approximately 0.2–0.5 µm and a width of about 5 µm (Figure 2.5). The tablets are closely packed and organized into a layer. The intercrystalline organic matrix separates the platelets and has a thickness of about 20 nm. All the platelets on the same layer are twin-related whether they share a boundary or not.[82] Studies on the biomineralization of nacre reveal that individual aragonite tablets are nucleated on the underlying matrix sheet and grow rapidly in a direction perpendicular to the shell surface (the crystallographic *c*-axis). Growth parallel to the lamina follows after the tablet has reached its maximum thickness. The growing crystal tablets are hexagons.[89]

Additionally, based on this nanostructure, it has long been established that a platelet diffracts as a single crystal.[88] The platelet was described as a "pseudo" single crystal, *i.e.* an organomineral composite made up of aragonite nano-grains all coherently oriented according to diffraction and embedded in a crystallized organic phase.[88] The identified elastic properties of the "inter-crystalline" organic matrix were found to be twice as high as those of the "intracrystalline" one.[87]

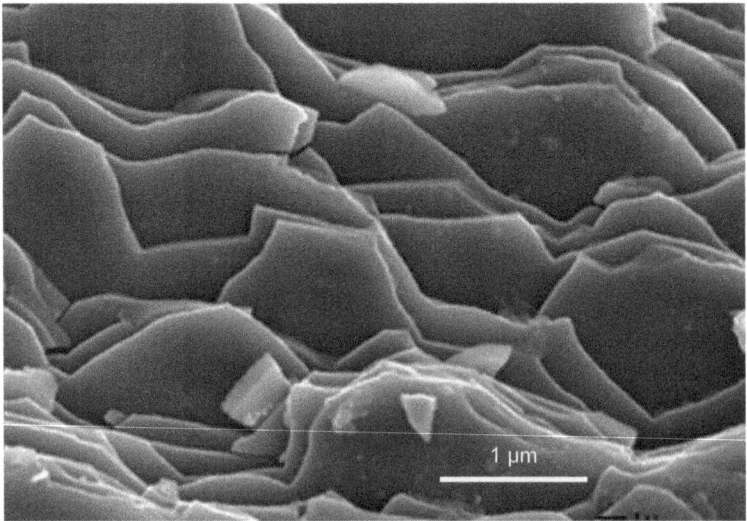

Figure 2.5   SEM image of hexagonal aragonite platelets with an intercrystalline organic matrix in-between.

2.2.3.1.3 Level 3: Platelet Layers and Interlamellar Organic Matrix. The platelet layers in nacre are parallel organized and sandwiched between interlamellar organic materials. The thickness of a platelet layer and an organic matrix layer are 500 ± 40 nm and 26 ± 5 nm, respectively.[90] This structure is the traditional "brick/mortar" structure, but the platelet layers are not smoothly arranged; there are some fine structures between the layers, as described below.

Additionally, interlocking of the platelets, which results from penetration of the platelet layers and rotation of the layers as they are stacked, is observed in nacre.[83] The "platelet interlocks", observed on the fracture surface by SEM, contribute an important toughening mechanism for nacre (see Figure 2.6) and the formation of platelet interlocks results from small rotation between layers of platelets. Between platelet layers, nanoscale mineral columns through the organic matrix layers were also observed by TEM. They are randomly distributed on the surface of the aragonite platelets and are called mineral bridges. Their average diameter is 46 ± 8 nm with a height of 26 nm.[90] Moreover, nanoscale asperities are also present on the surface of the platelets.[91] Observing the cross-section of a platelet layer more carefully, it was seen that the lamellae are parallel and roughly equally spaced (about 50–80 nm) from one another and that the spaces between the layers are filled with an electron-lucent organic material that shows no discernible regular structure.[92]

The interlamellar sheets between the platelet layers are composed of several layers of organic material. Thin layers of β-chitin are sandwiched between two thicker layers of silk fibroin-like proteins. The fibre axis of the β-chitin and silk proteins are perpendicular to each other and aligned with the *a*- and *b*-axes of the aragonite platelets, respectively. This well-defined spatial relationship

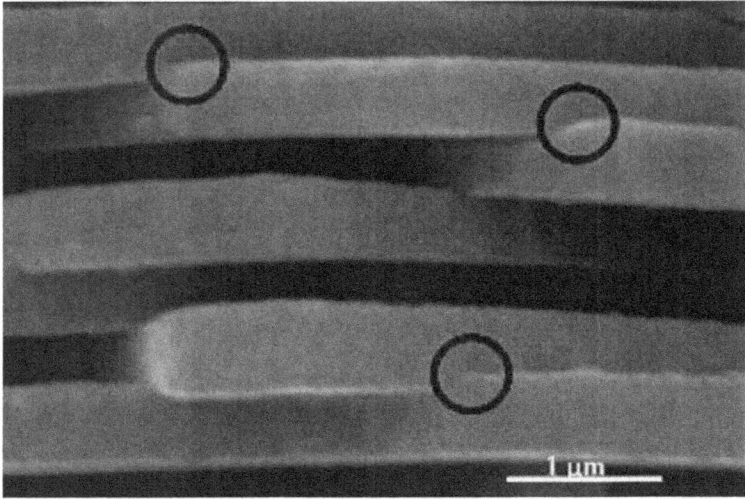

Figure 2.6   SEM image showing interlocks on the fracture surface of the platelets.[83]

between the substrate and overgrowth phase suggests an epitaxial mechanism of nucleation.[4,92] Furthermore, solid-state NMR investigations of nacre reveal the presence of an amorphous surface layer around the aragonite platelets, which has a thickness of 3–5 nm and which contains hydrogen carbonate groups.[93] Molecular modeling of the organic–mineral interactions has indicated the significant role of nonbonded interactions on the mechanical response of the organic material.[94,95]

2.2.3.1.4 Level 4: Lamellar Array Pattern. In the abalone shell, the inter-tablet boundaries form tessellated bands perpendicular to the lamellae boundaries; in other situations, such as in the pearl oyster, the inter-tablet boundaries are distributed randomly.[84] In fact, the crossed lamellar structure is the most widespread structure in mollusks, in which the lamellae are organized in more hierarchical levels.[96,97] For example, the crossed-lamellar layer of the *Strombus decorus persicus* seashell is composed of three sub-layers with differently oriented basic building blocks.[96] Each of the three crossed-lamellar sub-layers consists of the four orders of lamellar hierarchy, and each order of hierarchy is built of the lower-order blocks. The smallest building block of the crossed-lamellar layer, the 4th-order one, has a polygonal shape of approximate dimensions $100 \times 200 \times 100$ nm.[96]

2.2.3.1.5 Level 5: Mineral Mesolayers and Growth Bands. At the next hierarchical level, the structure is still "brick/mortar" like, in which the "brick" is the mineral mesolayer with a thickness of approximately 300 μm and the "mortar" is the organic material (so-called "growth bands") with a thickness of about 20 μm.[98] The periodic growth arrests create mesolayers that can also play an important role in the mechanical performance of nacre, acting as powerful crack deflectors.[99] The thin organic layers separating the meso-layers are thought to be formed upon shell growth in periods of less calcification.[84]

2.2.3.1.6 Level 6: Calcite Layer and Aragonite Layer. The longitudinal cross-section of nacre-containing shells, such as the abalone shell, displays two layers with distinct microstructures: a prismatic calcite layer (P) and an inner nacreous aragonite layer (N). Such an arrangement provides an optimal protective function of the shell:[82] the outer layer prevents penetration of the shell, while the nacreous layer is capable of dissipating mechanical energy through inelastic dissipation.[84]

Aragonite and calcite have very similar crystal structures; they also have the same composition. Both of them consist of alternating layers of calcium ions and carbonate ions perpendicular to the *c*-axis (in the *ab* plane). In aragonite, growth is preferred along the *c*-axis; thus, under conditions of normal temperature and pressure, aragonite forms as thin needles (acicular crystals) that do not generally grow into large crystals. Calcite, on the other hand, tends to form larger crystals, but they are very brittle and cleave easily along its {10.4} planes.[4]

2.2.3.1.7 Mechanical Advantages of the Hierarchical Structure. The basic order in the seashell build-up hierarchy may be of great importance to the superior mechanical properties of seashells owing to an enhanced ability for crack arrest at inter-lamellar boundaries.[96] Based on the hierarchical structure (Figure 2.7), toughening mechanisms at multiple length scales have been proposed to explain the significant improvement in performance. The negligible role of nano-asperities[100] and mineral bridges,[90,100–102] the significant role of interlocking,[103] the unfolding of protein molecules[104,105] and mineral–protein interaction,[94,95,106] as well as the brick morphology (waviness),[107] have been shown to significantly increase the toughness and fracture strength of nacre.

Additionally, each composition of nacre has a great influence on its properties. It was believed that the organic matrix was the key to nacre's fracture resistance.[108,109] Studies also show that loss of water in nacre did not influence the microstructure but caused a substantial decrease in stiffness. The nature of water in nacre itself has been investigated using FTIR.[110] An enhancement in toughness by three orders of magnitude and a two-fold increase in hardness have been reported for $CaCO_3$-built seashells compared with non-biogenic $CaCO_3$ crystals.[96]

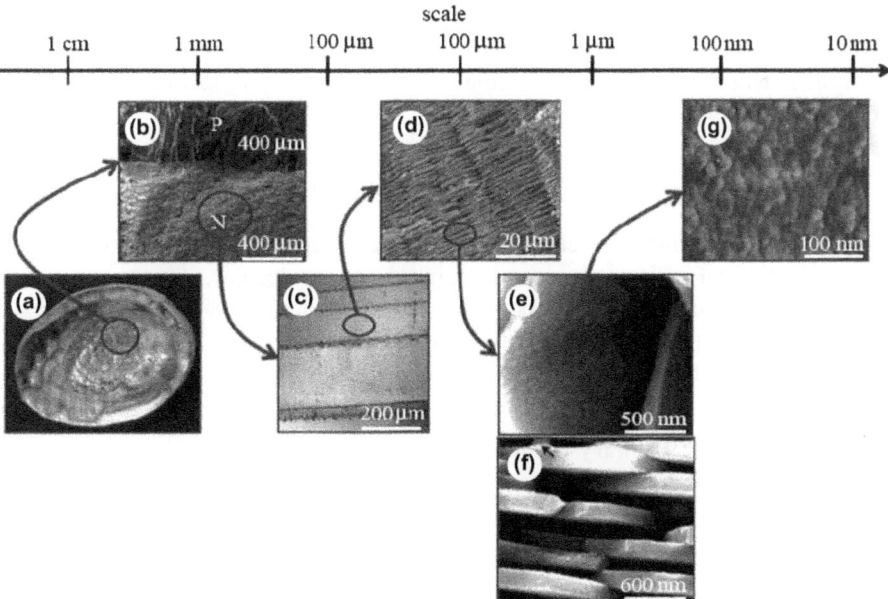

Figure 2.7 Hierarchical organization in seashells, showing at least six structural levels: (a) a bivalve shell; (b) the prismatic calcite layer (P) and nacre (N); (c)–(g) mesoscale, microscale and nanoscale structures.[84]

2.2.4   Spider Silk

Over the past 150 years, numerous studies have been conducted on spider silk and also many popular articles have been written about this amazing material. Spider silk can simultaneously exhibit a combination of tensile properties, including high breaking strength, high initial stiffness and high toughness;[111] especially the unusual combination of high strength and extensibility is a characteristic unavailable to date in synthetic materials and yet is attained in nature with a relatively simple protein processed from water.[112]

All spiders produce and use a variety of silks, but the orb-weaving araneid spiders appear to make the most diverse use of them and produce the orb-web.[113] Spiders produce at least seven different silks, each synthesized and spun by separate silk glands and spinnerets, and each with specific properties that appear to be optimized to perform different functional roles.[114] All of the silks are composed completely of protein and each undergoes an irreversible transition from a soluble to an insoluble form during processing,[115] but they have different mechanical properties because of different amino acid compositions and molecular organizations.[113]

Owing to its size and accessibility, major ampullate silk (produced by the major ampullate gland) has been the focus of most studies. The major ampullate silk, also called spider frame silk or dragline silk, constitutes the frame and radii of the orb-web as well as the dragline upon which the spider lowers itself. It is stronger per unit weight than high tensile steel, approaching the stiffness or strength of the super-strong polymeric material Kevlar, and also having an elasticity of up to 35%.[113] The exceptional mechanical properties of major ampullate silk have attracted much attention from scientists in various disciplines to investigate the molecular and structural origins of its mechanical properties.

2.2.4.1   Hierarchical Structure of Spider Silk

Nature is always instructively economical in its achievement of macroscopic material diversity through rearrangement of molecular, nanoscopic and microscopic building blocks. Like the previously described natural materials, spider silk also possesses a hierarchical structure (Figure 2.8) and here we present the five structural levels in detail.

2.2.4.1.1   Level 1: Primary Structure of the Proteins.   The primary structure of the proteins is the linear sequence of amino acids that form the molecule. The proteins are composed almost entirely of repetitive elements. Four types of shared amino acid motifs are observed in all of the proteins and they are recognized as amino acid sequence motifs: (1) GPGXX: GPGGX/GPGQQ; (2) Ala rich: poly-Ala/poly-Gly-Ala; (3) GGX; and (4) a "spacer" sequence with amino acids that do not conform to the typical amino acid composition of spider silks.[116] All members of the spider silk gene family that have been described at the protein sequence level share a limited set of amino acid motifs.[117]

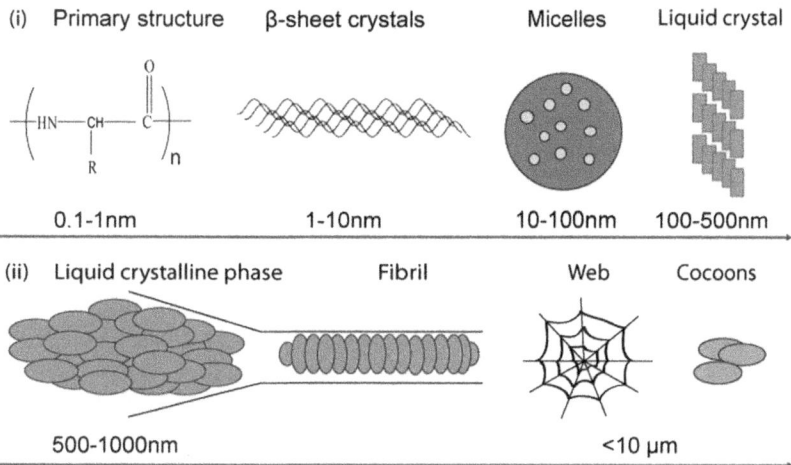

Figure 2.8   Hierarchical organization in a spider silk assembly. (i) Spider silk proteins consist of repeats of amino acid sequences as structural motifs that self-assemble into higher-level structures such as β-sheet crystals and elastic β-spirals. The β-sheet structures further assemble into soft micelles with the hydrophilic ends at the perimeter. With increasing protein concentration, micelles transform into metastable liquid crystalline structures. (ii) More ordered fibrils emerge from spinning ducts and form constructed webs or cocoons. (Adapted from Kluge *et al.*[128])

For example, spider major ampullate silk is composed of two proteins, major ampullate spidroin 1 (MaSp1) and major ampullate spidroin 2 (MaSp2). MaSp1 is further composed of poly-Ala/poly-Gly-Ala/GGX, where X typically stands for alanine, tyrosine, leucine or glutamine, and MaSp2 contains GPGGY/GPGQQ/poly-Ala.[115,118]

2.2.4.1.2   Level 2: Amino Acid Sequence Motifs and Secondary Structure.
Although it has been known for a long time that each of the seven glands produced proteins with unique amino acid compositions, it was only in the 1970s that the amino acid sequence motifs in spider fibroins were eluci-dated.[119] Further, it is only in the past few years that an understanding has emerged of the reasons for the unique mechanical properties that spider silk possesses. In particular, the proteins that comprise the silks and their sequen-ces have provided key information that relates directly to these properties.[114] Each amino acid motif will consistently construct specific secondary and tertiary structures in the fibre. The amino acids motifs are suggested to be structural modules, with large internal repeats flanked by shorter (~100 amino acid) terminal domains (N- and C-termini).[112]

Secondary structure describes the local spatial arrangement of main-chain atoms of any chosen segment of a polypeptide chain. The common types in proteins include α-helix, β-sheet and β-turn. Hayashi *et al.*[117] suggested that the different structures of the β-sheet regions could be the crystalline areas in the major ampullate silk fibre. With interlocks between adjacent chains, poly-Ala β-sheets have a higher binding energy than the poly-Gly-Ala regions, so that the tensile strength of major ampullate silk (poly-Ala) is greater than that of minor ampullate silk. Also, β-turn spirals made from periodically spaced proline residues impart extensibility to major ampullate and flagelliform silks. With at least 43 continuously linked β-turns in its spring-like spirals, flagelliform silk possess 200% extensibility compared to 35% extensibility of major ampullate silk, which has at most nine continuous β-turns.[117] All in all, sequence motifs such as polyalanine (polyA) and poly(alanine-glycine) (polyAG) (β-sheet-forming), GGX (3_1-helix), GXG (stiffness) and GPGXX (β-turn spiral) are key components in different silks whose relative positioning and arrangement are intimately tied with the end material properties.[112] Additionally, the C-terminal domain also has a role in the alignment of secondary structural features and acts as a molecular switch that controls fibre assembly.[120] Therefore, new materials with tailored properties are expected to be achieved by controlling the arrangement of these building block sequence motifs as well as terminal domains.

2.2.4.1.3 Level 3: Tertiary Structure: Crystal and Amorphous. Tertiary structure refers to the overall three-dimensional arrangement of all atoms in a protein. Guan proposed three different structural models for spider major ampullate silk (Figure 2.9).[121] In the first molecular model, nanocrystallites are dispersed in an amorphous matrix.[122] In the second model, highly oriented and weakly oriented β-sheets coexist.[123] In the third model, crystal β-sheet regions containing alanine (red lines) and glycine (blue lines) are

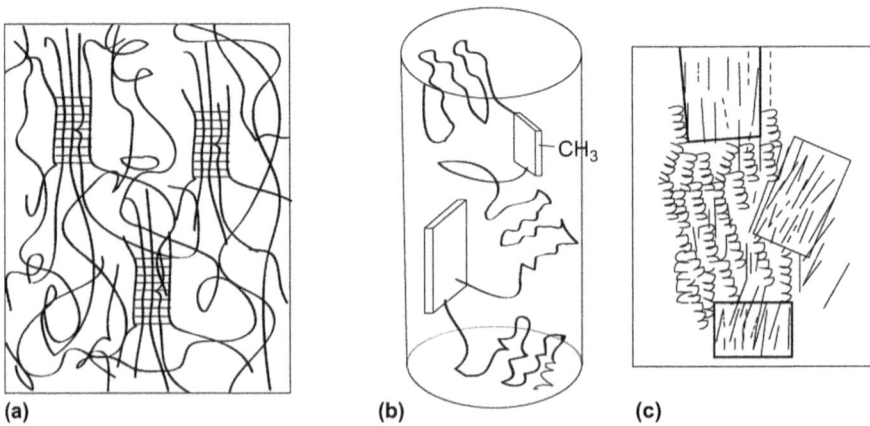

(a)   **(b)**   **(c)**

Figure 2.9   Different structural models proposed for spider major ampullate silk. (a) Nanocrystallites are dispersed in an amorphous matrix (proposed by Termonia). (b) Highly oriented and weakly oriented β-sheets coexist (proposed by Jelinski). (c) β-Sheet regions, containing alanine (solid lines) and glycine (dashed lines), interweaved with predominantly glycine-rich 3_1-helical parts (curls) (adapted from Guan[121]).

interweaved with the glycine-rich amorphous portion, which has pre-dominantly 3_1-helical parts (blue curls).[124]

It was also suggested that the spider silk fibres have a three-phase model which is composed of the well-oriented crystals and the isotropic amorphous material as well as the weakly oriented and partially ordered third phase.[125,126] As for the crystals, it is well established that the alternation of glycine with either alanine or serine causes this sequence spontaneously to form β-pleated sheet crystals through natural physical cross-linking. As the most heavily studied secondary structure of silks, crystalline β-sheets contribute to the high tensile strength of silk fibres. Wide-angle X-ray diffraction (WAXD) of major ampullate silk in *Nephila clavipes* indicates that the mean (minimum) crystal dimensions are approximately $2 \times 5 \times 7$ nm, the crystallinity is in the range 10–15%, and the crystals are strongly aligned with the fibre axis. However, Gosline *et al.* suggested that the total crystal content is higher than the 10–15% noted above, and predicted a crystal content of 20–25% based on mechanical tests.[127] In the semi-amorphous state, one-third is oriented while the rest is isotropic.[125] The semi-amorphous regions of silk are commonly made up of β-spirals and helical structures and thus provide silk with elasticity.[128] The amorphous chains, which interconnect the crystals, are estimated to be 16–20 amino acid residues long.[127] It was proposed that the poorly oriented crystallites may be important in effectively coupling the highly oriented crystalline domains and the amorphous regions to produce the exceptional material.[123]

Intermolecular forces are also used by major ampullate silk to self-assemble into nanocomposites composed of β-sheet nanocrystals imbedded in an amorphous matrix. Keten *et al.* used mechanical shearing simulations and illustrated that the nanoscale behavior of silk protein assemblies is controlled by the distinctly different secondary structure content and hydrogen bonding in the crystalline and semi-amorphous regions.[118] The fundamental stability criteria for the hydrogen-bonded interactions between water and the amide groups in protein chains was shown recently by using quantum mechanics simulations.[129] Although the primary, secondary and tertiary structures of spider silk have been known for decades, understanding the complete primary sequence, conformational structure of the repetitive amino acid motifs and their assembling mechanism has only just begun.[130,131]

2.2.4.1.4 Level 4: Liquid Crystalline Phase.   The liquid crystalline phase is an intermediate state during natural silk processing. Liquid crystalline spinning of spider silk offers desirable properties and makes it possible to efficiently spin a thread from molecules as large as silk proteins. Specifically, in the spider's gland and duct the molecules seem to form a nematic phase[132] or a cholesteric liquid crystalline phase[133] as the concentration is increased by evaporation of water. They form a substance that flows as a liquid but maintains some of the orientational order characteristics of a crystal, with the long axes of neighboring molecules aligned.[134] Observations reveal that the "rods" constituting the liquid crystalline phase of silk are not individual molecules or molecular segments, but instead arise at the supermolecular level. The rods are aggregates of essentially globular molecules, held together by non-covalent associations.[135]

Liquid crystalline spinning of spider silk stores protein dope molecules in a highly concentrated liquid crystalline state and prealigns the large molecules in the unspun dope, which may reduce the formation of defects, and then extends these in the spinning duct to form a supremely tough thread with minimal forces.[134]

2.2.4.1.5 Level 5: Fibril. After external drawdown of the liquid crystalline phase "spinning dope", the silk fibril forms. At the fibril level, in a cross-section the fibres have a circular profile, with a delicate triple-layered "coat–skin–core" structure in *Nephila* spider silk[136] that was observed as a skin–core duality structure.[137] This structure was also observed by AFM experiments.[138] The thin outer layer consists of higher electron density while the inner material has slightly lower electron density and numerous cavities of very low electron density.[136] Microfibrils enclosing fine channels might impart energy-dispersive properties by deflecting the tips of cracks forcing their way across the thread.[134]

2.2.4.1.6 Mechanical Advantages of the Hierarchical Structure. Spider major ampullate silk is one of the best biological materials known with respect to stiffness and strength (see Table 2.1). The web material is optimized to allow it to absorb a large amount of energy without breaking. The other major component of the orb-web, the viscid silk, is extremely extensible, breaking at extensions of 200% or more, and has a very low stiffness.[113] The hierarchical structure of the silk, *i.e.* the primary structure, the amino acid motif, the secondary structure, the tertiary structure and a skin–core duality structure in order, as well as the special liquid crystalline

Table 2.1 Mechanical properties of spider silks and some other fibre materials.

Material	Strength (MPa)	Elongation (%)
Major ampullate silk (dragline silk)[110]	4000	35
Minor ampullate silk[110]	1000	5
Flagelligorm silk (viscid silk)	1000	>200
Kevlar	4000	5
Rubber	1	600
E-glass	3445	—[a]
S2-glass	4890	—[a]
C fibre	6370	—[a]
SWNT	13 000–53 000	~23
Human hair	180	—[a]
Nylon 66	75	5

[a]Low values.

spinning process, is crucial in order for the silk to retain these exceptional mechanical properties.

Another unique feature of major ampullate silks is the supercontraction when exposed to water, and the process is reversible and repeatable. Super-contraction refers to the ability of the material contracting to a fraction of its original length in order to make the necessary space for the incoming solvent, and the result of the process is a swollen gel-like phase with a volume that can be 10 times larger than the dry network.[139] Unstrained silks can retract to about one-half of their original length in water. In this way the wetting-induced contraction of spider silk tightens the web whenever the humidity is very high and is able to restore the shape and tension of a slack web after deformation by precipitation, wind or prey.[140] Supercontraction was suggested to be the result of a rearrangement of the GPGXX motif within the silk proteins, where X represents one of a small subset of amino acids.[141] The mechanical properties change dramatically when the fibre is wet, owing to the formation of a hydrogen-bonded network in the amorphous phase and hydrophobic effects, as well as the fact that the pre-stress of the chains interconnecting the nanocrystals is released.[142,143]

2.3   Biomimetic Design and Fabrication

Mimicking biomimetic model structures has been extensively studied in the literature and many fabrication methods have been developed to duplicate the intricate structures of these biological systems. The characterization routes to the study of hierarchical structures as well as the duplication of these structures to achieve unique properties has been of much interest.

Attempts to mimic the nacre structure, primarily the brick and mortar architecture, are numerous. A self-assembly process that mimics the nano-laminated structure is reported in the literature that results in a 100-fold enhancement in properties using aragonite and biopolymers.[144] The flat pearl-like structure is also duplicated in silica and mica,[145] as is the layer-by-layer assembly.[146,147] The nacre structure has also inspired the design of hybrid materials based on natural polymers organized at the molecular level with inorganic solids such as clay and silicates.[148] Rapid assembly of inorganic tablets of materials such as talc has also been attempted, again inspired by nacre.[149] Nacre-like films have been designed using nanoclay sheets with soft polymer coatings with rapid self-assembly *via* simple processing routes.[150] Electrophoretic deposition techniques have been shown to be able to success-fully produce laminated high-quality structures and this technique has been used to make nacre-like structures using polymers and nanoclays.[151–153] Another strategy used is the template inhibition method that involves making films from solution, using soluble macromolecules anchored on insoluble matrices. This is called the template-inhibition strategy, and it consists of mineral deposition from solution onto a well-ordered two-dimensional struc-ture of a self-assembled film on solid or liquid substrates.[154] This techniques was also used to develop thin film crystals of $CaCO_3$ on chitosan.[155] Most of the current efforts on nacre mimicking have thus been targeted towards the laminated structures and the alternating hard and soft components.

Silk is considered as a biomimetic model due to a potential large range of mechanical properties, which are the result of both the semi-crystalline nanostructure and also the degree of hydration.[156] Silks are also being explored as materials for bio-microfluidics.[157] Spider silks have been investigated for a variety of biomaterial applications, such as substitutes for high-performance muscle,[158] and also several biomedical applications, including drug delivery, sutures and tissue engineering.[159–162] Further, silk-mimetic proteins have been fabricated using a variety of processing routes. These have included self-assembly of a number of silk-protein inspired copolymers, β-sheet-rich fibres or films from triblock copolymers, multiblock copolymers composed of poly(isoprene) and Ala5-spacer-Ala5, and brush copolymers from acrylonitrile and silk peptides.[163,164] An extensive review on composite materials based on silk proteins is reported by Hardy and Scheibel.[160]

Other important processing routes being investigated include the potential of genetically engineered polypeptides with tailored selectivity as means of self-assembly.[165–167] Often these biomimetics efforts, based on genetic engineering, result in potential for applications in distinctly different avenues, such as a recent innovation in textile materials based on engineered silk proteins.[168] These methodologies have a high potential to impact a large range of applications, from electronics to biomedicine.

Bone and teeth structures, although an inspiration, are not subjects of recent studies on replacement materials for these tissues; most studies for replacement of these hard tissues[169] are based on replacement of function rather than structure. Advances in tissue engineering[170] are likely to facilitate further progress, owing to advances in the knowledge of biomineralization routes as a result of biomimetic efforts.

2.4   Perspectives on Future of Biomimetic Technologies

Innovative techniques ranging from electrospinning, genetic pathways, spin coating, layer-by-layer deposition, dip coating, solvent evaporation, self-assembly and others are utilized to mimic many aspects of the organization of biological systems. Often, these attempts achieve a modest improvement in properties and are used for the design of composites for a variety of applications, ranging from photonic to biomedical. Other studies have evaluated the key roles of structural nuances of biological materials, such as the interlocks in nacre, on the properties. Hence an interest in the design of the next generation of composites that truly mimic biosystems would be of use for combination with modeling routes to identify key elements of the structure in biosystems that are particularly relevant to enhancement in properties. Current studies on modeling the properties of these hierarchical structures have also benefitted from advancements of computational resources. In addition, unique and advanced characterization tools such as high-resolution electron microscopy, photoacoustic spectroscopy and atomic force microscopy reveal unprecedented details of structure. A combination of modeling and characterization abilities are expected to strengthen efforts in biomimicking in the future.

References

1. P. Fratzl and R. Weinkamer, *Prog. Mater. Sci.*, 2007, **52**, 1263.
2. R. Lakes, *Nature*, 1993, **361**, 511.
3. S. Weiner and H. D. Wagner, *Annu. Rev. Mater. Sci.*, 1998, **28**, 271.
4. S. Weiner and L. Addadi, *J. Mater. Chem.*, 1997, **7**, 689.
5. D. R. Katti, S. M. Pradhan and K. S. Katti, *J. Biomech.*, 2010, **43**, 1723.
6. P. Fratzl, H. S. Gupta, E. P. Paschalis and P. Roschger, *J. Mater. Chem.*, 2004, **14**, 2115.
7. R. B. Martin and D. B. Burr, *Structure, Function, and Adaptation of Compact Bone*, Raven Press, New York, 1989.
8. W. I. Schmidt, *Naturwissenschaften*, 1936, 24.
9. V. Baranauskas, B. C. Vidal and N. A. Parizotto, *Appl. Biochem. Biotechnol.*, 1998, **69**, 91.
10. D. R. Baselt, J. P. Revel and J. D. Baldeschwieler, *Biophys. J.*, 1993, **65**, 2644.
11. L. M. Siperko and W. J. Landis, *J. Struct. Biol.*, 2001, **135**, 313.
12. S. Weiner and H. D. Wagner, *Annu. Rev. Mater. Sci.*, 1998, **28**, 271.
13. J. Currey, *The Mechanical Adaptions of Bones*, Princeton University Press, Princeton, 1984.
14. G. H. Bourne, *The Biochemistry and Physiology of Bone*, Academic Press, New York, 1972.
15. P. A. Timmins and J. C. Wall, *Calcif. Tissue Res.*, 1977, **23**, 1.
16. P. Fratzl, *Nat. Mater.*, 2008, **7**, 610.
17. M. M. Giraudguille, *Calcif. Tissue Int.*, 1988, **42**, 167.
18. S. M. Pradhan, D. R. Katti and K. S. Katti, *J. Nanomech. Micromech.*, 2011, **1**, 7.
19. M. M. Giraud-Guille, *Curr. Opin. Solid State Mater. Sci.*, 1998, **3**, 221.
20. K. E. Kadler, D. F. Holmes, J. A. Trotter and J. A. Chapman, *Biochem. J.*, 1996, **316**, 1.
21. E. P. Paschalis, K. Verdelis, S. B. Doty, A. L. Boskey, R. Mendelsohn and M. Yamauchi, *J. Bone Mineral Res.*, 2001, **16**, 1821.
22. M. L. Watson and R. A. Robinson, *Am. J. Anat.*, 1953, **93**, 25.
23. R. A. Robinson and M. L. Watson, *Anat. Rec.*, 1952, **114**, 383.
24. W. J. Landis, K. J. Hodgens, J. Arena, M. J. Song and B. F. McEwen, *Microsc. Res. Tech.*, 1996, **33**, 192.
25. S. Lees, K. S. Prostak, V. K. Ingle and K. Kjoller, *Calcif. Tissue Int.*, 1994, **55**, 180.
26. S. Lees and E. A. Page, *Connective Tissue Res.*, 1992, **28**, 263.
27. G. Marotti and S. Lees, *Calcif. Tissue Int.*, 1993, **53**, S47.

28. V. Ziv, I. Sabanay, T. Arad, W. Traub and S. Weiner, *Microsc. Res. Tech.*, 1996, **33**, 203.
29. H. J. Gao, B. H. Ji, I. L. Jager, E. Arzt and P. Fratzl, *Proc. Natl. Acad. Sci. U. S. A.*, 2003, **100**, 5597.
30. N. M. A. Board, *Hierarchical Structures in Biology as a Guide for New Materials Technology*, National Academy of Science, Washington, 1994.
31. M. L. K. Tate, *J. Biomech.*, 2011, **44**, 304.
32. A. Gautieri, A. Russo, S. Vesentini, A. Redaelli and M. J. Buehler, *J. Chem. Theory Comput.*, 2010, **6**, 1210.
33. D. K. Dubey and V. Tomar, *Mater. Sci. Eng., C*, 2009, **29**, 2133.
34. D. K. Dubey and V. Tomar, *Acta Biomater.*, 2009, **5**, 2704.
35. D. K. Dubey and V. Tomar, *J. Phys.: Condens. Matter*, 2009, **21**, 205103.
36. M. J. Buehler, *J. Comput. Theor. Nanosci.*, 2006, **3**, 603.
37. A. Fritsch and C. Hellmich, *J. Theor. Biol.*, 2007, **244**, 597.
38. J. Y. Rho, T. Y. Tsui and G. M. Pharr, *Biomaterials*, 1997, **18**, 1325.
39. R. Murugan and S. Ramakrishna, *Compos. Sci. Technol.*, 2005, **65**, 2385.
40. W. J. Landis, *Bone*, 1995, **16**, 533.
41. V. Ziv and S. Weiner, *Connective Tissue Res.*, 1994, **30**, 165.
42. B. H. Ji and H. J. Gao, *Compos. Sci. Technol.*, 2006, **66**, 1212.
43. B. H. Ji and H. J. Gao, *Mater. Sci. Eng., A*, 2004, **366**, 96.
44. B. H. Ji and H. J. Gao, *J. Mech. Phys. Solids*, 2004, **52**, 1963.
45. E. Pena, J. A. Pena and M. Doblare, *J. Biomech.*, 2008, **41**, 2659.
46. C. Hellmich, J. F. Barthelemy and L. Dormieux, *Eur. J. Mech., A: Solids*, 2004, **23**, 783.
47. J. L. Katz, A. Misra, P. Spencer, Y. Wang, S. Bumrerraj, T. Nomura, S. J. Eppell and M. Tabib-Azar, *Mater. Sci. Eng., C*, 2007, **27**, 450.
48. T. Nomura, E. Gold, M. P. Powers, S. Shingaki and J. L. Katz, *Dent. Mater.*, 2003, **19**, 167.
49. T. Siegmund, M. R. Allen and D. B. Burr, *J. Biomech.*, 2008, **41**, 1427.
50. R. Bhowmik, K. S. Katti and D. R. Katti, *J. Eng. Mech. Div., Am. Soc. Civ. Eng.*, 2009, **135**, 413.
51. R. Bhowmik, K. Katti and D. Katti, *J. Mater. Sci.*, 2007, **42**, 8795.
52. G. E. Fantner, T. Hassenkam, J. H. Kindt, J. C. Weaver, H. Birkedal, L. Pechenik, J. A. Cutroni, G. A. G. Cidade, G. D. Stucky, D. E. Morse and P. K. Hansma, *Nat. Mater.*, 2005, **4**, 612.
53. D. K. Dubey and V. Tomar, *J. Mat. Sci.: Mater. Med.*, 2010, **21**, 161.
54. D. K. Dubey and V. Tomar, *Ann. Biomed. Eng.*, 2010, **38**, 2040.
55. D. K. Dubey and V. Tomar, *Mech. Res. Commun.*, 2008, **35**, 24.
56. P. W. Lucas, *Dental Functional Morphology: How Teeth Work*, Cambridge University Press, Cambridge, 2004.
57. B. R. Lawn, J. J. W. Lee and H. Chai, *Annu. Rev. Mater. Res.*, 2010, **40**, 55.
58. M. C. Maas and E. R. Dumont, *Evol. Anthropol.*, 1999, **8**, 133.
59. C. Robinson, H. D. Briggs, P. J. Atkinson and J. A. Weatherell, *J. Dent. Res.*, 1979, **58**, 871.
60. M. L. Snead, D. H. Zhu, Y. P. Lei, S. N. White, C. M. Snead, W. Luo and M. L. Paine, *Mater. Sci. Eng., C*, 2006, **26**, 1296.

61. J. C. Elliott, *Dent. Enamel*, 1997, **205**, 54.
62. W. Vonkoenigswald and W. A. Clemens, *Scanning Microsc.*, 1992, **6**, 195.
63. K. Suzuki, T. Sakae and Y. Kozawa, *Connective Tissue Res.*, 1998, **38**, 113.
64. R. P. Apkarian, M. D. Gutekunst and D. C. Joy, *J. Electron Microsc. Tech.*, 1990, **14**, 70.
65. J. P. Simmer and A. G. Fincham, *Crit. Rev. Oral Biol. Med.*, 1995, **6**, 84.
66. C. Robinson, J. Kirkham, S. J. Brookes, W. A. Bonass and R. C. Shore, *Int. J. Dev. Biol.*, 1995, **39**, 145.
67. A. G. Fincham, J. Moradian-Oldak and J. P. Simmer, *J. Struct. Biol.*, 1999, **126**, 270.
68. J. D. Bartlett, B. Ganss, M. Goldberg, J. Moradian-Oldak, M. L. Paine, M. L. Snead, X. Wen, S. N. White and Y. L. Zhou, *Curr. Top. Dev. Biol.*, 2006, **74**, 57.
69. H. C. Margolis, E. Beniash and C. E. Fowler, *J. Dent. Res.*, 2006, **85**, 775.
70. D. Carlstrom, J. E. Glas and B. Angmar, *J. Ultrastruct. Res.*, 1963, **8**, 24.
71. J. H. Kinney, J. A. Pople, G. W. Marshall and S. J. Marshall, *Calcif. Tissue Int.*, 2001, **69**, 31.
72. E. Johansen, *J. Dent. Res.*, 1964, **43**, 1007.
73. E. Johansen and H. F. Parks, *Arch. Oral Biol.*, 1962, **7**, 185.
74. G. W. Marshall, S. J. Marshall, J. H. Kinney and M. Balooch, *J. Dent.*, 1997, **25**, 441.
75. J. H. Kinney, S. J. Marshall and G. W. Marshall, *Crit. Rev. Oral Biol. Med.*, 2003, **14**, 13.
76. P. Ungar, *Nature*, 2008, **452**, 703.
77. M. A. Meyers, P. Y. Chen, A. Y. M. Lin and Y. Seki, *Prog. Mater. Sci.*, 2008, **53**, 1.
78. M. Sarikaya, K. E. Gunnison, M. Yasrebi, D. L. Milius and I. A. Aksay, *Mat. Res. Soc. Symp. Proc.*, 1990, **174**, 109.
79. J. D. Currey, *Proc. R. Soc. London, Ser. B*, 1977, **196**, 443.
80. A. P. Jackson, J. F. V. Vincent and R. M. Turner, *Proc. R. Soc. London, Ser. B*, 1988, **234**, 415.
81. A. P. Jackson, J. F. V. Vincent and R. M. Turner, *J. Mater. Sci.*, 1990, **25**, 3173.
82. M. Sarikaya, *Microsc. Res. Tech.*, 1994, **27**, 360.
83. K. S. Katti, D. R. Katti, S. M. Pradhan and A. Bhosle, *J. Mater. Res.*, 2005, **20**, 1097.
84. G. M. Luz and J. F. Mano, *Philos. Trans. R. Soc. London, Ser. A*, 2009, **367**, 1587.
85. B. Mohanty, K. S. Katti, D. R. Katti and D. Verma, *J. Mater. Res.*, 2006, **21**, 2045.
86. X. D. Li, W. C. Chang, Y. J. Chao, R. Z. Wang and M. Chang, *Nano Lett.*, 2004, **4**, 613.
87. P. Stempfle, O. Pantale, M. Rousseau, E. Lopez and X. Bourrat, *Mater. Sci. Eng., C*, 2010, **30**, 715.
88. M. Rousseau, E. Lopez, P. Stempfle, M. Brendle, L. Franke, A. Guette, R. Naslain and X. Bourrat, *Biomaterials*, 2005, **26**, 6254.

89. L. Addadi, D. Joester, F. Nudelman and S. Weiner, *Chem.–Eur. J.*, 2006, **12**, 981.
90. F. Song and Y. L. Bai, *J. Mater. Res.*, 2003, **18**, 1741.
91. R. Z. Wang, Z. Suo, A. G. Evans, N. Yao and I. A. Aksay, *J. Mater. Res.*, 2001, **16**, 2485.
92. Y. Levi-Kalisman, G. Falini, L. Addadi and S. Weiner, *J. Struct. Biol.*, 2001, **135**, 8.
93. C. Jager and H. Colfen, *CrystEngComm*, 2007, **9**, 1237.
94. P. Ghosh, D. R. Katti and K. S. Katti, *J. Nanomater.*, 2008, 8.
95. P. Ghosh, D. R. Katti and K. S. Katti, *Biomacromolecules*, 2007, **8**, 851.
96. B. Pokroy and E. Zolotoyabko, *J. Mater. Chem.*, 2003, **13**, 682.
97. R. Menig, M. H. Meyers, M. A. Meyers and K. S. Vecchio, *Mater. Sci. Eng., A*, 2001, **297**, 203.
98. A. Lin and M. A. Meyers, *Mater. Sci. Eng., A*, 2005, **390**, 27.
99. R. Menig, M. H. Meyers, M. A. Meyers and K. S. Vecchio, *Acta Mater.*, 2000, **48**, 2383.
100. K. R. Katti, S. M. Pradhan and K. S. Katti, *Rev. Adv. Mater. Sci.*, 2004, **6**, 162.
101. K. Tushtev, M. Murck and G. Grathwohl, *Mater. Sci. Eng., C*, 2008, **28**, 1164.
102. D. R. Katti, K. S. Katti, J. M. Sopp and M. Sarikaya, *Comput. Theor. Polym. Sci.*, 2001, **11**, 397.
103. K. S. Katti and D. R. Katti, *Mater. Sci. Eng., C*, 2006, **26**, 1317.
104. B. L. Smith, T. E. Schaffer, M. Viani, J. B. Thompson, N. A. Frederick, J. Kindt, A. Belcher, G. D. Stucky, D. E. Morse and P. K. Hansma, *Nature*, 1999, **399**, 761.
105. P. Nukala and S. Simunovic, *Biomaterials*, 2005, **26**, 6087.
106. P. Ghosh, D. R. Katti and K. S. Katti, *Mater. Manuf. Process.*, 2006, **21**, 676.
107. H. D. Espinosa, A. L. Juster, F. J. Latourte, O. Y. Loh, D. Gregoire and P. D. Zavattieri, *Nat. Commun.*, 2011, **2**, 9.
108. F. Song and Y. L. Bai, *Acta Mech. Sin.*, 2001, **17**, 251.
109. N. M. Neves and J. F. Mano, *Mater. Sci. Eng., C*, 2005, **25**, 113.
110. D. Verma, K. Katti and D. Katti, *Spectrochim. Acta, Part A*, 2007, **67**, 784.
111. C. Viney, *J. Text. Inst.*, 2000, **91**, 2.
112. F. G. Omenetto and D. L. Kaplan, *Science*, 2010, **329**, 528.
113. J. M. Gosline, M. E. Demont and M. W. Denny, *Endeavour*, 1986, **10**, 37.
114. R. V. Lewis, *Chem. Rev.*, 2006, **106**, 3762.
115. M. Xu and R. V. Lewis, *Proc. Natl. Acad. Sci. U. S. A.*, 1990, **87**, 7120.
116. P. A. Guerette, D. G. Ginzinger, B. H. F. Weber and J. M. Gosline, *Science*, 1996, **272**, 112.
117. C. Y. Hayashi, N. H. Shipley and R. V. Lewis, *Int. J. Biol. Macromol.*, 1999, **24**, 271.
118. S. Keten and M. J. Buehler, *J. R. Soc., Interface*, 2010, **7**, 1709.
119. S. O. Anderson, *Comp. Biochem. Physiol.*, 1970, **35**, 705.

120. F. Hagn, L. Eisoldt, J. G. Hardy, C. Vendrely, M. Coles, T. Scheibel and H. Kessler, *Nature*, 2010, **465**, 239.
121. Z. B. Guan, *Polym. Int.*, 2007, **56**, 467.
122. Y. Termonia, *Macromolecules*, 1994, **27**, 7378.
123. A. H. Simmons, C. A. Michal and L. W. Jelinski, *Science*, 1996, **271**, 84.
124. J. D. van Beek, S. Hess, F. Vollrath and B. H. Meier, *Proc. Natl. Acad. Sci. U. S. A.*, 2002, **99**, 10266.
125. D. T. Grubb and L. W. Jelinski, *Macromolecules*, 1997, **30**, 2860.
126. Z. Yang, D. T. Grubb and L. W. Jelinski, *Macromolecules*, 1997, **30**, 8254.
127. J. M. Gosline, P. A. Guerette, C. S. Ortlepp and K. N. Savage, *J. Exp. Biol.*, 1999, **202**, 3295.
128. J. A. Kluge, U. Rabotyagova, G. G. Leisk and D. L. Kaplan, *Trends Biotechnol.*, 2008, **26**, 244.
129. D. Porter and F. Vollrath, *Soft Matter*, 2008, **4**, 328.
130. J. E. Trancik, J. T. Czernuszka, D. J. H. Cockayne and C. Viney, *Polymer*, 2005, **46**, 5225.
131. T. Izdebski, P. Akhenblit, J. E. Jenkins, J. L. Yarger and G. P. Holland, *Biomacromolecules*, 2010, **11**, 168.
132. K. Kerkam, C. Viney, D. Kaplan and S. Lombardi, *Nature*, 1991, **349**, 596.
133. P. J. Willcox, S. P. Gido, W. Muller and D. L. Kaplan, *Macromolecules*, 1996, **29**, 5106.
134. F. Vollrath and D. P. Knight, *Nature*, 2001, **410**, 541.
135. D. Kaplan, W. W. Adams, B. Farmer and C. Viney, *Silk Polymers: Materials Science and Biotechnology*, American Chemical Society, Washington, 1993.
136. S. Frische, A. B. Maunsbach and F. Vollrath, *J. Microsc. (Oxford, U. K.)*, 1998, **189**, 64.
137. R. W. Work, *Trans. Am. Microsc. Soc.*, 1984, **103**, 113.
138. S. F. Y. Li, A. J. McGhie and S. L. Tang, *Biophys. J.*, 1994, **66**, 1209.
139. J. D. van Beek, J. Kummerlen, F. Vollrath and B. H. Meier, *Int. J. Biol. Macromol.*, 1999, **24**, 173.
140. R. W. Work, *J. Exp. Biol.*, 1985, **118**, 379.
141. C. Boutry and T. A. Blackledge, *J. Exp. Biol.*, 2010, **213**, 3505.
142. R. Ene, P. Papadopoulos and F. Kremer, *Polymer*, 2010, **51**, 4784.
143. R. Ene, P. Papadopoulos and F. Kremer, *Soft Matter*, 2009, **5**, 4568.
144. A. Sellinger, P. M. Weiss, A. Nguyen, Y. F. Lu, R. A. Assink, W. L. Gong and C. J. Brinker, *Nature*, 1998, **394**, 256.
145. M. Fritz, A. M. Belcher, M. Radmacher, D. A. Walters, P. K. Hansma, G. D. Stucky, D. E. Morse and S. Mann, *Nature*, 1994, **371**, 49.
146. P. Podsiadlo, A. K. Kaushik, B. S. Shim, A. Agarwal, Z. Y. Tang, A. M. Waas, E. M. Arruda and N. A. Kotov, *J. Phys. Chem. B*, 2008, **112**, 14359.
147. P. Podsiadlo, S. Paternel, J. M. Rouillard, Z. F. Zhang, J. Lee, J. W. Lee, L. Gulari and N. A. Kotov, *Langmuir*, 2005, **21**, 11915.
148. E. R. Ruiz-Hitzky, M. Darder and P. Aranda, *J. Mater. Chem.*, 2005, **15**, 3650.

149. N. Almqvist, N. H. Thomson, B. L. Smith, G. D. Stucky, D. E. Morse and P. K. Hansma, *Mater. Sci. Eng., C*, 1999, **7**, 37.
150. A. Walther, I. Bjurhager, J. M. Malho, J. Pere, J. Ruokolainen, L. A. Berglund and O. Ikkala, *Nano Lett.*, 2010, **10**, 2742.
151. B. Long, C. A. Wang, Y. Huang and J. L. Sun, *Rare Metal Mater. Eng.*, 2007, **36**, 844.
152. X. Y. Wang, Y. M. Du and J. W. Luo, *Nanotechnology*, 2008, 19.
153. B. Long, C. A. Wang, W. Lin, Y. Huang and J. L. Sun, *Compos. Sci. Technol.*, 2007, **67**, 2770.
154. C. M. Li and D. L. Kaplan, *Curr. Opin. Solid State Mater. Sci.*, 2003, **7**, 265.
155. T. Kato and T. Amamiya, *Chem. Lett.*, 1999, 199.
156. D. Porter and F. Vollrath, *Adv. Mater.*, 2009, **21**, 487.
157. P. Domachuk, K. Tsioris, F. G. Omenetto and D. L. Kaplan, *Adv. Mater.*, 2010, **22**, 249.
158. I. Agnarsson, A. Dhinojwala, V. Sahni and T. A. Blackledge, *J. Exp. Biol.*, 2009, **212**, 1989.
159. J. G. Hardy and T. R. Scheibel, *Biochem. Soc. Trans.*, 2009, **37**, 677.
160. J. G. Hardy and T. R. Scheibel, *Prog. Polym. Sci.*, 2010, **35**, 1093.
161. H. J. Kim, U. J. Kim, H. S. Kim, C. M. Li, M. Wada, G. G. Leisk and D. L. Kaplan, *Bone*, 2008, **42**, 1226.
162. B. D. Lawrence, J. K. Marchant, M. A. Pindrus, F. G. Omenetto and D. L. Kaplan, *Biomaterials*, 2009, **30**, 1299.
163. J. M. Smeenk, M. B. J. Otten, J. Thies, D. A. Tirrell, H. G. Stunnenberg and J. C. M. van Hest, *Angew. Chem. Int. Ed.*, 2005, **44**, 1968.
164. J. M. Smeenk, P. Schon, M. B. J. Otten, S. Speller, H. G. Stunnenberg and J. C. M. van Hest, *Macromolecules*, 2006, **39**, 2989.
165. C. Tamerler and M. Sarikaya, *Philos. Trans. R. Soc. London, Ser. A*, 2009, **367**, 1705.
166. C. Tamerler and M. Sarikaya, *Acta Biomater.*, 2007, **3**, 289.
167. M. Sarikaya, C. Tamerler, A. K. Y. Jen, K. Schulten and F. Baneyx, *Nat. Mater.*, 2003, **2**, 577.
168. V. Rossbach, P. Patanathabutr and J. Wichitwechkarn, *Fibers Polym.*, 2003, **4**, 8.
169. K. S. Katti, *Colloids Surf., B*, 2004, **39**, 133.
170. R. Langer and J. P. Vacanti, *Science*, 1993, **260**, 920.

CHAPTER 3

Natural Fibres as Composite Reinforcement Materials: Description and New Sources

KARINE CHARLET

Clermont Université, IFMA, Institut Pascal, BP10448, 63000 Clermont Ferrand, France;
CNRS, UMR 6602, Institut Pascal, 63177 Aubière Cedex, France
Email: karine.charlet@ifma.fr

3.1 Introduction

Traditionally used in textile industries, natural vegetable fibres are increasingly used or pinpointed as having potential in effective reinforcement materials of polymer matrices. They exhibit several advantages, which explains the growing interest of composite manufacturers (mainly in automobile and building industries), wishing to replace the more commonly used glass fibres.[1-3] A recent survey revealed that about 90% of the fibres used as composite reinforcement in Europe were glass fibres and, at most, only 3% were natural fibres.[4] Another research study claims that almost 40 million tons of natural cellulose fibres could be obtained from already available agricultural by-products, while the current total fibre consumption in the world is about 70 million tonnes.[5] Moreover, it has been estimated that substituting synthetic fibres by natural fibres in automotive composite parts would reduce the material weight by 30% and their cost by 20%.[6] According to the Canadian manager of the national program on bio-products, the replacement of only 3.5 kg of glass fibres by natural fibres in each car produced in North America would reduce greenhouse

RSC Green Chemistry No. 16
Natural Polymers, Volume 1: Composites
Edited by Maya J John and Thomas Sabu
© The Royal Society of Chemistry 2012
Published by the Royal Society of Chemistry, www.rsc.org

gas emissions by at least 500 million tonnes per year.[7] This strategy is encouraged by many governments worldwide through new legislation that deals with the end-of-life handling of materials. For example, in Europe, the deposition fraction of a vehicle is to be limited to 5% of its weight by 2015.[8] In 2006, already 19 European countries had managed to reuse or recycle 80% of a car.[9] This promotes the use of natural and degradable fibres in composite materials instead of synthetic and inert ones. This substitution is likely to be feasible at a reasonable cost provided the processes and equipment used to produce fibre reinforced composites would be the same. However, these environmentally friendly materials also present some drawbacks that still limit their use.

In this chapter, we will only focus on vegetable fibres (except wood), as protein fibres will be discussed in another part. After a brief description of the benefits and the difficulties brought about by the use of such fibres in composite materials, a survey carried out on new sources of cellulosic fibres is presented. This gathers information on natural fibres for which investigations as composite reinforcements began only a few years ago and already reveal promising characteristics (isora, vakka, artichoke, celery, switchgrass, *etc.*). This part is presented as a comparison of these fibres with those of more "classical" fibres, *i.e.* fibres that have been under study for a couple of years (flax, hemp, sisal, jute, bamboo, ramie, cotton, *etc.*). The objective is to show that the field of research of suitable, available and effective fibres expected to improve the properties of polymers is far from being limited.

3.2 Advantages and Drawbacks of Natural Fibres as Composite Reinforcement Materials

The first advantage of natural fibres is that they are available worldwide, depending on the species concerned.[10] In North America, the main produced fibres come from the growing of cotton, hemp, flax, sorghum and switchgrass, whereas in South America, sisal, abaca, curaua, pineapple, bagasse and ramie are the most harvested vegetable fibres. Some African countries are well-known for the production of kenaf, okra, sisal or rhectophyllum fibres. In Europe, hemp and flax are the most developed forms of non-edible plants, but also sorghum and china reed. Asia, thanks to the large variety of climates it offers, gathers several types of vegetable fibre harvests: bamboo, abaca, sisal, kenaf, ramie, flax, hemp, coir, jute, cotton, isora, vakka, okra and china reed are available more or less locally in this part of the world. The worldwide extent of vegetable fibre growing enables their cost to be relatively low (Table 3.1).

In parallel to these fibres, by-products resulting from traditional harvests or even weeds have been pinpointed as possible local sources of cellulose fibres, such as hop stems, nettle stems, cornstalks, artichoke stems, cotton stalks, soybean straws, velvetleaf stems, *etc.* The inclusion in composites of fibres that are currently considered as waste products brings the advantage of limiting the required disposal areas and of giving value to products that were previously worth almost nothing. Table 3.1 summarizes the production data from various survey reports on vegetable fibres.[10–19]

The added value that can result from the use and exploitation of such natural fibres in polymer reinforcement is not only economic; it may also encourage the development of local enterprises (and thus, of rural areas) that possess the know-how in terms of extraction and treatment of these fibres. It may also generate a (sometimes new) agricultural-based economy, especially when dealing with by-product promotion.[20]

According to many parameters, such as the species or the position in the plant, natural fibres generally exhibit large aspect ratios (defined as length/ diameter) that can vary from a few score, notably for by-products, to several thousands for nettle, ramie or flax fibres (Table 3.2).[15,21–45] This particularity has an important influence on the mechanical properties of the derived composites,[46,47] as they provide a quasi-continuous reinforcement within the composites: this enables a more homogeneous charge transfer between the matrix and the fibres and hinders possible crack propagation. It is generally considered that the minimum fibre aspect ratio needed to achieve optimum composite properties must range between 100 and 200.[48] These relatively long fibres can also be cut to a defined length to ease the processing of composites, such as injection.

Owing to the lightness of their constituents (mainly sugar polymers) and to their microstructure, the density of natural fibres is relatively low (circa 1.4–1.5) compared, for example, to that of E-glass fibres (Table 3.3).[11,15,16,21–23,33,36,39–41,49–62]

Table 3.1 Production and cost of natural fibres and of glass fibres (approximate values for raw material, largely dependent on period, country and stock size).

Fibre	Production (Mt)	Year	Cost ($/kg)[a]	Ref.
Abaca	0.07	2007	1.5–2.5	11,12
Coir	1.0	2005	0.2–0.4	13
Cotton	21.2	2003	1.5–2.2	10
Curaua	0.00015	2003	n.a.	14
E-glass	2.2	2003	1.2–1.8	10,11,15
Flax	0.75	2004	0.5–1.5	10,11
Flax	1.0	2007	n.a.	12
Hemp	0.08	2004	0.6–1.8	10,11
Hop	0.1	2007	n.a.	12
Jute	2.77	2004	0.35	10,11,16
Pineapple (Brazil)	0.02	2000s	n.a.	15
Ramie	0.27	2004	1.5–2.4	10
Rice straw	10	2006	n.a.	17
Sisal	0.31	2003	0.6	10,11
Sisal	0.23	2007	n.a.	13
Switchgrass	n.a.	1999	0.06–0.08	18
Vegetable fibres[b]	5.0	2003	0.2–1	10,15
Wheat straw	0.6	2004	n.a.	19

[a]n.a. = not available.
[b]Except cotton, wood and straws.

Table 3.2   Dimensions (mean value and/or [range]) and aspect ratio of natural fibres.

Fibre	Diameter, d (μm)	Length, L (mm)	Aspect ratio (L/d)	Ref.
Abaca	20 [6–46]	[2–12]	350	21,22
Artichoke	–	[100–160]	–	23
Bamboo	14 [7–40]	2.7 [1.5–4.4]	190	22,24
Banana	[11–81]	[0.9–5.5]	75	15,21,22
Bark of cotton stalk	[10–20]	[1–2]	50–200	25–27
Celery	260	–	–	28
Coir	[10–24]	[0.3–1.2]	45	15,21,22,29,30
Cornstalk	27	0.8	30	31
Cotton	20 [10–45]	20 [2–64]	1000	15,21,22,29
Curaua	60 [20–130]	–	–	32,33
Flax	19 [5–76]	33 [4–140]	1700	21,22,24,34
Hemp	25 [3–51]	25 [8.3–55]	1000	10,24
Hop stems	16.5	2	120	35
Isora	10.1	1	100	36
Jute	20 [5–30]	2 [1–6]	100	15,22,24
Kapok	19 [10–35]	19 [7–35]	1000	22,24
Kenaf	21 [12–50]	5 [1.5–11]	240	21,22,24
Miscanthus	13.7	0.552	40	37
Nettle	20 [10–126]	50 [2–87]	2500	22,38,39
Pineapple leaf	20 [7–80]	8 [3–10]	400	22,24,40
Ramie	50 [5–126]	120 [40–260]	2400	15,21,22,24
Rhectophyllum	174 [100–250]	6 [5–8]	35	41
Rice straw	350	1.53	6	17
Sisal	20 [4–50]	3 [0.5–8]	150	15,21,22,24
Sorghum	[8–15]	1.4 [0.4–3.4]	130	24
Soybean straw	15.6	1.5	100	42
Switchgrass	20 [10–30]	0.24	120	43,44
Velvet leaf	11.4	0.94	82	45

Actually, cellulose, their main constituent, has a specific weight around $1.56 \, \mathrm{g \, cm^{-3}}$, which varies according to the relative humidity.[63–65] This slight drop in density between the main constituent and the fibre can be explained by the presence of lighter impurities (such as waxes and oils[64]), but especially by their porosity mainly due to the lumen, an inner cavity that contains the cytoplasm during the fibre life but which becomes empty when dead. This is why we can find in literature either "bulk" or "apparent" density, which is a global measurement of the fibre specific weight, and "real" or "absolute" density, which only takes into account the weight of the cell walls. The presence of the lumen is also responsible for the excellent properties of natural fibres as thermal and acoustic insulators.[15,66–69] This lightness brings about advantages not only during the fabrication stage of the derived composites, especially during transportation, but also during their life cycle; their use, for example as an automotive part, may decrease the weight of the structure and thus reduce fuel consumption.

The low density and particular microstructure of natural fibres are known to be mainly responsible for their relatively good specific

Table 3.3 Specific weights of natural fibres and of glass fibres (apparent densities in brackets).

Fibre	Specific weight, ρ (g cm^{-3})	Ref.
Abaca	1.4–1.5 (1.1–1.2)	11,21,22,49
Artichoke	1.58 (1.21)	23
Bamboo	0.6–1.5	22,50
Banana	1.3–1.35 (0.7)	21,22,50
Celery	1.5 (0.45–0.80)	51
Coir	1.15–1.5	11,22,52,53
Cotton	1.5–1.6	11,22,52–54
Curaua	1.4 (0.92–1.1)	22,32,33,55,56
Date	0.96–0.99	50
E-glass	2.5–2.6	11,15,16,52,53,57
Elephant grass	0.82–1.08	58
Flax	1.4–1.54 (1.38)	11,16,21,22,52–54,57
Hemp	1.4–1.6	11,16,22,52,54,57
Isora	1.35	36
Jute	1.3–1.52 (1.23)	11,16,21,22,52–54,57
Kenaf	1.2–1.4	22,57
Luffa	(0.92)	59
Miscanthus	1.4	60
Nettle	(0.72)	39
Pineapple leaf	1.44	40
Ramie	1.50–1.56 (1.44)	11,21,22,52,54
Rhectophyllum	0.95	41
Sisal	1.33–1.5 (1.2)	11,21,22,50,52,53,61
Switchgrass	0.74	62
Vakka	0.81	50

mechanical properties, sometimes higher than those of glass fibres (Table 3.4).[11,15,16,21–23,25,28,31,32,35–42,45,50–58,70–72] Globally, the specific tensile strength of plant fibres is between 1600 and 2950 MPa cm^3 g^{-1}, while their specific tensile modulus varies from 10 to 130 GPa cm^3 g^{-1}.[15] Actually, the elastic modulus of cellulose is about 140 GPa,[73] while that of hemicelluloses has been estimated at 8 GPa;[74] both varying according to the relative humidity. It is also well known that lignin increases the fibre structure stiffness by acting as a compatibilizer between cellulose and hemicelluloses.[66] Note that, in Table 3.4, some mechanical data have been obtained by testing elementary fibres and others by testing bundles of fibres. This is why the specific strength and Young's modulus are expressed either in MPa cm^3 g^{-1} (for ultimate fibres) or in g denier^{-1} (for bundles).

The varying contents of the constituents in the fibres, gathered in Table 3.5,[15,22,23,25,29,31,32,35,36,39,41–43,45,59–61,72,75–83] are one of the explanations for the different and scattered mechanical properties among and within the different species. Actually, it is understandable, amongst other points, that the higher the cellulosic content, the stiffer the fibre. Moreover, the particular arrangement of the cellulose microfibrils, embedded under a helicoidal shape in the amorphous matrix made of hemicelluloses, lignin and pectins, is partly responsible for the deformation mode of natural fibres. Usually, the larger the angle of the microfibrils with the fibre axis, called the microfibril angle

Table 3.4 Specific mechanical properties of natural fibres or bundles and of E-glass fibres.

Fibre	Strength (MPa cm³ g⁻¹)	Failure strain (%)	Young's modulus (GPa cm³ g⁻¹)	Ref.
Abaca	414	4	34	11,22
Artichoke	50–220	–	3–11	23
Bamboo	383–552	1.4–1.7	22–39	22,50
Banana	450	3.4–4	13–15	22,50
Celery	23–167	2–5.7	1–5	28,51
Coir	145–174	15–46	3–5	11,22,28,52,53
Cotton	265–290	3–10	5–8	11,22,52,53
Curaua	360–1000	3–4.3	8.4–36	22,32,55,56
Date (base)	478	24	2	50
Date (leaf)	312	2.7	11	50
E-glass	940–1350	2.5–3.4	28–30	11,15,16,52,53,57
Elephant grass	226–272	2.5–2.8	9–10	58
Flax	240–1070	1.2–3.3	26–76	11,16,22,52,53,57,70
Hemp	210–1264	1.6–3	24–50	11,16,22,52,53,57
Isora	370–440	5–6	13–15	36
Jute	270–650	1.2–2.0	7–39	11,16,22,52,53,57
Kapok	41	2	2	22
Kénaf	538	3	12–42	22,57
Miscanthus	652	–	42.5	37
Nettle	900–2200	1.7–2.1	53–121	22,38,57
Pineapple leaf	287–1130	1.6–2	24–57	22,40
Ramie	330–610	2–3.8	29–43	11,22,52,53
Rhectophyllum	588	27.5	6.1	41
Sisal	340–530	2–5.5	6–30	11,22,50,52,53
Switchgrass	60–88	–	6	71,72
Vakka	678	3.5	20	50

Fibre	Strength (g denier⁻¹)	Failure strain (%)	Young's modulus (g denier⁻¹)	Ref.
Bark of cotton stalk	2.9	3.0	144	25
Coir	–	16	50	21
Cornstalk	2.2	2.2	127	31
Cotton	3.5	8.0	44	54
Flax	6.0	2–3	215–300	21,54
Hemp	6.3	1–6	225–285	22,54
Hop stems	4.1	3.3	160	35
Jute	4.2	1.2–1.5	200–350	22,54
Nettle	0.22–0.57	2.3–2.6	–	39
Ramie	7.0	4–4.4	160–170	22,54
Sisal	–	2–3	290	22
Soybean straw	2.7	3.9	94	42
Velvet leaf	2.9	2.5	194	45

(Table 3.6[19,22,25,31,35,36,38,41,42,45,57,59,79,84–95]), the higher the failure strain, since the fibrils will be able to twist when stretched.[96,97] Nevertheless, this is not a universal rule since, in some fibres, microfibrils may lie in different directions in the cell walls. For example, in cotton, the microfibril angle is largely dependent

Table 3.5 Biochemical composition of natural fibres, in % (w/w) on dry fibres.

Fibre	Cellulose	Hemicelluloses	Lignin	Ash	Ref.
Abaca	56–70.2	15–21.7	5.6–10	3	22,29,75
Artichoke	75.3	–	4.3	2.2	23
Bamboo	26–43	15–26	21–31	1.7–5	22,29
Banana	60–65	15	5–10	1.2	15,22
Bark of cotton stalk[a]	79	–	14	0.5	25
Coir	35–44	0.2–15.4	33–45	–	15,75
Cornstalk[a]	81	–	8.4	–	31
Cotton	90–95	1–6.3	0.7–1.6	1.4	15,22,29,75,76
Curaua	73–73.6	10–20	1.5–7.5	1.0	22,32,77
Date (leaf)	33.5	28.5	26.5	6.5	78
Flax	62–72	17–18.5	2–5	1.5	15,22,75
Hemp	63–78	5–16	3–6	1.5	22,75,79
Hop stem[a]	84	–	6	2.0	35
Isora	74.8	–	23.0	1.0	36
Jute	59–71	13.3–15	10–13	0.4	15,22,75
Kenaf	37–57	18–26	15–21	2–5	22,29
Luffa	63	19.4	11.2	0.4	59
Miscanthus	43.1	26.7	22.1	3.9	60
Nettle	79–80	10–12.5	0.5–3.8	–	22,39
Okra	60–70	15–20	5–10	–	80
Pineapple	80–83	12.3–17	3.5–12	–	15,22,81
Ramie	72–85	14	0.5–0.7	0.3	15,22
Rice straw	44.3	33.5	20.4	–	82
Rhectophyllum	68.2	16.0	15.6	–	41
Sisal	60–78	12–13	8–14	0.3	15,22,61,75
Soybean straw[a]	85	–	11.8	1.0	42
Straw (rice, wheat)	30–50	25–30	15–20	5–10	29
Switchgrass	29–43	26–36	12–24	2–5	43,72,83
Velvet leaf[a]	69	–	17	3.2	45

[a]Samples treated with sodium hydroxide.

on the maturity of the plant (from 1° to 45°),[22,31,76,98] whereas bamboo fibres possess several concentric layers (more than other bast fibres), *i.e.* several microfibril angles.[99,100]

Another advantage linked to the lightness and the good mechanical properties of natural fibres is that their use as composite reinforcement requires less polymer matrices. Actually, to obtain the same properties with natural fibre reinforced plastics as with glass fibre reinforced plastics, a simple law-of-mixture says that the volume fraction of natural fibres has to be higher than that of synthetic fibres. This reduces the required volume fraction of the polymer matrix, which is a positive thing when dealing with a petrochemical-based product.

The plants from which the fibres described in the following are taken have the advantage of being renewable, generally quite rapidly. Except for some species such as bamboo or china reed, which reach maturity only after several years, vegetable fibres are often extracted from annual plants. This particularity, compared for example to wood fibres, acts in favour of the very current concern of sustainable development.

Table 3.6 Microfibril angle and crystallinity of cellulose in natural fibres.

Fibre	Microfibril angle (°)	Crystallinity (%)	Ref.
Abaca	22.5	52	84,85
Banana	10–12	45–55	19,22,85,86
Bark of cotton stalk	15	47	25
Coir	30–49	25–33	19,22,84,87
Cornstalk	10.9	52	31
Curaua	–	76	22
Flax	5–11	70	22,84,88–90
Hemp	2–7.5	60–88	22,57,79,84,91
Hop stems	8	44	35
Isora	20–26	–	36
Jute	7–10	78	22,57,92
Luffa	–	59	59
Nettle	3	–	38
Pineapple	6–18	44–75	19,22
Ramie	3–12	61–70	22,84,85,88,93
Rhectophyllum	40	–	41
Rice straw	–	40	19
Sisal	10–25	55–71	22,57,94,95
Soybean straw	12	47	42
Switchgrass	–	51	31
Velvet leaf	18	38	45
Wheat straw	–	55–65	19

Furthermore, vegetable fibres are known to be CO_2 neutral, since their composting or combustion does not release into the atmosphere their excess in carbon dioxide (captured through photosynthesis to produce the sugars of their skeleton).[66] Their production and their extraction generally require low amounts of energy, according to the species considered, compared to the fabrication of synthetic fibres which consumes high levels of energy (mainly through heating).

Last but not least, natural fibres are less abrasive to mixing compared with traditional glass fibres.[14,33,66] This leads to advantages when dealing with processing or recycling of natural fibre-based composites.[91]

However, vegetable fibres exhibit several drawbacks that are more or less handicaps for their further development as polymer reinforcements. First of all, as can be seen in the tables which gather the fibre properties (Tables 3.2–3.7), there is generally a broad dispersion of the values. This is true not only for mechanical properties and dimensions, but also for the composition contents. Some works have already underlined this aspect of the fibres and explained its origin by the numerous variations undergone by the fibres during their growth (humidity, sunlight, wind, *etc.*). Other sources of property scattering are the level of maturity, the position in the plant and the extraction processes.[29,66,101,102] This results in a large variation in fibre quality according to global parameters, even if the methods used to measure them are often the same. For example, the determination of the cellulose, hemicelluloses and lignin contents is almost always obtained using the Van Soest and Wine method developed in 1967.[103] Nevertheless, the cellulose contents often vary by several percent for a given species (see Table 3.5). Concerning the mechanical properties, attempts have been made to calculate a property that could help compare the fibres properly (such as a calculated E-modulus),[104] even if this does not really reduce the scattering. Both the fibre stiffness and strength are directly linked with their growth environment, the former being more dependent on the cellulose content and quality, and the latter on structural defects (generally called kink bands). Since enzymatic "accidents" can occur during the formation of the long polymeric chains, the dispersion in the biochemical contents and thus in the morphological parameters cannot be controlled (and so, limited) during cell life. As far as flaws generated by the extraction processes are concerned, they can be reduced both in amount and in dangerousness, as proved by several studies comparing the properties of fibres extracted by hand with those of fibres extracted mechanically: a large improvement in strength was highlighted, but no reduction in scattering.[105]

With the aim of improving the reliability of the properties of fibres, several authors have used statistical analysis, mainly the Weibull equation, to refine their description.[80,106–110] Whereas this kind of numerical method can help in understanding the effect of fibre length on its strength and modulus, it does not limit the range of properties.

Unlike synthetic fibres that are standardized and whose properties are generally very accurate, natural fibres suffer from their lack of reliability. This is particularly incapacitating when trying to model the fibre or the derived composite behaviour. As composite manufacturers expect reproducible mechanical characteristics, the only way to ensure an effective quality level is to take large security coefficients on the fibre properties, which in turn reduces their attractiveness and prevents full exploitation of their advantages.

Another important characteristic of vegetable fibres is that they absorb water very easily compared with most synthetic fibres (Table 3.7).[19,21–23,25,29,31,32,35,42,45,50,57,60,72,80,83,111,112] This is likely to be due not only to the strongly polarized hydroxyl groups present at the surface of the natural fibres, but also to the non-crystalline (and less-oriented) regions present in their cell walls. Actually, as seen in Table 3.6, these non crystalline parts can reach 60% of the fibre volume. They mainly comprise lignin and hemicelluloses but also some cellulose. However, only the latter two are decisive in relation to moisture behaviour since they are highly hydrophilic whereas lignin is hydrophobic.[21]

The first problem generated by this property is a noticeable dimensional instability, especially in the transverse direction of the fibre. Values as high as 45% of volume swelling have been reported for jute and abaca, but they generally range between 30% and 40%.[21,22,54] This can raise problems, not only during the processing stage, but also during the life of the derived composite. Actually, during fabrication, low humidity levels may cause a non-uniform dispersion of the fibres within the matrix.[66] Moreover, once the composite is processed, a small change in relative humidity of the atmosphere can lead to fibre/matrix decohesion, *i.e.* initiating the collapse of the structure. To reduce

Table 3.7 Moisture regain for some natural fibres (21 °C, 65% RH).

Fibre	Moisture regain (%)	Ref.
Abaca	5–14	21,22,29
Artichoke	5	23
Bamboo	9–10	50
Banana	9–15	19,21,22,29,50
Bark of cotton stalk	8.8	25
Coir	10–13	19,21,29
Cornstalk	7.9	31
Cotton	7–25	21,22,29
Curaua	9–12	32,111
Date	9–11	50
Flax	7–12	21,22,29,57
Hemp	6–12	21,22,29,57
Hop stems	8.3	35
Isora	5–8	112
Jute	8.5–17	21,22,29,57
Kenaf	9.5–17	22,31,45
Miscanthus	11	60
Nettle	11–17	22
Okra	8	80
Pineapple	10–13	19,22,29
Ramie	7.5–17	22,29
Rice straw	6.5	19
Sisal	10–22	21,22,29,50
Soybean straw	11.2	42
Switchgrass	6–12	72,83
Vakka	12	50
Velvetleaf	9.8	45
Wheat straw	10	19

the sensitivity of the natural fibres to moisture, some surface treatments have been successfully applied and enabled a relative stabilization of the reinforcement (see Chapter 8). Nevertheless, moisture does not play in favour of developing such eco-composites, especially in applications where the variation in humidity is high.

In addition, natural fibres are highly temperature-dependent: their properties, especially the mechanical ones, drop dramatically when the temperature increases. Previous studies reported a temperature threshold for natural fibre use as low as 200 °C, since the fibres show damage above that.[48,66,80,113–116] Generally, degradation is almost complete around 350 °C, as shown by thermogravimetric analysis for okra, artichoke, curaua, kenaf, jute, isora, luffa, sisal, bamboo and hemp fibres.[17,32,59,80,112,117,118] This low thermal stability becomes a problem when processing composites. Actually, natural fibre composites are mainly processed by thermoforming, which requires relatively high temperatures (according to the thermal and rheological properties of the polymer) to ensure low viscosity and good impregnation and/or dispersion of the reinforcement. This acts as an important ageing factor of both the polymer and the fibres, and causes premature damage in the derived composite.

When dealing with long fibre composites, fibre entanglement within the composite may occur, especially during processing. This is particularly true of natural fibres, due to the porosity present within the bundles, *i.e.* between fibres. Madsen and Lilholt[119,120] showed that, according to the processing conditions and the species considered, there was an optimal fibre weight content in the derived composite, since increasing this content also increases the porosity and thus reduces the mechanical properties. Their studies proved that porosity has to be taken into account, particularly when designing plant fibre-based rather than synthetic fibre-based composites. Moreover, this particularity may prevent plant fibre-based composites from reaching the same mechanical properties as highly concentrated glass fibre reinforced composites, which goes against eco-composites development.

At the scale of the composites, the interfaces between the natural fibres and the polymer matrix are generally weak, and many treatments have been applied to the fibres to enhance compatibility between these materials and the mechanical properties of the derived composites. More details about that aspect will be found in Chapter 8.

Focusing on vegetable fibres, they are often naturally occurring in the shape of bundles of fibres. This implies that another natural material enters the fibre composition in the form of sandwiching cement (Figure 3.1). In the example of flax, several studies underlined the fact that the middle lamella was a weak link within a bundle,[121] and it has recently been shown that the interfacial strength between two fibres was at least twice as low as the interfacial strength between elementary fibres and classical polymer matrices.[122] As a consequence, this

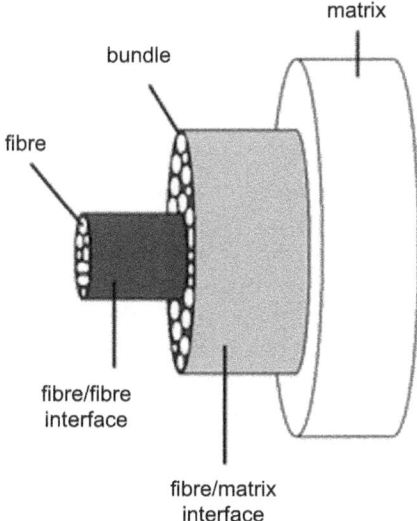

Figure 3.1 Schematic representation of the interfaces within a natural fibre-based composite.

cement, which is generally made of pectins, lignin and waxes in contents that differ according to the species considered,[123,124] contributes significantly to mechanical weakness. An understanding of the mechanical behaviour of this phase would help to enhance the reliability of derived composites.

Added to the other drawbacks of natural fibres, their rot sensitivity and their not-so-pleasant odour have also already been underlined.[66,101] Concerning this last point, Müssig *et al.* highlighted the role of pectins in this problem, based on previous studies.[22,125] They showed that the elimination of pectins from hemp bundles increased the odour intensity caused mainly by mildew contamination or overheating, since the fibre surfaces available for these gaseous reactions were more numerous. They also proved that enzymatic treatment of bundles led to a reduction of the odour intensity, especially at low temperatures, and that, after a few months, this intensity returned to its initial level. Although synthetic fibres do not exhibit this kind of problem, more investigations must be carried out to convince potential users that odour should not get in the way of natural fibre composites development.

A more economic problem that could be posed when dealing with plant-based composites is their long term supply. The recent controversies around the use of fields to produce plants for other applications (mainly for bioethanol), rather than for feeding people, may raise the question of investing in new technologies to produce eco-composites, since geopolitical aspects are likely to intervene in the debate. This may also curb the industrial enthusiasm in shifting to more environmentally friendly materials, even if this shift is largely encouraged by several governments.

3.3   Description of New Sources of Vegetable Fibres

The fibres whose properties are described in this chapter come from different sources. They are mainly extracted either from leaves, from stems or from seeds, and some plants even possess several types of fibres, such as cotton: the most exploited cotton fibres come from their seeds, but others can be found in the bark of their stems. The most common cellulosic fibres (flax, hemp, ramie, cotton, jute, kenaf) used to strengthen polymers have already been described elsewhere;[21,29,57,66] thus we will focus here on fibres that gave rise to recent interest as composite reinforcement materials (Figure 3.2).

Among the new sources of fibres, isora fibres, which fill in the bark of *Helicteres isora* plants, have gained a relatively great importance due to their good mechanical properties. This plant is mainly harvested in East and South India. The fibres are extracted from the plant by retting and look like jute,[36] but fibre production is still at a local level. Elementary isora fibres are, as in many other vegetable plants, gathered in bundles that appear relatively porous, which is an advantage when combined in a polymer matrix (Figure 3.3a).[36,112] When added to natural rubber in a 1:1 proportion, isora fibres enhance the stiffness of the matrix by almost 30%.[36] Several chemical treatments applied to these fibres (acetylation, benzoylation, silane treatment) changed the morphology of the

Figure 3.2 Images of plants considered as new sources of cellulose fibres.

fibres (Figure 3.3b–f).[36,112] Thus the dissolution of superficial fatty acids and lignin components makes the fibres thinner and the pores clearer, enhancing the ability to graft within a polymer matrix and improving its mechanical properties.[126] For example, the addition of randomly oriented triton-treated isora fibres to an unsaturated polyester resin was shown to increase the tensile strength and modulus by about 40% and 140%, respectively, and the flexural strength and modulus by about 50% and 85%, respectively. The explanation given for these improvements was the decrease in hydrophilicity brought about

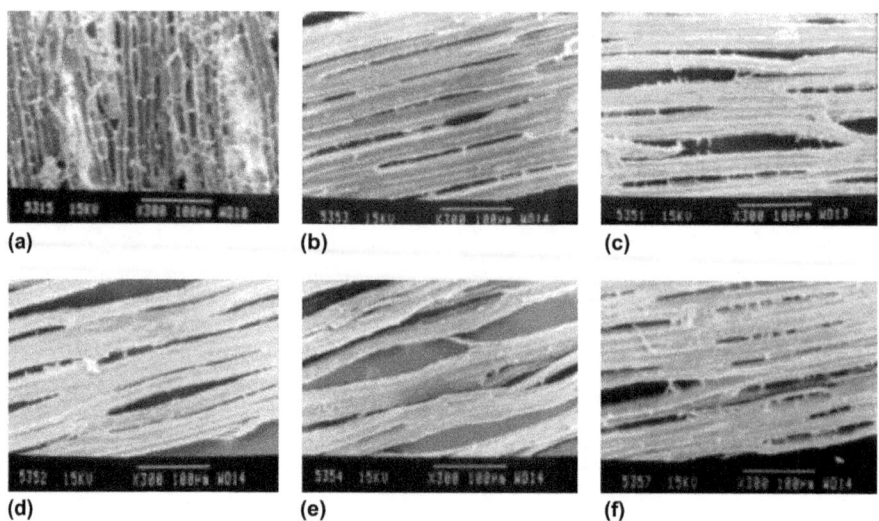

Figure 3.3 SEM photographs of (a) untreated, (b) alkali-treated, (c) acylated, (d) benzoylated, (e) toluene diisocyanate-treated and (f) silane-treated isora fibres.[36,112]

by the different treatments, which increases the compatibility of the fibres with the polyester matrix as well as their dispersion within the polymer.

Switchgrass (*Panicum virgatum*) is a perennial North American native plant that requires little water and few nutriments but leads to high productivity.[83] It possesses fibres in its leaves and stems whose mechanical properties and chemical composition are interesting enough to foresee various applications.[127] For example, owing to their relatively low lignin content, these fibres are quite easily pulped and thus attractive for paper making industries. Nevertheless, since switchgrass receives interest mainly as a possible source of bioethanol, few studies have been considering the possibility of using these fibres in polymer composites. One study revealed that the incorporation of 30 wt% of switchgrass fibres to a polypropylene matrix led to an increase by a factor of about 2.5 in the polymer flexural modulus.[43]

China reed plant, also called *Miscanthus*, originally comes from Japan, China and Korea, and is now also harvested in Europe (Figure 3.4).[37,128] It takes at least three years for the plant to mature and to grow several meters high.[129] At that stage, a third of the volume of its stem is made of relatively strong and straight fibres that can compete, in terms of mechanical properties, with more traditional vegetable fibres.[37] For example, it has recently been shown that composites made of polypropylene and 30 wt% of China reed fibres or hemp fibres had the same value of Young' modulus;[128] the addition of such fibres to the polymer led to increased stiffness by a factor of 2.5. Further enhancement of the composite properties was achieved by treating the China reed fibres with polypropylene-grafted maleic anhydride (PP-*g*-MA): an increase of 12% and

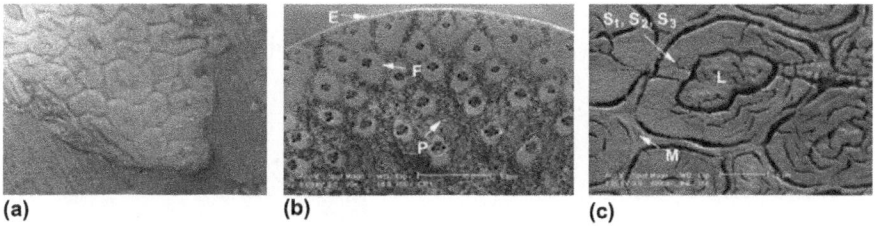

Figure 3.4   Cross-sections of China reed fibres (F, fibre; P, parenchyma cells; E, epidermis; L, lumen; M, middle lamella; S_1, S_2, S_3, sub-layers).[37,128]

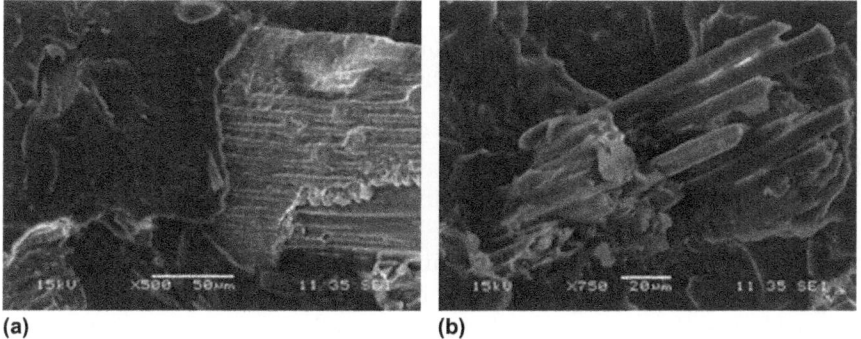

Figure 3.5   SEM micrographs of tensile fracture of PP/30% China reed fibre composite: (a) untreated fibres; (b) fibres treated with 2% PP-*g*-MA.[128]

30% of the tensile modulus and the tensile strength, respectively, could be obtained. Actually, this treatment enables better cohesion between the fibres and the matrix, as can be seen on composite rupture micrographs (Figure 3.5):[128] whereas the untreated fibres are mainly pulled out of the matrix before the composite failure, the treated fibres are well bonded to the polymer and the rupture of the composite occurs by fibre breakage as well as fibre slipping. Attempts have also been made to use *Miscanthus* fibres to reinforce a poly(lactic acid) (PLA) matrix: the results of tensile tests showed that this addition had a positive impact on Young's modulus (value increased by almost 50%) but a negative impact on strength (decrease of 40%), with or without addition of compatibilizer.[128] However, added to a cellulose diacetate (CDA) matrix, China reed fibres increased notably both the stiffness and the strength of the polymer: for a fibre volume fraction of 40%, the composite's Young's modulus and strength were found to be respectively 4 and 1.2 times higher than the properties of the pure polymer.[37] Very similar to China reed plant, elephant grass (*Pennisetum purpureum*), originally from Africa, possesses fibres in its leaves that can be extracted either by chemical or by mechanical means. The chemical method has been shown to lead to stronger polyester/elephant grass

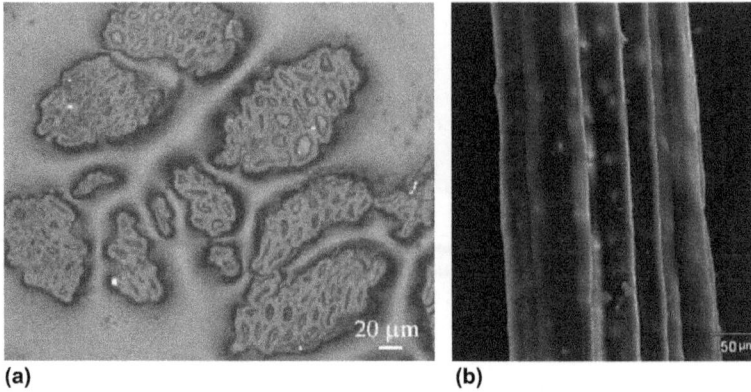

(a)   **(b)**

Figure 3.6   (a) Optical micrograph showing a cross-section of bundles of okra fibres. (b) SEM micrograph of a longitudinal view of a bundle of okra fibres.[80]

composites than the mechanical extraction, with an increase of about 45% of the tensile strength and modulus.[58] Moreover, even the mechanically treated elephant grass fibres led to an increase of the plain polyester matrix strength and modulus by a factor of 1.5, for a fibre fraction of 31 vol%.[58] These results were comparable to bamboo/polyester composites.

Okra fibres (*Abelmoschus esculentus*), also called "lady's finger" fibres, are extracted from the stems of a 2 m high plant that mainly grows in Egypt and Southern Asia. Although it is cropped for its fruits, the mechanical properties of its bark fibres are similar to those of the more traditional natural fibres used in composite materials.[80] Their colour varies from whitish to yellowish, depending on the UV radiation received by the plant. Like many other bark fibres, they are gathered in bundles of several polygonal elementary fibres, whose thickness varies along the fibre (Figure 3.6).[80] Recently, various chemical treatments have been applied to okra fibres (notably bleaching and acetylation), aimed at reducing their hydrophilic character.[130] While the degradation of the treated fibres due to exposure to water was greatly reduced compared to the raw fibres, their mechanical properties dramatically dropped after any of these treatments. For example, the Young's modulus of acetylated or bleached okra fibres was divided by 3 or 4, respectively. However, the effect of adding treated or untreated okra fibres to polymers has not yet been studied in terms of mechanical improvements, but it is expected that such fibres would constitute effective composite reinforcement materials.

Curaua fibres (*Ananas erectifolius*) are mainly produced in Amazon regions. The decortication of the leaves provides stiff, strong and smooth bundles of fibres (Figure 3.7).[32,56] Investigations carried out in the last decade on these fibres as composite reinforcements have already aroused strong interest from automobile companies.[33,131] Actually, a recent study has revealed that the use of curaua fibres to replace E-glass fibres in certain car parts could lead to economic, environmental and social benefits, since, for example, they cost 50%

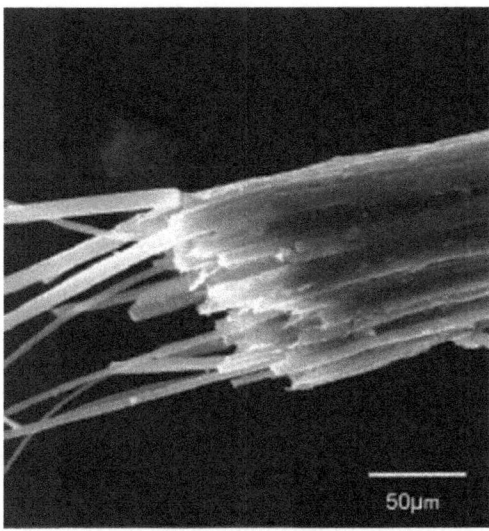

Figure 3.7   Fracture surface of a bundle of curaua fibres.[56]

less than glass fibres.[55] Nevertheless, an effort still has to be made to lighten the natural fibre composites in order to implement them at a higher level in car industries. Moreover, alkali treatment has been applied to curaua fibres to enhance the properties of a cornstarch-based polymer:[56] this treatment led to a drop in the fibres' properties but also to an improvement in the toughness of the derived composite. This was attributed to better interfacial bonding induced by the alkali treatment as well as a reduction of the scattering of fibre strength values. Also, the addition of 40 wt% of curaua fibres to a polyamide-6 matrix brought about an increase in the polymer stiffness by a factor of 3, which was further improved when the fibres were alkali-treated before composite processing.[132]

Royal palm (*Roystonea regia*), also known as vakka, is an Indian and Central American tree that produces light and strong external fibres.[50] These fibres, extracted from the ripened leaves of the tree, have been recently investigated as polyester reinforcements.[133] The addition of 37 vol% of fibres led to an increase in the polymer tensile strength and stiffness by a factor of 3, which is higher than the reinforcement efficiency of other natural fibres such as sisal and banana (for the same fibre volume fraction). The flexural properties were also shown to be improved by the presence of vakka fibres, especially the flexural modulus which increased by a factor of 1.3 for a fibre fraction of 39 vol%. More surprisingly, the vakka fibre-based composites were shown to exhibit an increase in their dielectric strength when the fibre fraction increased;[133] this acts in favour of a possible use of such composites in electrical insulation applications.

The ripe fruit of the *Luffa cylindrica*, called sponge-gourd, possesses fibres entangled around the beehive-like structure of the fruit nucleus. This plant is traditionally harvested in Africa, in Asia and in Central and South America, either for its edible young fruit or for its mature fruit that can be used as bath sponges.[59,134] Added to the low density of these fibres, their mat structure makes them worthwhile fibres for polymer reinforcement since they can be used directly to process composites without the need, for example, to be woven beforehand. Among the few studies already carried out yet on *Luffa* fibre-based composites, one highlighted the beneficial effect of these fibres on the stiffness of a polypropylene matrix (increase of 40% for a fibre fraction of 15 wt%), in spite of a reduction in the composite tensile strength compared to the pure matrix strength.[135] Another study confirmed this lack of mechanical improvement in a polyester matrix reinforced with *Luffa* fibres, but underlined the modification to the fracture mode of the polymer induced by this addition (from a fragile one to a more controlled one), making these fibres attractive in hybrid composites.[136]

Date (*Phoenix sylvestris*) is a palm tree from the South East of Asia that possesses fibres both in the stems of its leaves and in the netted structure that surrounds the base of the tree (called amplexicaul fibres). The first type of fibre is extracted by beating the leaves followed by water retting, whereas the second type is obtained by carding the dried netted structure. Even if they differ a lot in terms of stiffness, these two types of fibres are relatively light and strong and could be used to strengthen parts of polymer composites.[50] Added to an epoxy matrix, date fibres extracted from leaves increased the flexural modulus by about 50% and even by 90% when the fibres were treated beforehand with acetic and maleic anhydrides.[78] This enhancement was much lower in the case of an unsaturated polyester matrix (13%) and was even null with treated fibres. Moreover, date palm fibres were shown to enhance the thermal and mechanical stabilities of a polypropylene matrix submitted to severe weathering conditions.[137] This is a great advantage when dealing with outdoor uses of composites.

Hop (*Humulus lupulus*) is a plant mainly produced in the USA, in Germany and in Ethiopia. It belongs to the same family as hemp (*Cannabaceae*) and is thus a potential source of cellulose fibres, since its stem, although generally considered as a by-product, possesses external and internal fibres.[35] The surface of a hop stem is irregular and coated with numerous deposits of non-cellulosic substances, most of which are removed during fibre extraction, leading to smooth and clean bundles of fibres (Figure 3.8). A recent study revealed that the addition of fibres extracted from the outer bark of a hop stem to a polypropylene matrix led to better mechanical properties than jute fibres, for the same composite density.[138]

Rhectophyllum fibres (*Rhectophyllum camerunense*) are plant fibres that can be extracted from the long air roots of a common creeper native to the Gulf of Guinea.[41] These bundles of fibres are very long and porous and exhibit a very low density. Their high cellulose content and good mean specific properties are likely to make them attractive as reinforcement materials for composites, even if they have not yet been tested in this role.

Like nettle, velvetleaf is often considered as a weed since this plant (*Abutilon theophrasti*) thrives to the detriment of other crops. This invasiveness, added to

(a)   **(b)**

Figure 3.8   SEM images of (a) a hop stem and (b) a bundle of hop fibres.[35]

mechanical properties very close to those of more common natural fibres, as well as the recent improvement of fibre processing, acts in favour of using these fibres as polymer reinforcements.[45] Nevertheless, no study has yet been carried out to determine the properties of the derived composites.

Also considered as by-products, corn and artichokes stems, cotton stalks, corn and rice husks, rice and soybean straws have all been shown to contain the same kind of ligno-cellulosic fibres as those mentioned above.[19,23,25,31,42,82,139] Moreover, as revealed by SEM images (Figure 3.9), the bundles extracted from these products also have the same type of morphology as the traditional plant bundles, *i.e.* rough elementary fibres glued together in scores by non-cellulosic components. They are thus likely to provide effective composite reinforcement, although the effect of most of them has not yet been characterized. The main conclusion of the studies carried out on these by-products is that their wide availability and their low cost compensate for their relatively low mechanical properties. What can also be noted is the similar shape between the fibres extracted from some of these by-products and cotton fibres, with several convolutions along the length of elementary fibres (Figure 3.10).[25,42,45,140]

Among the different forms of plant waste, rice straws have been largely studied these last few years as potential polymer reinforcements. For example, incorporating 50 wt% of rice straw fibres into a high-density polyethylene matrix was shown to increase the storage modulus, tensile strength and impact strength of the polymer by about 150, 50 and 25%, respectively.[139] Another example concerns rice straw/polyester composites made of 40 vol% of fibres: their flexural strength and modulus were 20 and 54% higher than those of the plain matrix.[141] More recently, composites made of poly(lactic acid) and 20 wt% rice straw fibres treated with 8 wt% poly(butyl acrylate) were tensile tested and exhibited a tensile strength 25% higher than that of the neat matrix.[142] Added to waste tyres, rice straw fibres led to composites that presented higher flexural properties and acoustic insulation than wood particle boards, giving a foretaste of these cheap natural fibres as effective reinforcements of recycled polymers in insulation boards.[143]

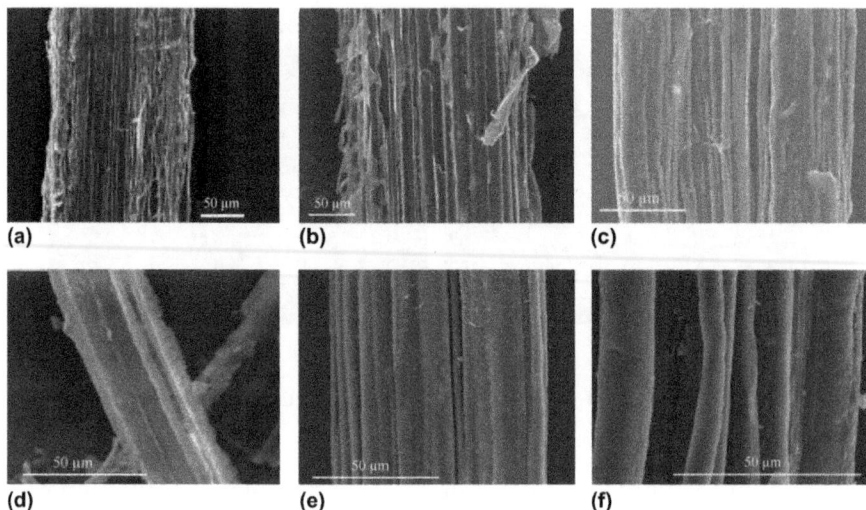

Figure 3.9   SEM images of untreated bundles of fibres extracted from (a) artichoke stem,[23] (b) corn stalk,[31] (c) corn husk,[19] (d) rice straw,[82] (e) soybean straw[42] and (f) cotton stalk.[25]

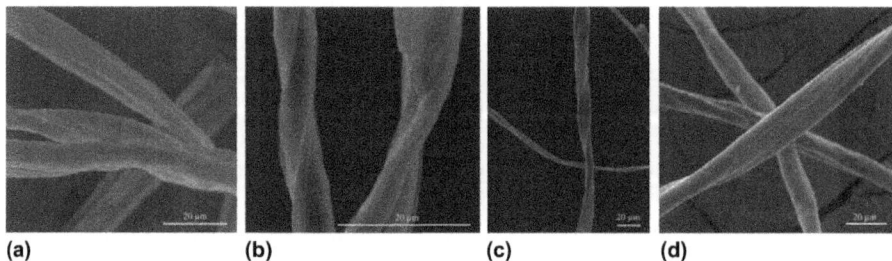

Figure 3.10   SEM images of elementary fibres of (a) soybean straw,[42] (b) cotton stalk[25] and (c) velvet leaf,[45] having the same convolutions as seen in (d) cotton.[140]

3.4   Concluding Remarks

Despite the numerous hindrances related to natural fibres, notably the strong dependence of their mechanical behaviour on temperature and humidity and their lack of reliability, promising developments of plant fibre composites are expected since ecological concerns are growing relatively fast in many domains. Current studies tend, among others, to limit the influence of external conditions on the composites' stability in order to enhance the reliance of industrialists on such materials, while other investigations aim at increasing their mechanical properties to improve their use and popularity among composite users. One particular objective is to substitute synthetic fibres, mainly E-glass fibres, by natural fibres, which could help in improving the carbon footprint and the energy balance for the production of such eco-composites.

Improvements in the reliability of natural fibre based composites, *i.e.* reduction of the fibres properties ranges, could be achieved by different means, such as a better fibre selection. This could be done, for example, by improving the fibre extraction processes, by selecting varieties or even by modifying plants genetically in order to obtain almost identical and defect-free fibres, for a given harvesting area. Nevertheless, this last idea, still under debate in relation to other natural products (especially food resources), is far from being the most "environmentally friendly" solution.

Hopefully, as the number of scientific papers dealing with the characterization of new plant fibres for polymer reinforcement is increasing every month, the development of natural fibre-based composites is expected to increase, especially at a local level, thanks to the large variety of species available worldwide. Likewise, the use of materials previously considered as by-products or waste, to strengthen or stiffen polymeric matrices, is a fashionable, popular and certainly profitable way to respond to government expectations in composite applications.

References

1. P. Wambua, J. Ivens and I. Verpoest, *Compos. Sci. Technol.*, 2003, **63**, 1259.
2. M. Pervaiz and M. M. Sain, *Resour., Conserv. Recycl.*, 2003, **39**, 325.
3. S. V. Joshi, L. T. Drzal, A. K. Mohanty and S. Arora, *Composites, Part A*, 2004, **35**, 371.
4. E. Witten, *The Composites Market in Europe: Market Developments, Challenges, and Opportunities*, survey report, Industrievereinigung Verstärkte Kunststoffe (AVK), Frankfurt, 2008.
5. S. Huda, N. Reddy, D. Karst, W. Xu, W. Yang and Y. Yang, *J. Biobased Mater. Bioenergy*, 2007, **1**, 177.
6. W. Fung and M. Hardcastle, *Textiles in Automotive Engineering*, Woodhead, Cambridge, 2001.
7. http://www.nrc-cnrc.gc.ca/fra/dimensions/numero3/lin.html; accessed 1 September 2011.
8. Directive 2000/53/EC of the European Parliament and the Council on end-of-life vehicles, no. 32000L0053, September 2000.
9. Report from the Commission to the Council, the European Parliament, the European Economic and Social Committee, and the Committee of Regions on the implementation of Directive 2000/53/EC on end-of-life vehicles for the period 2005–2008, no. 52009DC0635, December 2009.
10. ADEME, *Etude de Marché des Nouvelles Utilisations des Fibres Végétales*, survey report, Ernst and Young, 2005.
11. S. W. Beckwith, *Compos. Fabr.*, 2003, **19**, 12.

12. FAO Commodities and Trade Division, 2008, www.fao.org/es/esc/ common/ecg/323/en/SB_JUNE08.pdf, accessed on 1 September 2011.
13. S. Piotrowski and M. Carus, in *Industrial Applications of Natural Fibres*, ed. J. Müssig, Wiley, Chichester, 2010, p. 73.
14. A. L. Leão, S. M. Sartor and J. C. Carashi, *Mol. Cryst. Liq. Cryst.*, 2006, **448**, 161.
15. K. G. Satyanarayana, G. G. C. Arizaga and F. Wypych, *Prog. Polym. Sci.*, 2009, **34**, 982.
16. C. Scarponi, *JEC Compos. Mag.*, 2009, **46**, 46.
17. F. Yao, Q. Wu, Y. Lei, W. Guo and Y. Xu, *Polym. Degrad. Stab.*, 2008, **93**, 90.
18. P. Girouard, M. Walsh and D. Becker, in *Proceedings of the 4th Biomass Conference of the Americas*, Oakland, California, Aug. 29th–Sept. 2nd 1999, p. 85.
19. N. Reddy and Y. Yang, *Trends Biotechnol.*, 2005, **23**, 22.
20. I. V. Rijswijk and W. D. Brouwer, in *Natural Polymers and Composites*, ed. L. H. C. Mattoso, A. Leao and E. Frollini, Sao Carlos, Brazil, 2002.
21. S. K. Batra, in *Handbook of Fiber Chemistry*, ed. M. Lewin and E. M. Pearce, Dekker, New York, 1998, p. 505.
22. J. Müssig, H. Fischer, N. Graupner and A. Drieling, in *Industrial Applications of Natural Fibres*, ed. J. Müssig, Wiley, Chichester, 2010, p. 269.
23. V. Fiore, A. Valenza and G. Di Bella, *Compos. Sci. Technol.*, 2011, **71**, 1138.
24. M. S. Ilvessalo-Pfäffli, *Fibre Atlas: Identification of Papermaking Fibers*, Springer, Berlin, 1993.
25. N. Reddy and Y. Yang, *Bioresour. Technol.*, 2009, **100**, 3563.
26. T. M. Saleh and S. A. El-Meadawy, *Egypt. J. Chem.*, 1972, **15**, 361.
27. Y. Fahmy and H. Ibrahim, *Cell. Chem. Technol.*, 1976, **10**, 723.
28. B. Bakri and S. J. Eichhorn, *Cellulose*, 2010, **17**, 1.
29. R. M. Rowell, in *Properties and Performance of Natural-Fibre Composites*, ed. K. L. Pickering, Woodhead, Cambridge, 2008.
30. G. Leson, in *Proceedings of the International Coir Convention, Common Fund for Commodities*, Colombo, Sri Lanka, 13–14 June 2002.
31. N. Reddy and Y. Yang, *Polymer*, 2005, **46**, 5494.
32. M. A. S. Spinacé, C. S. Lambert, K. K. G. Fermoselli and M. A. De Paoli, *Carbohydr. Polym.*, 2009, **77**, 47.
33. J. R. Araujo, B. Mano, G. M. Teixeira, M. A. S. Spinacé and M.-A. De Paoli, *Compos. Sci. Technol.*, 2010, **70**, 1637.
34. K. Charlet, J. P. Jernot, S. Eve, M. Gomina and J. Bréard, *Carbohydr. Polym.*, 2010, **82**, 54.
35. N. Reddy and Y. Yang, *Carbohydr. Polym.*, 2009, **77**, 898.
36. M. Lovely and R. Joseph, *J. Appl. Polym. Sci.*, 2007, **103**, 1640.
37. L. Lundquist, B. Marque, P. O. Hagstrand, Y. Leterrier and J. A. E. Manson, *Compos. Sci. Technol.*, 2003, **63**, 137.
38. E. Bodros and C. Baley, *Mater. Lett.*, 2008, **62**, 2143.

39. L. Bacci, S. Baronti, S. Predieri and N. di Virgilio, *Ind. Crops Prod.*, 2009, **29**, 480.
40. S. Mishra, A. K. Mohanty, L. T. Drzal, M. Misra, S. Parija, S. K. Nayak and S. S. Tripathy, *Compos. Sci. Technol.*, 2003, **63**, 1377.
41. A. Béakou, R. Ntenga, J. Lepetit, J. A. Atéba and L. O. Ayina, *Composites, Part A*, 2008, **39**, 67.
42. N. Reddy and Y. Yang, *Bioresour. Technol.*, 2009, **100**, 3593.
43. M. J. A. Van Den Oever, H. W. Elbersen, E. R. P. Keijers, R. J. A. Gosselink and B. De Klerk-Engels, *J. Mater. Sci.*, 2003, **38**, 3697.
44. K. Goel, T. Radiotis, R. Eisner, G. Sherson and J. Li, *Pulp Paper Canada*, 2000, **101**, 41.
45. N. Reddy and Y. Yang, *Bioresour. Technol.*, 2008, **99**, 2449.
46. A. Kelly and W.R. Tyson, *J. Mech. Phys. Solids*, 1965, **13**, 329.
47. B. Jiang, C. Liu, C. Zhang, B. Wang and Z. Wang, *Composites, Part B*, 2007, **38**, 24.
48. M. J. John and R. D. Anandjiwala, *Polym. Compos.*, 2008, **29**, 187.
49. I. A. Ansari, G. C. East and D. J. Johnson, *J. Text. Inst.*, 2001, **92**, 331.
50. K. M. M. Rao and K. M. Rao, *Compos. Struct.*, 2007, **77**, 288.
51. C. Santulli, C. Caneva, I. M. De Rosa and F. Sarasini, in *Proceedings of the International Conference on Innovative Natural Fibre Composites for Industrial Applications*, Rome, Italy, 10–13 October 2007.
52. A. Beukers, in *Lightness, the Inevitable Renaissance of Minimum Energy Structures*, ed. E. Van Hinte, 010 Publishers, Rotterdam, 1999, p. 72.
53. S. J. Eichhorn, C. A. Baillie, N. Zafeiropoulos, L. Y. Mwaikambo, M. P. Ansell, A. Dufresne, K. M. Entwistle, P. J. Herrera-France, G. C. Escamilla, L. Groom, M. Hughes, C. Hill, T. G. Rials and P. M. Wild, *J. Mater. Sci.*, 2001, **36**, 2107.
54. M. S. Rahman, in *Industrial Applications of Natural Fibres*, ed. J. Müssig, Wiley, Chichester, 2010, p. 135.
55. R. Zah, R. Hischier, A. L. Leao and I. Braun, *J. Clean. Prod.*, 2007, **15**, 1032.
56. A. Gomes, T. Matsuo, K. Goda and J. Ohgi, *Composites., Part A*, 2007, **38**, 1811.
57. J. Summerscales, N. P. J. Dissanayake, A. S. Virk and W. Hall, *Composites, Part A*, 2010, **41**, 1329.
58. K. M. M. Rao, A. V. R. Prasad, M. N. V. R. Babu, K. M. Rao and A. V. S. S. K. S. Gupta, *J. Mater. Sci.*, 2007, **42**, 3266.
59. V. O. A. Tanobe, T. H. D. Sydenstricker, M. Munaro and S. C. Amico, *Polym. Test.*, 2005, **24**, 474.
60. P. Visser and V. Pignatelli, in *Miscanthus for Energy and Fibre*, ed. M. B. Jones and M. Walsh, Cromwell Press, London, 2001, p. 109.
61. R. D. Anandjiwala and M. John, in *Industrial Applications of Natural Fibres*, ed. J. Müssig, Wiley, Chichester, 2010, p. 181.
62. J. Ai and U. Tschirner, *Bioresour. Technol.*, 2010, **101**, 215.
63. S. V. Skaven-Haug, in *Proceedings of the 4th International Peat Congress*, Espoo, Finland, 1972.

64. L. Y. Mwaikambo and M. P. Ansell, *J. Mater. Sci. Lett.*, 2001, **20**, 2095.
65. C. C. Sun, *Int. J. Pharm.*, 2008, **346**, 93.
66. M. J. John and S. Thomas, *Carbohydr. Polym.*, 2008, **71**, 343.
67. I. Frydrych, G. Dziworska and J. Bilska, *Fibres Text. East. Eur.*, 2002, **10**, 40.
68. T. Ashour, H. Wieland, H. Georg, F. J. Bockisch and W. Wu, *Mater. Des.*, 2010, **31**, 4676.
69. Y. Li, Y. Luo and S. Han, *J. Biobased Mater. Bioenergy*, 2010, **4**, 164.
70. K. Charlet, C. Baley, C. Morvan, J. P. Jernot, M. Gomina and J. Bréard, *Composites, Part A*, 2007, **38**, 1912.
71. K. N. Law, B. V. Kokta and C. B. Mao, *Bioresour. Technol.*, 2001, **77**, 1.
72. K. Kaack and K. U. Schwarz, and P. E. Brander, *Ind. Crops Prod.*, 2003, **17**, 131.
73. I. Sakurada, Y. Nukushina and T. Ito, *J. Polym. Sci.*, 1962, **57**, 651.
74. W. J. Cousins, *Wood Sci. Technol.*, 1978, **12**, 161.
75. J. E. G. Van Dam, Wet Processing of Coir, *CFC/FAO Techno-economic Manual*, 2002, no. 6, 1.
76. P. J. Wakelyn, N. R. Bertoniere, A. D. French, S. H. Zeronian, T. P. Nevell, D. P. Thibodeaux, E. J. Blanchard, T. A. Calamari, B. A. Triplett, C. K. Bragg, C. M. Welch, J. D. Timpa, W. E. Franklin, R. M. Reinhardt, T. L. Vigo, in *Handbook of Fiber Chemistry*, ed. M. Lewin and E. M. Pearce, Dekker, New York, 1998, vol. 15, p. 577.
77. D. Behrens, *Curauá-Faser-eine Pflanzenfaser als Konstruktionswerkstoff?*, Köster, Berlin, 1999.
78. H. Kaddami, A. Dufresne, B. Khelifi, A. Bendahou, M. Taourirte, M. Raihane, N. Issartel, H. Sautereau, J. F. Gérard and N. Sami, *Composites, Part A*, 2006, **37**, 1413.
79. A. Thygesen, J. Oddershede, H. Lilholt, A. B. Thomsen and K. Stahl, *Cellulose*, 2005, **12**, 563.
80. I. M. De Rosa, J. M. Kenny, D. Puglia, C. Santulli and F. Sarasini, *Compos. Sci. Technol.*, 2010, **70**, 116.
81. B. M. Cherian, A. L. Leao, S. F. de Souza, S. Thomas, L. A. Pothan and M. Kottaisamy, *Carbohydr. Polym.*, 2010, **81**, 720.
82. X. Chen, J. Yu, Z. Zhang and C. Lu, *Carbohydr. Polym.*, 2011, **85**, 245.
83. D. R. Keshwani and J. J. Cheng, *Bioresour. Technol.*, 2009, **100**, 1515.
84. A. Chakravarty and J. W. S. Hearle, *J. Text. Inst.*, 1967, **58**, 651.
85. L. Y. Mwaikambo, *AJST*, 2006, **7**, 120.
86. A. G. Kulkarni, K. G. Satyanarayana, P. K. Rohatgi and K. Vijayan, *J. Mater. Sci.*, 1983, **18**, 2290.
87. D. S. Varma, M. Varma and I. K. Varma, *Text. Res. J.*, 1984, **54**, 827.
88. R. Meredith, *Mechanical Properties of Textile Fibers*, Interscience, New York, 1956.
89. C. Baley, *Composites, Part A*, 2002, **33**, 939.
90. G. C. Davies and D. M. Bruce, *Text. Res. J.*, 1998, **68**, 623.
91. A. K. Bledski and J. Gassan, *Prog. Polym. Sci.*, 1999, **24**, 221.
92. E. T. N. Bisanda and M. P. Ansell, *J. Mater. Sci.*, 1992, **27**, 1690.

93. Y. Nishiyama and T. Okano, *J. Wood Sci.*, 1998, **44**, 310.
94. P. Wilson, *HFRS*, FAO report, Rome, 1971, no. 8.
95. N. Chand, S. Sood, P. K. Rohatgi and K. G. Satyanarayana, *J. Sci. Ind. Res.*, 1984, **43**, 489.
96. L. Y. Mwaikambo and M. P. Ansell, *J. Appl. Polym. Sci.*, 2002, **84**, 2222.
97. X. Hu and Y. J. Hsieh, *J. Polym. Sci., Polym. Phys.*, 1996, **34**, 1451.
98. V. A. Krakhmalev and A. A. Paiziev, *Cellulose*, 2006, **13**, 45.
99. N. Parameswaran and W. Liese, *Wood Sci. Technol.*, 1976, **10**, 231.
100. C. S. Gritsch, aand R. J. Murphy, *Ann. Bot. (Oxford, UK)*, 2005, **95**, 619.
101. S. Thomas and L.A. Pothan, in *Natural Fibre Reinforced Polymer Composites,* ed. S. Thomas and L. A. Pothan, Old City Publishing, Philadelphia, 2009, p. 3.
102. K. Charlet, J. P. Jernot, J. Bréard and M. Gomina, *Ind. Crops Prod.*, 2010, **32**, 220.
103. P. J. Van Soest and R. H. Wine, *J. Assoc. Off. Anal. Chem.*, 1967, **50**, 50.
104. A. Nechwatal, K. P. Mieck and T. Reusmann, *Compos. Sci. Technol.*, 2003, **63**, 1273.
105. H. L. Bos, PhD thesis, Eindhoven, Netherlands, 2004.
106. T. Dooley, T. S. Creasy and A. Cuellar, *Composites, Part A*, 2000, **31**, 1255.
107. N. E. Zafeiropoulos and C. A. Baillie, *Composites, Part A*, 2007, **38**, 629.
108. K. L. Pickering, G. W. Beckermann, S. N. Alam and N. J. Foreman, *Composites. Part A*, 2007, **38**, 461.
109. F. A. Silva, N. Chawla and R. D. de Toledo Filho, *Compos. Sci. Technol.*, 2008, **68**, 3438.
110. N. Defoirdt, S. Biswas, L. De Vriese, L. Q. N. Tran, J. Van Acker, Q. Ahsan, L. Gorbatikh, A. Van Vuure and I. Verpoest, *Composites. Part A*, 2010, **41**, 588.
111. E. M. F. Aquino, L. P. S. Sarmento, W. Oliveira and R. V. Silva, *J. Reinf. Plast. Compos.*, 2007, **26**, 219.
112. K. U. Joseph, J. Rani and M. Lovely, in *Proceedings of EcoComp2003*, London, September 2003.
113. B. Wielage, T. Lampke, G. Marx, K. Nestler and D. Starke, *Thermochim. Acta*, 1999, **337**, 169.
114. J. Gassan and A. K. Bledzki, *J. Appl. Polym. Sci.*, 2001, **82**, 1417.
115. S. Garkhail, PhD Thesis, University of London, 2002.
116. K. Van de Velde and P. Kiekens, *J Appl. Polym. Sci.*, 2002, **83**, 2634.
117. G. W. Beckermann and K. L. Pickering, Composites, *Part A*, 2008, **39**, 979.
118. C. Albano, J. Gonzalez, M. Ichazo and D. Kaiser, *Polym. Degrad. Stab.*, 1999, **66**, 179.
119. B. Madsen and H. Lilholt, *Compos. Sci. Technol.*, 2003, **63**, 1265.
120. B.Madsen and H.Lilholt, in *Proceedings of the 16th International Conference on Composite Materials*, Kyoto, Japan, July 2007.
121. G. Romhany, J. Karger-Kocsis and T. Czigany, *J. Appl. Polym. Sci.*, 2003, **90**, 3638.
122. K. Charlet and A. Béakou, *Int. J. Adhes. Adhes.*, 2011, **31**, 875.

123. S. L. Bai, C. M. L. Wu and Y. M. Mai, *Adv. Compos. Lett.*, 1999, **8**, 13.
124. C. Morvan, C. Andème-Onzighi, R. Girault, D. S. Himmelsbach, A. Driouich and D. E. Akin, *Plant Physiol. Biochem.*, 2003, **41**, 935.
125. H. Fischer, H. Gerardi, D. Knittel and V. Antonov, in *Proceedings of the 15th International Conference STRUTEX*, Liberec, Czech Republic, 2008, p. 331.
126. M. K. Joshy, M. Lovely and R. Joseph, *Int. J. Polym. Mater.*, 2009, **58**, 2.
127. N. Reddy and Y. Yang, *Biotechnol. Bioeng.*, 2007, **97**, 1021.
128. A. Bourmaud and S. Pimbert, *Composites, Part A*, 2008, **39**, 1444.
129. F. Tröger, G. Wegener and C. Seemann, *Ind. Crops Prod.*, 1998, **8**, 113.
130. I. M. De Rosa, J. M. Kenny, M. Maniruzzaman, M. Moniruzzaman, M. Monti, D. Puglia, C. Santulli and F. Sarasini, *Compos. Sci. Technol.*, 2011, **71**, 246.
131. B. Mano, J. R. Araújo, M. A. S. Spinacé and M.-A. De Paoli, *Compos. Sci. Technol.*, 2010, **70**, 29.
132. P. A. Santos, M. A. S. Spinacé, K. K. G. Fermoselli and M.-A. De Paoli, *Composites, Part A*, 2007, **38**, 2404.
133. K. M. M. Rao, K. M. Rao and A. V. R. Prasad, *Mater. Des.*, 2010, **31**, 508.
134. J. L. Guimarães, E. Frollini, C. G. da Silva, F. Wypych and K. G. Satyanarayana, *Ind. Crops Prod.*, 2009, **30**, 407.
135. H. Demir, U. Atikler, D. Balköse and F. Tıhmınlıoğlu, *Composites, Part A*, 2006, **37**, 447.
136. C. A. Boynard and J. R. M. D'Almeida, *Polym.-Plast. Technol. Eng.*, 2000, **39**, 489.
137. B. F. Abu-Sharkh and H. Hamid, *Polym. Degrad. Stab.*, 2004, **85**, 967.
138. Y. Zou, N. Reddy and Y. Yang, *J. Appl. Polym. Sci.*, 2010, **116**, 2366.
139. F. Yao, Q. Wu, Y. Lei and Y. Xu, *Ind. Crops Prod.*, 2008, **28**, 63.
140. O. Carmody, R. Frost, Y. Xi and S. Kokot, *Surf. Sci.*, 2007, **601**, 2066.
141. A. V. R. Prasad, K. M. M. Rao, M. A. Kumar and K. M. Rao, *Indian J. Fibre Text. Res.*, 2006, **31**, 335.
142. L. Qin, J. Qiu, M. Liu, S. Ding, L. Shao, S. Lü, G. Zhang, Y. Zhao and X. Fu, *Chem. Eng. J.*, 2011, **166**, 772.
143. H. S. Yang, D. J. Kim, Y. K. Lee, H. J. Kim, J. Y. Jeon and C. W. Kang, *Bioresour. Technol.*, 2004, **95**, 61.

CHAPTER 4

Relation between Structural Anisotropy in Natural Fibres and Mechanical Properties in Composites

ELESSANDRA DA ROSA ZAVAREZE*[a] AND
ALVARO RENATO GUERRA DIAS[b]

[a] School of Food Engineering, Federal University of Pampa, 96413-170,
Bagé, Brazil; [b] Department of Agroindustrial Science and Technology,
Federal University of Pelotas, 96010-900, Pelotas, Brazil
*Email: elessandrad@yahoo.com.br

4.1 Introduction

As a result of the increasing demand for environmentally friendly materials and the desire to reduce the cost of traditional fibres (carbon, glass and aramid) and reinforced petroleum-based composites, new bio-based composites have been developed.[1] Researchers have begun to focus their attention on natural fibre composites, which are composed of natural or synthetic resins, reinforced with natural fibres. The natural fibre component may be wood, sisal, flax, hemp, coconut, cotton, flax, jute, banana leaf fibres, bamboo, wheat straw or other fibrous material. Natural fibres can be divided into vegetable, animal and mineral fibres. The main polymers involved in the composition of vegetable fibres are cellulose, hemicelluloses, lignin and pectin, while fibres of animal origin consist of proteins (hair, silk, wool).[2]

RSC Green Chemistry No. 16
Natural Polymers, Volume 1: Composites
Edited by Maya J John and Thomas Sabu
© The Royal Society of Chemistry 2012
Published by the Royal Society of Chemistry, www.rsc.org

Polymer matrix composites using natural plant fibres as reinforcement have attracted much interest in recent years. This is because these natural fibres can be harvested from renewable resources, possess long aspect ratios for efficient stress transfer, and certain aspects of their mechanical properties are comparable to those of existing inorganic glass fibres.[3] Thermoplastics are alternative matrix materials that possess a number of clear advantages over thermoset matrices. These include enhanced shelf-life, post-forming, toughness and recyclability. The physical and chemical nature of a fibre surface defines its compatibility and adhesion with a polymer, as well as determining such characteristics as surface free energy, wettability and the possibility of interaction with the matrix or coupling agents. Natural fibres offer many advantages such as energy efficiency, low cost, low density, high toughness, acceptable specific strength and bio-degradability.[4,5] However, they do suffer from a few limitations because bio-fibres are hydrophilic in nature and thus are less compatible with relatively hydrophobic polymer matrices. The hydrophilic nature of bio-fibres is also responsible for the water absorption characteristics of biocomposites, thus limiting their application. Because the mechanical properties of composites are related to the compatibility and interaction between their components, the improvement of the interface and interphase interactions in natural fibre/polyester composites is essential.[6]

Fibres serve as reinforcement in composites and show high tensile strength and stiffness, while the matrix holds the fibres together, transmits the shear forces and also functions as a coating. The selection of suitable fibres is determined by the required values of stiffness and tensile strength of a composite. Further criteria for choosing suitable reinforcing fibres include elongation at failure, thermal stability, adhesion of the fibres and matrix, dynamic behaviour, long-term behaviour and price and processing costs.[2] The mechanical properties of fibres determine the stiffness and tensile strength of a composite. Depending on the fibre orientation, the materials behaviour of composites can be divided into quasi-isotropic (with all short fibres randomly orientated, and no privileged direction of mechanical properties), anisotropic (with all fibres orientated in one or more directions with corresponding mechanical properties) or orthotropic (fibres orientated mainly in two directions orthogonal to each other and showing corresponding materials behaviour).[2,7] Owing to the anisotropy in composites reinforced with natural fibres, some properties modify the fibre orientation, *i.e.* the mechanical properties depend on the direction in which force is applied. The fibres adopt a particular orientation in the process, which is responsible for the increased stiffness in the direction parallel to the flow. Materials can be classified as either isotropic or anisotropic. Isotropic materials have the same material properties in all directions, and anisotropic materials have different properties along different directions with respect to a specified point in a material.[7]

The mechanical properties of composite materials depend on the structure in a complex way. The mechanical properties of natural fibres are affected by the variability in plants and by the processing stage and damage sustained during processing; thus, there is a large distribution in observed mechanical properties.[8,9] The mechanical performance of composites is mainly dependent upon the properties of the matrix and reinforcement material and the interaction between the matrix and reinforcement. Moreover, the chemical composition, crystallinity, surface properties, diameter, cross-sectional shape, length, strength and stiffness vary from fibre to fibre. Natural fibres can improve the mechanical properties of composites, particularly thermoplastic products. Fibre-reinforced plastic composites are used in several structural applications and fields such as construction, aerospace, automotive parts, sports equipment, machinery and the food industry.[10]

Many researchers have studied the effect of natural fibres on the properties of reinforced composites.[2,3,6,8,11–24] This chapter will review the properties of natural fibres and composites, emphasizing the co-relation between the structural anisotropy in natural fibres and the mechanical properties in composites. The importance and applications of natural fibre composites will also be discussed.

4.2 Natural Fibres

Natural fibres can be further subdivided into vegetable, animal and mineral fibres. Mineral fibres are generally used in very small amounts, often due to factors that affect human health during extraction. Vegetable fibres (*e.g.*, cotton, flax, hemp, jute) are composed of cellulose, while fibres of animal origin consist of proteins (*e.g.*, hair, silk, wool). The surfaces of natural fibres are rough and uneven and provide good adhesion to the matrix in a composite structure. When determining the properties of natural fibres, one has to keep in mind that one is dealing with natural products with properties that are strongly influenced by their natural growth environment, such as temperature, humidity, the composition of the soil, strength of its fibres, density, *etc.* Additionally, the way the plants are harvested and processed results in a variation in properties.[25]

Natural fibres are rounded, elongated structures with hollow cross-sections spread over the entire plant and are classified according to the anatomic origin of the stem, leaf, wood and surface fibres. Stem fibres occur in the phloem of, for example, jute, ramie, flax and palm fibres. Banana is also extracted from the pseudo-stem of the plant. Leaf fibres are extracted from the leaves of plants such as sisal, curaua, pineapple and palm. Wood fibres can be subdivided into two types, hardwood and softwood, and these have different properties. Wood is a hydrophilic porous composite of cellulose, lignin and hemicellulose polymers that are rich in functional groups such as hydroxyls, which readily interact with water molecules by hydrogen bonding.[26] Surface fibres form a protective layer on the stems, leaves, fruits and seeds of plants such as açaí, coconut and cotton. Vegetable fibres are composed primarily of cellulose, hemicellulose, lignin and small amounts of pectin, inorganic salts, nitrogenous substances and natural dyes, which constitute what is called the soluble fraction. Cellulose, the largest fractional component in vegetable fibres, is a semicrystalline polysaccharide made up of D-glucosidic bonds. The large number of hydroxyl groups in cellulose (three in each repeating unit) imparts hydrophilic properties to natural fibres. Hemicellulose is strongly bound to the cellulose fibrils, presumably by hydrogen bonds. Hemicellulose polymers are branched and fully amorphous and have a significantly lower molar mass than cellulose. Because of its open structure containing many hydroxyl and acetyl groups, hemicellulose is partly soluble in water and is hygroscopic. Lignins are amorphous, highly complex, mainly aromatic polymers with phenylpropane units, but have the lowest water sorption of natural fibre components.[27,28]

A fibre has a length that is much greater than its diameter. The length-to-diameter ratio is known as the aspect ratio, which can vary greatly. Continuous fibres have high aspect ratios, whereas discontinuous fibres have low aspect ratios. Continuous-fibre composites normally have a preferred orientation, while discontinuous fibres generally have a random orientation.[7] Each individual fibre cell has a complex structure formed by cell walls that surround the lumen. The walls consist of layers formed by clusters of microfibrils in plants that grow in a spiral shape with different angular orientations. The main component of microfibrils is cellulose; cellulose microfibrils are interconnected through a network of hemicellulose molecules. Another important constituent of the walls is lignin, a primarily hydrophobic substance that forms an impregnated layer near the surface, which is responsible for support. Microfibrils are composed of micelles, which are crystals of cellulose molecules arranged parallel to different allotropic structures.

4.3   Sources of Natural Fibres

Plant fibre sources include cotton, flax, hemp, sisal, jute, kenaf, henequen, corn and coconut. Fibres from mineral sources include asbestos fibres. Asbestos is the generic commercial designation for a group of naturally occurring mineral silicate fibres of the serpentine and amphibole series. These include the serpentine mineral chrysotile (white asbestos) and the five amphibole minerals: actinolite, amosite (brown asbestos), anthophyllite, crocidolite (blue asbestos) and tremolite. Asbestos consist of short, discontinuous fibres of high length/diameter ratio and high surface area. Fibres from animal sources include silk and wool. Fibres from animal sources include wool, a fibre formed in the fleece of sheep or similarly hairy animals (*e.g.*, alpacas, llamas, vicunas, yaks, camels, cashmere goats, mohair goats and angora rabbits) produced in many places around the world. After shearing, the wool is washed, carded and dyed before being woven or knitted into fabric. Wool is flexible and absorbs moisture, making it cool in summer and warm in winter. Silk is another fibre from animal sources that is produced from the cocoons of the silkworm. Silk absorbs moisture, dries quickly, will not shrink, is easily dyed, retains its shape and has a natural shimmer.

Natural fibres exhibit a variation in properties due to environmental conditions. There are many types of vegetable fibres, which are the most important and most used fibres in several industrial applications.

4.3.1   Cotton Fibre

Worldwide, cotton is one of the most important fibres used in the textile industry. Hand picking of cotton fruits are highly labour-intensive, which is often carried out by machine. In many parts of the world, however, picking is still carried out by hand. The seeds, dead leaves and other debris are removed by ginning. The cotton can absorb up to 20% of its dry weight in moisture without feeling wet and is also a good heat conductor.

4.3.2   Jute Fibre

According to the International Jute Study Group,[29] jute is one of the most well-known vegetable fibres, which is mostly grown in countries such as India, Bangladesh, China, Nepal and Thailand. Together, they produce about 95% of the global production of jute fibres. Like those of synthetic fibre composites, the mechanical properties of the final products made of jute fibres also depend on the individual properties of the matrix, fibre and the nature of the interface between them.[19] The fibres of jute are extracted from the ribbon of the stem. These fibres are obtained by successively retting in water, beating, stripping the fibre from the core and drying. Owing to its short fibre length, jute is the weakest stem fibre, although it withstands rotting very easily. It is used as a packaging material and in carpet backing, ropes, yarns and wall decorations.[25]

4.3.3   Flax Fibre

Flax fibre is readily marketed in a variety of industries, including textiles, composites and specialty paper. The individual fibres are formed in bundles that encircle the core tissue. Flax must undergo the process of retting to separate the fibre from woody cells, which are termed shives and constitute the major waste component of flax fibres.[30] When ripe, flax plants are pulled from the ground rather than cut, to avoid a loss in fibre length from the stubble left in the field. Flax is a strong fibre and experiences an increase in strength under wet conditions and can absorb 20% moisture without feeling wet. Flax has good heat conducting properties and is hard wearing and durable. Flax is used in the production of linen, ropes and sacks.

4.3.4   Ramie Fibre

Ramie produces some of the strongest and longest plant fibres, which are lustrous with an almost silky appearance. Fibres from ramie are used in clothing fabrics, industrial packaging, twines, cordages, canvas, car outfits, *etc.* Recently, ramie fibres were successfully tested in fibre-reinforced composites.[31]

4.3.5   Hemp Fibre

Hemp fibres are present in bundles that are as long as the stems, which can easily be peeled off the xylem surface by hand or machine. The fresh stem consists of a hollow cylinder of 1–5 mm thick xylem covered by 10–50 μm cambium, 100–300 μm cortex, 20–100 μm epidermis and 2–5 μm cuticle.[14] A hemp yarn is strong and has the highest resistance against water of all natural fibres; however, it should not be creased excessively, to avoid breakage. Fibres from hemp stems have been widely used in the production of cords and clothing and have potential as a reinforcement material in polymer/matrix composites.

4.3.6   Sisal Fibre

Sisal fibre is one of the most widely used natural fibres and is very easily cultivated. Sisal fibre is a hard fibre extracted from the leaves of the sisal plant (*Agave sisalana*), which originated in Mexico and is now mainly cultivated in East Africa, Brazil, Haiti, India and Indonesia. A sisal plant produces about 200–250 leaves, each of which contains 1000–1200 fibre bundles composed of 4% fibre, 0.75% cuticle, 8% dry matter and 87.25% water.[32] Sisal produces sturdy and strong fibres that are very well resistant against moisture and heat. It is mainly used in ropes, mats, carpets and as a cement reinforcement.[25] The characteristics of the sisal fibres depend on the properties of the individual constituents, the fibrillar structure and the lamellae matrix. Sisal fibres are composed of numerous elongated fusiform fibre cells that taper towards each end. The fibre cells are linked together by middle lamellae, which consist of hemicellulose, lignin and pectin.

4.3.7   Henequen Fibre

Henequen fibre (*Agave fourcroydes*) is composed of approximately 77% cellulose, 4–8% hemicelluloses, 13% lignin and 2–6% pectin and waxes by weight. Henequen has been used extensively to make ropes, carpets and cordages for a long period of time.[33] The structural arrangement of lignin and cellulose inside the fibre is based fundamentally on lignin, which acts as a cementing matrix for the cellulose component and has its own separate structure. The cellulose fibre is the rigid part of the fibre and the mechanical properties of henequen fibre such as stiffness and strength are mainly influenced by its structural arrangement.[34]

4.3.8   Coir Fibre

Coir is a natural fibre extracted from the husk of coconut and used in products such as floor mats, doormats, brushes and mattresses. Coconut fibre is obtained from the husk of the fruit of the coconut palm. The fruits are dehusked on a spike, and after retting, the fibres are extracted from the husk by beating and washing. The fibres are strong, light and can easily withstand heat and salt water. After nine months of growth, the nuts are still green and contain white fibres, which can be used for the production of yarn, rope and fishing nets. After 12 months of growth, the fibres are brown and can be used for brushes and mattresses.[25]

4.4 Composites

4.4.1 Biopolymer Composites

A composite material can be defined as a combination of two or more materials that results in better properties than those of the individual components used alone. Unlike metallic alloys, each material retains its individual chemical, physical and mechanical properties. The two constituents behave as a reinforcement and a matrix, respectively. The main advantages of composite materials are their high strength and stiffness combined with low density, compared with bulk materials, allowing for a weight reduction in the finished product.[7]

Biodegradable polymers are designed to decompose upon disposal by the action of living organisms. Extraordinary progress has been made in the development of practical processes and products using polymers such as starch, cellulose and lactic acid.[35] Biopolymer composites are becoming increasingly employed to replace oil-based components.[36] Understanding their mechanical properties is essential to guarantee high performance during their lifetime.[13] The main impediment to the development of biopolymers is their low mechanical performance and high sensitivity to environmental conditions (temperature, humidity).[37]

According to Guessasma and Bassir,[38] the use of biopolymer-based composites such as starchy materials to replace oil-based materials is a challenging task that has led to several recent research developments. The improvement in the mechanical performance of biocomposites is a key prerequisite to increase the potential of their use as materials in packaging and single-use components, among other applications. The improvement in the mechanical performance of thermomoulded glassy biocomposites is correlated to the basic understanding of the relationship between the composites' elastic properties and microstructural attributes. Recently, the tailoring of the elastic properties of such composites has been proved to be dependent on the composites' intrinsic properties, the spatial arrangement of phases and interface behaviour.[36] Starch is an inexpensive, annually renewable material derived from corn and other crops. The biodegradation of starch products recycles atmospheric CO_2 trapped by starch-producing plants. The starches contain amylose and amylopectin at ratios that vary with starch source. This variation provides a natural mechanism for regulating the material properties of starch.[35]

Biopolymers such as cellulose, starch and proteins can also be modified to produce composites. Chemically modified plant cellulose is used in a remarkably diverse set of applications. For example, cellulose acetate is used in many common applications, including toothbrush handles and adhesive tape backing.[35] Kadokawa *et al.*[16] studied the properties of cellulose/starch composite fibrous material by reconstitution from a mixture. These authors indicate that the crystalline structures of the cellulose and starch were largely disrupted in the gel, and the scanning electron microscopy and X-ray diffraction data of the obtained material show a compatibilized fibrous structure. The cellulose and starch are composed of the same unit structure, *i.e.* a glucose unit, but are linked through different glycosidic bonds, such as β-$(1\rightarrow4)$ for cellulose and α-$(1\rightarrow4)$ and α-$(1\rightarrow6)$ for starch. The roles of cellulose and starch in nature are completely different; the former is a structural material and the latter is an energy storage material. Cellulose is the most abundant organic substance on Earth and, thus, a very important renewable resource that has a number of traditional applications, including its use in furniture, clothing and medical products.[39]

Kumar and Zhang[17] studied the properties of aligned ramie-fibre-reinforced and soy protein composites and described that the alignments of fibres along different directions had a significant effect on the mechanical properties of the soy protein/ramie fibre composites; moreover, the soy protein/ramie fibre composites with vertically aligned fibres exhibited the highest mechanical properties. The arylation of the soy protein/ramie fibre composites led to a significant improvement in water resistance and elastic modulus. The mechanical properties of the protein composites are proportional to the volume fraction of fibres as well as the orientation of the fibres. To improve the properties of soy protein isolate materials, a series of modified protein composites such as soy protein isolate thermoplastic reinforced with chitin whiskers,[40] montmorillonite exfoliated soy protein isolate nanocomposites[41] and water-induced hydrophobic soy protein isolate materials[42] have been reported.

4.4.2 Thermoplastic Composites

Thermoplastics are polymers that require heat to make them processable. After cooling, such materials retain their shape. In addition, these polymers may be reheated and reformed, often without significant changes in their properties.[43] Thermoplastic polymers constitute an important class of materials with a wide variety of applications. Thermoplastic composites offer some substantial advantages over thermoset composites. Owing to the higher toughness of the matrix, they offer a higher impact resistance. Manufacturing cycle times consisting of melting the matrix, shaping and consolidation by cooling are significantly shorter than those for thermosets, which require a time-consuming curing step. The main disadvantage of thermoplastic composites is the need for high processing temperatures and pressures, caused by the high melt viscosity of the matrix. In addition, proper impregnation of the fibres at the micron level has proven difficult and often results in products with a locally high void content. One way to improve fibre impregnation is to bring the matrix and the fibres in more intimate contact before the final moulding step or, in other words, to reduce the required flow length of the polymer matrix.[44,45] According to Riedel and Nickel,[2] the impregnation of fibres in a composite structure, particularly thermoplastic resins, depends on the viscosity of the matrix, which can be influenced by temperature.

Polymers of natural origin (*e.g.*, starch and cellulose) must be modified either physically or chemically to make them suitable for processing as thermoplastic resins. For example, the structure of starch can be made thermoplastic with adjuvants such as glycerol and water. The physical, chemical, mechanical and thermal properties of biopolymers are also influenced by physical and chemical modifications. The esterification of hydroxyl groups on side chains is preferred to convert cellulose into a thermoplastic material that retains its cellulose chain structure.[2] Starch polymers have been applied in several areas, including in starches used in composites, building materials and polymer blends. The main drawbacks of thermoplastic starches are their brittleness, moisture absorption and low mechanical properties. The incorporation of natural fibres is an alternative to improve their thermal and mechanical properties and preserve their biodegradability. Lignocellulosic fibres can be used for the development of new composites because of their good mechanical and physical properties.[23]

Ma *et al.*[46] reported on the properties of thermoplastic starch composites with micro winceyette fibre as reinforcement. The introduction of micro winceyette fibre improved the tensile strength, water resistance and thermal stability of thermoplastic starch because of the good adhesion between the starch and fibre. With the increase in fibre content from 0 to 20%, the initial tensile strength was trebled up to 15.16 MPa, while the elongation was reduced from 105 to 19%. The Young's modulus of thermoplastic starch behaved analogously to the tensile strength as a consequence of introducing fibres. The reinforcement effect was gradually weakened with the increase in water content, but at high water content ($>30\%$) both the fibre and water content had no effect on the tensile strength. A considerable increase in tensile strength indicated that thermoplastic starch was well suited as the matrix for natural cellulose fibres. These authors reported that this was due to the remarkable intrinsic adhesion of the fibre/matrix interface caused by the chemical similarity of starch and the cellulose fibre.

4.4.3 Natural Fibre Composites

Today, natural fibres constitute an interesting alternative in composite technology. The use of fibres such as flax, hemp, jute or sisal in this industry to date is small, in part because the availability of a durable semi-finished product with consistently good quality is often a problem. The natural fibres are embedded in a biopolymer matrix system, the function of which is to hold the fibres together. This provides and stabilizes the shape of the composite structure, transmits the shear forces between the mechanically high-quality fibres, and protects them against radiation and other aggressive media.[2]

The properties of composites are strongly dependent on the properties of their constituent materials, their distribution and the interaction between them. The composite properties may be the volume fraction sum of the properties of the constituents, or the constituents may interact in a synergistic way, resulting in improved or better properties. Apart from the nature of the constituent materials, the geometrical properties of the reinforcement, such as shape, size and size distribution, influence the properties of the composite to a great extent. The concentration distribution and orientation of the reinforcement also affect the properties. The shape of the discontinuous phase, which may by spherical, cylindrical or rectangular cross-sanctioned prisms or platelets, the size distribution, which controls the texture of the material, and volume fraction determine the interfacial area; this area plays an important role in determining the extent of the interaction between the reinforcement and the matrix.[18]

Although natural fibres can offer the resulting composites many advantages, typically, polar fibres have inherently low compatibility with non-polar polymer matrices, especially hydrocarbon matrices such as polypropylene and polyethylene.[47] Xie *et al.*[48] reported that this incompatibility may cause problems during composite processing and for the material properties. Hydrogen bonds may form between the hydrophilic fibres, and thus the fibres tend to agglomerate into bundles and unevenly distribute throughout the non-polar polymer matrix during compounding processing.

Research and development have shown that these aspects can be improved considerably. Because natural fibres are cheap and have a better stiffness per weight than does glass, which results in lighter components, interest in natural fibres has grown. Moreover, the environmental impact is smaller because the natural fibre can be thermally recycled and fibres come from a renewable resource. Their moderate mechanical properties keep fibres from being used in high-tech applications, but for many reasons they can compete with glass fibres. The advantages are as follows: (1) low specific weight, which results in a higher specific strength and stiffness than glass; (2) it is a renewable resource and their production requires little energy; (3) they can be produced with low investment at low cost; (4) the feature friendly processing, with no wear of tooling or skin irritation; (5) thermal recycling is possible, whereas glass causes problems in combustion furnaces; and (6) good thermal and acoustic insulating properties. The disadvantages are as follows: (1) lower strength properties, particularly impact strength; (2) variable quality, depending on unpredictable influences such as weather; (3) moisture absorption, which causes swelling of the fibres; (4) restricted maximum processing temperature; (5) lower durability, though fibre treatments can improve this considerably; (6) poor fire resistance; and (7) the price can fluctuate by harvest results or agricultural politics.[25]

The properties of composite materials depend very much upon their structure. Composites differ from homogeneous materials in that considerable control can be exerted over larger-scale structures and, hence, over the desired properties. In particular, the properties of a composite material depend upon the shape of the heterogeneities, upon the volume fraction occupied by them and upon the interface among the constituents. The shapes of the heterogeneities in a composite material are classified as follows. The principal inclusion shape categories are: (1) the particle, with no long dimension; (2) the fibre, with one long dimension; and (3) the platelet or lamina, with two long dimensions. The inclusions may vary in size and shape within a category. For example, particulate inclusions may be spherical, ellipsoidal, polyhedral or irregular. If one phase consists of voids, filled with air or liquid, the material is referred to as a cellular solid. If the cells are polygonal, the material is a honeycomb; if the cells are polyhedral, it is a foam. In each composite structure we may moreover make the distinction between random orientation and pre-ferred orientation.[49]

Natural fibres are being widely used in fibre-reinforced polymer composites. A common feature of natural fibres is a much higher variability of mechanical properties. This necessitates study of the strength distribution of the fibres and efficient experimental methods for its determination. Common fibre reinforced composites are composed of fibres and a matrix. The fibres are the reinforce-ment and the main source of strength, while the matrix glues all the fibres together in shape and transfers stresses between the reinforcing fibres. The fibres carry the loads along their longitudinal directions. Sometimes, filler might be added to smooth the manufacturing process, impact special properties to the composites and/or to reduce the product cost.[18] Fibre-reinforced com-posites require a moderate adhesion between the matrix and fibre. Good adhesion between the two phases provides good mechanical strength by effi-cient transfer of load from the matrix to the fibre, but the material becomes brittle. Poor adherence results in low mechanical strength, but the energy absorbed increases the fracture energy dissipation during the detachment of the fibre. Researchers have investigated applications from several natural fibres, such as hemp,[14] flax,[9] sisal,[12,21] oil palm,[10] jute[15] and cotton fibres[20,50] to improve the properties of composites.

4.4.4   Anisotropy

The material behaviour of composites depends on the fibre orientation; when the material is anisotropic, all fibres are orientated in one or more directions, with corresponding mechanical properties.[2,7] The anisotropy of materials is defined as the variation in the material's physical response to the applied stress along different specimen axes. The local average fibre direction and the degree of anisotropy can be evaluated based on parameters related to the distribution of the different phases in a representative volume, without the need to analyse and/or reconstruct the geometry of each single fibre.[51] The fibre length dis-tribution in a material sample can be easily determined by separating the fibres form the matrix by burning or hydrolysis and observing them using an optical microscope, although further fibre breakage during extraction cannot be excluded. The measurement of fibre orientation is more difficult because fibres are dispersed in the matrix and cannot be separated from the matrix without altering their orientation.[48] Consequently, simplifying assumptions with respect to fibre orientation are usually made to apply these mechanical models, and fibre orientation factors are derived implicitly from tests rather than evaluated on the basis of fibre angles measurement.[52] Lillie *et al.*[53] also reported that it is important to determine the inherent anisotropy due to certain structural features of an unstressed material, such as an unequal partitioning of ortho-gonal fibres and induced anisotropy due to superimposed stresses on a non-linear material in which elastic moduli depend on loading conditions.

Anisotropic properties can be obtained by the addition of aligned organic or inorganic fibrous fillers. The incorporation of short fibres has caused many technical difficulties, such as an increase in the viscosity of the blend system, machine wear problems and increased energy consumption.[27] The fibre length distribution and fibre orientation distribution in composites play an important role in determining the composite mechanical properties. Owing to the partial fibre orientation, short-fibre composites can exhibit more or less anisotropy with respect to their mechanical properties. This inevitability of anisotropy in the mechanical properties of short-fibre reinforced composites can lead to severe deterioration in the total performance of the composites when their unfavourable direction is highly loaded or under attack by the environment.[54] Michalowski and Cermák[55] tested three distributions of fibres: random orientation, all fibres in the vertical direction and all fibres in the horizontal direction. These authors reported that the contribution of the fibres to the composite strength was the largest when they were placed in the direction of largest extension of the composite (here, horizontal). Vertical fibres in triaxial testing were subjected to compression; they had an adverse effect on the initial stiffness of the composite and they did not contribute to any strength increase. Specimens with a random distribution of fibres exhibited a smaller increase in strength than those with horizontal fibres because a portion of randomly distributed fibres was subjected to compression.

Fibre-reinforced composite materials typically exhibit anisotropy. That is, some properties vary depending upon which geometric axis or plane they are measured along. For a composite to be isotropic with respect to a specific property, such as Young's modulus, all reinforcing elements, whether fibres or particles, must be randomly oriented. This is not easily achieved for discontinuous fibres because most processing methods tend to impart a certain orientation to the fibres. Continuous fibres in the form of sheets are usually used to deliberately align the composite anisotropy along a particular direction that is known to be the principally loaded axis or plane. Honda and Narita[56] focused their study on locally anisotropic structures of fibrous laminated composites and reported that they are often found as parts of natural compounds and that the bony pelvis of humans is composed of inorganic calcium phosphates and organic collagen fibres. The inorganic component contributes to stiffness and strength, while organic fibres provide toughness. The organic fibres form curvilinear shapes, which may be the optimum shape to sustain external forces. In other words, the natural compounds have locally anisotropic properties distributed optimally throughout the compound to perform more effectively than simple anisotropic materials.

To develop structure–property relationships in composite systems, it is necessary to know the properties of the fibre and matrix constituents. Much research indicates that natural fibres exhibit complicated anisotropic structure. Natural fibres are often composed of millimetre-long microtubules generally oriented along the fibre direction. These microtubules are in turn composed of a variety of substances, including cellulose, lignin and hemicelluloses. The cellulose forms crystalline fibrils bound to one another by non-crystalline cellulose that spiral within the microtubules along the general direction of the natural fibre axis. Models indicate that fibre stiffness is influenced by the spiral angle of the crystalline fibrils as well as the concentration of the non-cellulosic substances. These structural parameters vary between the different types of natural fibres, likely accounting for some of the variations in reported fibre properties.[32,57]

Demirci *et al.*[58] reported that the random orientation of fibres is the main source of mechanical anisotropy, and the orientation distribution function of the fibrous matrix can be used to analyse the structure's direction-dependent behaviour. Besides, a practical tool is developed to determine an orientation distribution function and anisotropic parameters based on its images obtained with scanning electron microscopy or X-ray micro-computed tomography regardless of its planar density.

4.4.5 Mechanical Properties

Tensile strength, tensile modulus and elongation at break of the composites vary as the properties and composition of the fibre and characteristics of the matrix of the composite. The specific mechanical properties of natural fibres are important when used in composites. When comparing the tenacity and elongation at failure of both natural and synthetic fibres, hemp, flax and ramie fibres can compete with glass fibres, which serve as a reference because of their major importance in composite technology.[2] According to John and Thomas,[1] the primary effects of biofibre reinforcement on the mechanical properties of natural rubber composites include increased modulus, increased strength with good bonding at high fibre concentrations, decreased elongation at failure, greatly improved creep resistance over particulate-filled rubber, increased hardness and a substantial improvement in cut, tear and puncture resistance.

Both the long-term and short-term mechanical performance of a natural fibre composite depend on the properties of the constituents, their relative volume fractions and on their microscopic architecture, for example, the fibre orientation distribution, the fibre length distribution, the fibre shape, the size of lumen and whether it is filled with resin or not. Other aspects of equally high importance are variable fibre parameters such as surface morphology and surface composition.[59] The mechanical properties of a fibre-reinforced polymer composite depend not only on the properties of the constituents but also on the properties of the region surrounding the fibre, known as the interphase. Stress transfer from the matrix to the fibre takes place at such an interphase and, therefore, it is important to characterize its properties to better understand the performance of the composite. Sreekala *et al.*[10] studied the tensile strength, tensile modulus and elongation at break of the hybrid composites of oil palm fibre hybridized with glass fibre to achieve superior mechanical performance. They reported that hybrid composites are materials created by combining two or more different types of fibres in a common matrix. These authors also reported that the mechanical performance of composites is mainly dependent upon the properties of the matrix and the reinforcement and the interaction between the matrix and the reinforcement.

The mechanical properties of natural fibre composites have been studied previously by many authors. Oksman et al.[11] investigated how poly(lactic acid) acts as a matrix material for natural fibre composites using flax fibre, and reported that the mechanical properties of poly(lactic acid) and flax fibre composites are promising. The composite strength was about 50% better compared to similar polypropylene/flax fibre composites, which are used today in many industrial applications. The stiffness of poly(lactic acid) was increased from 3.4 to 8.4 GPa with an addition of 30% flax fibres. Venkateshwaran et al.[24] studied the addition of sisal fibre to improve the mechanical properties of the banana fibre/epoxy composite and reported that the addition of sisal up to 50% increases the tensile strength, flexural strength and impact strength around 16, 4 and 35%, respectively.

The plant fibre structure can be thought of as a composite material with the stiff and strong cellulose microfibrils embedded in a hemicellulose/lignin matrix. However, the composite structure in plant fibres is rather complex (e.g., two-phase matrix and cell wall layers). Moreover, plant fibres are part of a larger biological system, i.e. the plants, with a long evolutionary history, and their properties have therefore been highly optimized with respect to the functional requirements of plants. Thus, the study of plant fibre mechanical properties is not just an assessment of the reinforcement potential of plant fibres in man-made composites, but might as well provide insight into the form and function of a sophisticated composite material.

A number of structural aspects serve to restrain the practical attainable tensile properties of plant fibres, e.g. the degree of cellulose crystallinity, the microfibril angle and the cellulose content. Table 4.1 presents typical reported tensile properties and microfibril angles of different types of plant fibres. Stiffness and ultimate stress of hemp fibres have been reported in the ranges 30–60 GPa and 300–800 MPa, respectively. It can be seen in Table 4.1 that in particular the measured ultimate stress of plant fibres is much below the theoretical estimates. This might be explained by the presence of fibre defects, which has been shown to affect the failure mechanisms in plant fibres.[60] The measured fibre stiffness more closely reflects the theoretical estimates. The observed large variation in the measured tensile properties is a typical trait for materials with a natural origin, but some variation is also added by the experimental testing procedure. The measurement of tensile properties of single

Table 4.1 Microfibril angle and tensile properties of different plant fibres.[a]

Plant fibre	Microfibril angle (°)	Stiffness (GPa)	Ultimate stress (MPa)
Hemp	6	30–60	300–800
Flax	6–10	50–70	500–900
Jute	8	20–55	200–500
Sisal	10–25	9–22	100–800
Softwood	3–50	10–50	100–170

[a]Data for microfibril angles are from Gassan et al.,[61] Anagnost et al.,[62] Lilholt and Lawther[63] and Madsen.[64]

plant fibres is not a simple task, and problems are especially related to fibre gripping and the determination of the fibre cross-sectional area.

4.4.6 Effect of Anisotropy in Natural Fibres on Mechanical Properties

The mechanical properties of an anisotropic material are a function of orientation. An anisotropic material has properties that vary with direction within the material. This directional dependence is observed for other material properties such as ultimate strength, Poisson's ratio and thermal expansion coefficient. Material property data, including fibre concentration, matrix properties and properties of the composite in the principal material directions, are incorporated into micro-mechanical models to estimate the fibre properties. The determination of composite properties along the principal material directions is, however, somewhat more involved. The mechanics concerning coordinate system transformations can be applied to arrive at the well-known relationship between off-axis Young's modulus and the principal properties of a 2-D composite ply.[57] Young's modulus describes tensile elasticity or the tendency of an object to deform along an axis when opposing forces are applied along that axis; it is defined as the ratio of tensile stress to tensile strain. It is often referred to simply as the elastic modulus.

A broad spectrum of mechanical properties is provided by the selection of reinforcement types, style, form and proportion in combination with a different matrix. The directionality of strength in a composite can be greatly influenced by substituting longitudinal reinforcement by random mats or directional fabrics. The absolute value of the specific property desired depends on the fibre type chosen: glass, carbon, aramid, organic or natural fibres. The mechanical properties of composite plates reinforced by curvilinear fibres strongly depend on the fibre shapes, and these composites exhibit non-uniform stiffness and anisotropy. A very important feature of natural fibre composites is that the mechanical properties of both the fibre and polymer matrix are time-dependent.[56]

Neagu *et al.*[65] evaluated a quantitative analytical–experimental method to determine the stiffness properties of wood fibre from macroscopic tensile tests of composites. This method can be divided into three parts: (1) measurement of the Young's moduli, fibre volume fraction and fibre orientation distribution in the composites; (2) modelling of the Young's moduli in terms of the anisotropic elastic constants of the fibres using a laminate analogy and a micromechanical model; and (3) inverse modelling to estimate the elastic properties of the fibres from composite stiffness. The analytical–experimental method was employed to estimate the longitudinal Young's modulus of the fibres to evaluate their potential reinforcement in composites. A correlation between lignin content and fibre longitudinal Young's modulus was observed, and an optimal lignin content range at which fibre stiffness attains a maximum was identified for softwood kraft fibres.

Michalowski and Cermák[55] reported that the distribution of fibre orientation in practical applications is clearly anisotropic and that the use of existing isotropic models leads to inaccurate predictions of the strength gain attributed to fibres. For cases in which the predominant load is perpendicular to the preferred plane of fibre orientation, isotropic models will, in general, underestimate the utility of the fibres. These authors reported an increase in the strength of fibre-reinforced soils due to anisotropic hardening. They reported that when fibre-reinforced soil is subjected to large strains, the variation of fibre orientation distribution produces an anisotropic increase in strength. According to Lakes,[49] anisotropic composites offer superior strength and stiffness in comparison with isotropic ones. Material properties along a certain direction are enhanced at the expense of properties along other directions. It is sensible, therefore, to use anisotropic composite materials only if the direction of the applied stress is known in advance.

Ku *et al.*[66] reported that the effect of hemp fibre content and anisotropy can be examined on the basis of the tensile properties of the resultant composite materials. The tensile strength, with fibres along the perpendicular direction, tended to decrease with increasing hemp fibre content (a maximum decrease of 34% at 70% of hemp). The tensile strength of composites with fibres oriented along the parallel direction showed a different trend, and a maximum value was found with increasing fibre loading. These authors also reported that the tensile strength of composites with fibres oriented along the perpendicular direction was 20–40% lower than that of composites with fibres oriented along the parallel direction. Because the fibres lay perpendicular to the direction of load, they could not act as load-bearing elements in the composite matrix structure and became potential defects that could cause failure. They described that the best tensile properties were found in specimens cut from the composite sheets parallel to the direction of carding.

John *et al.*[12] studied the influence of fibre loading and fibre ratio on mechanical properties. They reported that the longitudinal orientation of fibres results in maximum tensile strength, and as the angle of orientation of fibres increases, the tensile strength decreases. When fibres are longitudinally oriented, the fibres are aligned along the direction of force and the fibres transfer stress uniformly. When transversely oriented, the fibres are aligned perpendicular to the direction of load and cannot take part in stress transfer.

The inclination angle relative to the fibre axis for cellulose microfibrils is a key parameter to assess fibre tensile properties. The microfibril angle can be determined by different techniques such as X-ray diffraction[67] and by fungal decomposition of the non-cellulose constituents in the cell wall.[68] Madsen[64] reported that both the strain and stress at which a composite fractures decrease when the fibre axis is inclined to increasing angles with respect to the test direction. This author evaluated the composite tensile strength and reported that there was a four-fold decrease when the inclination angle to the fibre orientation was increased to 26°, whereas the composite stiffness only decreased by a factor of two.

Tensile properties of plant fibre composites compared to glass fibre composites are presented in Table 4.2. Madsen[64] reported that the composites with

Table 4.2   Tensile properties of plant fibre composites. For means of comparison, reported properties of glass fibre composites are included.

Fibre type	Fibre orientation	Stiffness (GPa)	Ultimate stress (MPa)	Ref.
Hemp	Random	2.4–2.7	33–37	70
	Aligned	27.6	277	64
Flax	Random	3.4	36–39	71
	Aligned	28.7	288	64
Jute	Random	5.2	40–61	72
	Aligned	27.2	225	73
Wood	Random	4.2	28	74
	Aligned	4.2	52	74
Glass	Random	5.4	77	75
	Aligned	45.0	1020	76

a random fibre orientation possess moderate tensile properties with stiffness below 6 GPa and ultimate stress below 60 MPa. Table 4.2 shows that if the fibres are aligned, the tensile properties are considerably improved. Thus, (partial) alignment of plant fibres would be an alternative and cleaner approach to improve composite properties, instead of treatment of fibres with chemical additives. According to Table 4.2, the glass fibre composites are superior to plant fibre composites irrespective of fibre orientation, and moreover, that ultimate stress in particular is larger for glass fibre composites. Thus, Table 4.2 reveals the current status of plant fibre composites where stiffness is acceptable, but ultimate stress needs to be somewhat improved. The lower weight of plant fibre composites is usually used as an argument to compensate for their lower properties. The lower density of plant fibres in relation to glass fibres (about 1.5 *vs.* 2.6 g cm^{-3}) is advantageous with respect to a lower composite component weight. If tensile stiffness is selected as the key mechanical parameter, and component stiffness is predetermined, it is anticipated that a composite component based on plant fibres would possess the lowest weight. However, this implies that the stiffness of plant fibres is similar to the stiffness of glass fibres, which in general is not the case. The stiffness of glass fibres is reported to be about 75 GPa.[69] In the study reported by Madsen[64] the estimated largest value of plant fibre stiffness was about 65 GPa. Because of the difference in fibre stiffness, the anticipations of composite component weight are not that obvious. The difference in stiffness between plant and glass fibres is partly equalized in composites with a random fibre orientation.

The structural anisotropy of natural fibres can be estimated through a correlation with the mechanical properties of composites. The relationship between the molecular and supramolecular structures of regenerated cellulose fibres and their mechanical properties in the direction of the fibre axis has been studied by Eichhorn *et al.*[60] In the case of well-oriented cellulose fibres, the fibre compliance is related to the orientation parameter $\langle \sin^2 \phi \rangle_E$ by eqn (4.1):

$$\frac{1}{E} = \frac{1}{e_c} + \frac{\langle \sin^2 \phi \rangle_E}{2g} \qquad (4.1)$$

Where E is the modulus of elasticity in fibre direction, e_c is the chain modulus and g is the average shear modulus between adjacent chains.[77] The strain orientation parameter $\langle\sin^2\phi\rangle_E$ represents the average orientation of all polymer chains at zero strain. This orientation parameter is equivalent to the orientation parameter $\langle\sin^2\phi\rangle$ (eqn 4.2), which is the second moment of the molecular orientation distribution, only in well-oriented fibres with a Gaussian distribution of chain orientation:[78]

$$\langle\sin^2\phi\rangle = \frac{2}{3}\left(1 - \frac{\Delta n}{\Delta n_{max}}\right) \qquad (4.2)$$

In eqn (4.2), Δn is the measured birefringence and Δn_{max} is the maximum birefringence of cellulose. The validity of the model shown in eqn (4.1) for well-oriented regenerated cellulose fibres was confirmed by Kong and Eichhorn.[79] However, in order to fully understand the behaviour of cellulose fibre reinforcement in polymer composites, it is also necessary to study the mechanical fibre properties transverse to the fibre direction. Gindl *et al.*[77] studied the changing ratio of the modulus of elasticity of regenerated cellulose fibres parallel and transverse to the fibre direction, independent of their degree of preferred molecular orientation. By fitting an empirical mathematical relationship to the experimental data, the modulus of elasticity transverse to the direction of the cellulose chain was inferred, and an estimate of the anisotropy (*i.e.*, the ratio of the modulus of elasticity parallel and transverse to the fibre direction) of a regenerated cellulose fibre was made from the degree of preferred orientation of its constituent cellulose chains. In this study an empirical model was developed, which allowed making an estimate of the anisotropy of regenerated cellulose fibres based on their degree of preferred orientation expressed by birefringence. A mathematical function of the form was used and the parameters a, b and c were determined by least-squares fitting (eqn 4.3). The calculated anisotropy of regenerated cellulose fibres characterized in this study ranged from 3.1 for textile viscose fibre up to 15.3 for the experimental Bocell fibre.

$$E = a + b + e^{(c\times\Delta n)} \qquad (4.3)$$

4.5 Applications of Natural Fibre Composites

Natural fibre-containing composites are more environmentally friendly and are used in transportation (automobiles, railway coaches, aerospace), military applications, building and construction industries (ceiling panelling, partition boards), packaging, consumer products, handicraft articles, fancy goods, *etc.* Natural fibre composites are also used in the building and construction industries as panels for partitions and false ceilings, partition boards, wall, floor, window and door frames, roof tiles, and mobile or pre-fabricated buildings, which can be used in times of natural calamities such as floods, cyclones and earthquakes. Regarding storage devices, composites can be used in post-boxes, grain storage silos and bio-gas containers. Fibre composites can also be applied in furniture, electric devices, lampshades, suitcases, helmets and toys.

According to Tudu,[18] natural fibres are used in the automotive industry because of their low density, which may lead to a weight reduction of 10–30%, acceptable mechanical properties, good acoustic properties, favourable processing properties, such as low wear on tools, possibilities for new production technologies and materials, favourable accident performance, high stability and less splintering, favourable ecobalance during vehicle operation due to weight savings, occupational health benefits compared to glass fibres during production, lack of off-gassing of toxic compounds (in contrast to phenol resin bonded wood and recycled cotton fibre parts), reduced fogging behaviour and cost effectiveness, both for the fibres and the applied technologies.

Natural fibres have replaced synthetic fibres as reinforcement in various matrices. Natural fibre composites can effectively be used as substitutes for wood and in various other technical applications, *e.g.* automotive parts. Textiles, ropes, canvas and also paper can be made of local natural fibres, such as flax and hemp. The number of applications of natural fibre composites has multiplied in recent years owing to their biodegradability and high specific properties. The use of natural fibres in composite applications is being pursued extensively throughout the world. Consequently, natural fibre composite materials are being used to produce many components in the automotive sector. These materials are based largely on polypropylene or polyester matrices, which incorporate fibres such as flax, hemp or jute.[43]

Owing to their light weight, high strength-to-weight ratio, corrosion resistance and other attractive properties, natural fibre-based composites are becoming important in building and civil engineering fields as panels for partitions and false ceilings, partition boards, walls, floors, window and door frames, roof tiles and in mobile or pre-fabricated buildings. Despite their usefulness in practice, synthetic fibre-based composites are difficult to recycle after their designated service life. Moreover, natural fibre-based composites are environmentally friendly to a large extent. Bamboo is a very well-known and popular construction material, particularly in bamboo-rich regions. Sisal fibres obtained from the leaves of sisal plants have also proven to be very suitable reinforcement material in various polymeric matrices.[80] Composites made from soy oil-based resin and cellulose fibres have been used in roof structures. Recycled paper was previously tested in composite sheets and structural unit beams and was found to produce the required stiffness and strength required for roof construction.[81]

4.6 Final Considerations

Based on various economic and environmental considerations, the commercialization of composites reinforced with natural fibres should continue to increase. New environmental regulations and societal concern have triggered the search for new products and processes that are environmentally friendly. The incorporation of natural fibres into composite materials can reduce the dependence on non-renewable materials. On the basis of economic and envir-onmental considerations, the use of natural fibres will continue to increase in markets owing to advantages of its characteristics of biodegrability. Studies on specific properties of fibres, such as composition, molecular structure, source, orientation, length, anisotropy and their effect on mechanical properties, are important for applications in composites. Mechanical properties such as tensile strength, Young's modulus and elongation at break of natural fibre composites are a function of orientation of the fibres in the matrix. Therefore, the struc-tural anisotropy of fibres is related to the mechanical properties.

References

1. M. J. John and S. Thomas, *Carbohydr. Polym.*, 2008, **71**, 343.
2. U. Riedel and J. Nickel, *Applications of Natural Fiber Composites for Constructive Parts in Aerospace, Automobiles, and Other Areas*, ed. A. Steinbüchel, Wiley, New York, 2004.
3. K. L. Fung, X. S. Xing, R. Y. Li and Y.-W. Mai, *Compos. Sci. Technol.*, 2003, **63**, 1255.
4. K. Mohanty, M. Misra and G. Hinrichsen, *Macromol. Mater. Eng.*, 2000, **276**, 1.
5. A. K. Mohanty, M. Misra and L. T. Drzal, *J. Polym. Environ.*, 2002, **10**, 19.
6. G. Mehta, A. K. Mohanty, M. Misra and L. T. Drzal, *J. Mater. Sci.*, 2004, **39**, 2961.
7. F. C. Campbell, *Structural Composite Materials*, ASM International, Materials Park, OH, 2010.
8. H. L. Bos, M. J. A. Oever and O. C. J. J. Peters, *J. Mater. Sci.*, 2002, **37**, 1683.
9. J. Andersons, E. Sparniņš, R. Joffe and L. Wallström, *Compos. Sci. Technol.*, 2005, **65**, 693.
10. M. S. Sreekala, J. George, M. G. Kumaran and S. Thomas, *Compos. Sci. Technol.*, 2002, **62**, 339.
11. K. Oksman, M. Skrifvars and J.-F. Selin, *Compos. Sci. Technol.*, 2003, **63**, 1317.
12. M. J. John, S. Thomas and K. T. Varughese, *Compos. Sci. Technol.*, 2004, **64**, 955.
13. M. D. H. Beg, K. L. Pickering and S. J. Weal, *Mater Sci. Eng. A*, 2005, **412**, 7.
14. A. Thygesen, *Properties of Hemp Fibre Polymer Composites: An Optimi-sation of Fibre Properties Using Novel Defibration Methods and Fibre Characterization*, Thesis, University of Denmark, 2006.
15. K. S. Ahmed and S. Vijayarangan, *J. Appl. Polym. Sci.*, 2007, **104**, 2650.
16. J.-I. Kadokawa, M.-A. Murakami, A. Takegawa and Y. Kaneko, *Carbohydr. Polym.*, 2009, **75**, 180.
17. R. Kumar and L. Zhang, *Compos. Sci. Technol.*, 2009, **69**, 555.

18. P. Tudu, *Processing and Characterization of Natural Fiber Reinforced Polymer Composites*, Thesis, National Institute of Technology Rourkela, 2009.
19. C. Alves, P. M. C. Ferrão, A. J. Silva, L. G. Reis, M. Freitas, L. B. Rodrigues and D. E. Alves, *J. Cleaner Prod.*, 2010, **18**, 313.
20. S. G. Bajwa, D. S. Bajwa, G. Holt, T. Coffelt and F. Nakayama, *Ind. Crops Prod.*, 2011, **33**, 747.
21. A. C. H. Barreto, D. S. Rosa, P. B. A. Fechine and S. E. Mazzetto, *Composites, Part A*, 2011, **42**, 492.
22. V. Lopresto, C. Leone and I. Iorio, *Composites, Part B*, 2011, **42**, 717.
23. M. E. Vallejos, A. A. S. Curvelo, E. M. Teixeira, F. M. Mendes, A. J. F. Carvalho, F. E. Felissia and M. C. Area, *Ind. Crops Prod.*, 2011, **33**, 739.
24. N. Venkateshwaran, A. Elayaperumal, A. Alavudeen and M. Thiruchitrambalam, *Mater. Des.*, 2011, **32**, 4017.
25. K. V. Rijswijk, W. D. Brouwer and A. Beukers, *Application of Natural Fibre Composites in the Development of Rural Societies*, Delft, Delft University of Technology, 2001, p. 47.
26. N. Ayrilmis, S. Jarusombuti, V. Fueangvivat and P. Bauchongkol, *Polym. Degrad. Stab.*, 2011, **96**, 818.
27. Y. Li, Y. Iwakura, K. Nakayama and H. Shimizu, *Compos. Sci. Technol.*, 2007, **67**, 2886.
28. M. A. S. Spinacé, C. S. Lambert, K. K. G. Fermoselli and K. A. Paoli, *Carbohydr. Polym.*, 2009, **77**, 47.
29. International Jute Study Group, *Jute, Kenaf & Roselle Plants*, 2005; available at http://www.jute.org/plant.htm.
30. D. E. Akin, R. B. Dodd and J. A. Foulk, *Ind. Crops Prod.*, 2005, **21**, 369.
31. F. Bruhlmann, M. Leupin, K. H. Erismann and A. Fiechter, *J. Biotechnol.*, 2000, **76**, 43.
32. Y. Li, Y.-W. Mai and L. Ye, *Compos. Sci. Technol.*, 2000, **60**, 2037.
33. M. Aguilar-Vega and C. A. Cruz-Ramos, *J. Appl. Polym. Sci.*, 1995, **56**, 1245.
34. M. N. Cazaurang-Martinez, P. J. Herrera-Franco, P. I. Gonzalez-Chi and M. Aguilar-Vega, *J. Appl. Polym. Sci.*, 1991, **43**, 749.
35. R. A. Gross and B. Kalra, *Science*, 2002, **298**, 803.
36. S. Rjafiallah, S. Guessasma and D. Lourdin, *Composites, Part A*, 2009, **40**, 130.
37. M. Gáspár, Z. Benkő, G. Dogossy, K. Réczey and T. Czigány, *Polym. Degrad. Stab.*, 2005, **90**, 563.
38. S. Guessasma and D. H. Bassir, *Mech. Mater.*, 2010, **42**, 344.
39. D. Klemm, B. Heublein, H. P. Fink and A. Bohn, *Angew. Chem.*, 2005, **44**, 3358.
40. Y. Lu, L. Weng and L. Zhang, *Biomacromolecules*, 2004, **5**, 1046.
41. P. Chen and L. Zhang, *Biomacromolecules*, 2006, **7**, 1700.
42. R. Kumar and L. Zhang, *Biomacromolecules*, 2008, **9**, 2430.
43. S. Taj, M. A. Munawar and S. U. Khan, *Proc. Pak. Acad. Sci.*, 2007, **44**, 129.

44. D. Stavrov and H. E. N. Bersee, *Composites, Part A*, 2005, **36**, 39.
45. K. V. Rijswijk and H. E. N. Bersee, *Composites, Part A*, 2007, **38**, 666.
46. X. Ma, J. Yu and J. F. Kennedy, *Carbohydr. Polym.*, 2005, **62**, 19.
47. G. Cantero, A. Arbeliaz, R. Liano-Ponte and I. Mondargon, *Compos. Sci. Technol.*, 2003, **63**, 1247.
48. Y. Xie, C. A. S. Hill, Z. Xiao, H. Militz and C. Mai, *Composites, Part A*, 2010, **41**, 806.
49. R. Lakes, in *The Biomedical Engineering Handbook*, ed. J. D. Bronzino, CRC, Boca Raton, FL, 2nd edn., 2000, p. 2896.
50. G. J. Holt, T. A. Coffelt, P. Chow and F. S. Nakayama, *Int. J. Mater. Prod. Tech.*, 2009, **36**, 104.
51. A. Bernasconi, F. Cosmi and D. Dreossi, *Compos. Sci. Technol.*, 2008, **68**, 2574.
52. J. L. Thomason, *Compos. Sci. Technol.*, 2001, **61**, 16.
53. M. A. Lillie, R. E. Shadwick and J. M. Gosline, *J. Biomech.*, 2010, **43**, 2070.
54. B. Lauke and S.-Y. Fu, *Compos. Sci. Technol.*, 1999, **59**, 699.
55. R. L. Michalowski and J. Cermák, *Comput. Geotech.*, 2002, **29**, 279.
56. S. Honda and Y. Narita, *Compos. Struct.*, 2011, **93**, 902.
57. F. R. Cichocki and J. L. Thomason, *Compos. Sci. Technol.*, 2002, **62**, 669.
58. E. Demirci, M. Acar, B. Pourdeyhimi and V. V. Silberschmidt, *Compos. Mater. Sci.*, 2012, **52**, 157.
59. E. Marklund, *Modeling the Mechanical Performance of Natural Fiber Composites*, Thesis, Lulea University of Technology, 2007.
60. S. J. Eichhorn, R. J. Young, R. J. Davies and C. Riekel, *Polymer*, 2003, **44**, 5901.
61. J. Gassan, A. Chate and A. K. Bledzki, *J. Mater. Sci.*, 2001, **36**, 3715.
62. S. E. Anagnost, R. E. Mark and R. B. Hanna, *Wood Fiber Sci.*, 2002, **34**, 337.
63. H. Lilholt and J. M. Lawther, in *Comprehensive Composite Materials*, ed. A. Kelly and C. Zweben, Pergamon Press, New York, 2000, vol. 1, p. 303.
64. B. Madsen, *Properties of Plant Fibre Yarn Polymer Composites: An Experimental Study*, Thesis, Technical University of Denmark, 2004.
65. R. C. Neagu, E. K. Gamstedt and F. Berthold, *J. Compos. Mater.*, 2006, **40**, 663.
66. H. Ku, H. Wang, N. Pattarachaiyakoop and M. Trada, *Composites, Part B*, 2011, **42**, 856.
67. K. M. Entwistle and N. J. Terrill, *J. Mater. Sci.*, 2000, **35**, 1675.
68. S. E. Anagnost, R. E. Mark and R. B. Hanna, *Wood Fiber Sci.*, 2002, **34**, 337.
69. D. Hull and T. W. Clyne, *An Introduction to Composite Materials*, Cambridge University Press, Cambridge, 2nd edn., 1996.
70. S. Mishra, J. B. Naik and Y. P. Patil, *Compos. Sci. Technol.*, 2000, **60**, 1729.
71. P. R. Hornsby, E. Hinrichsen and K. Tarverdi, *J. Mater. Sci.*, 1997, **32**, 1009.
72. T. L. Andersen and D. Plackett, WO 02/064670 A1, 2002.

73. P. J. Roe and M. P. Ansell, *J. Mater. Sci.*, 1985, **20**, 4015.
74. C. M. Clemons, in *Proceedings of the 3rd International Wood and Natural Fibre Composites Symposium*, Kassel, Germany, September 2000, p. 1.
75. K. Oksman, *Appl. Compos. Mater.*, 2000, **7**, 403.
76. E. K. Gamstedt, L. A. Berglund and T. Peijs, *Compos. Sci. Technol.*, 1999, **59**, 759.
77. W. Gindl, M. Reifferscheid, R.-B. Adusumalli, H. Weber, T. Roder, H. Sixta and T. Schoberl, *Polymer*, 2008, **49**, 792.
78. M. G. Northolt, H. Boerstoel, H. Maatman, R. Huisman, J. Veurink and H. Elzerman, *Polymer*, 2001, **42**, 8249.
79. K. Kong and S. J. Eichhorn, *Polymer*, 2005, **46**, 6380.
80. A. Rai and C. N. Jha, *Express Text*, 2004, **1**, 6.
81. M. A. Dweib, B. Hu, H. W. Shenton and R. P. Wool, *Compos. Struct.*, 2006, **74**, 379.

CHAPTER 5

Flame Retardant Characteristics of Natural Fibre Composites

BALJINDER K. KANDOLA

Institute for Materials Research and Innovation, University of Bolton, Bolton, BL3 5AB, UK
Email: B.Kandola@bolton.ac.uk

5.1 Introduction

Composites are advanced engineering materials created by the synthetic assembly of two (a matrix and a reinforcing element) or more (other additives) components in order to obtain specific properties. The reinforcing component provides mechanical strength and stiffness to the composite, whereas the matrix helps in holding the reinforcement in the required shape and form. Since the mechanical properties are governed by the reinforcing component, its choice is crucial and depends upon the end-product requirements. The composite construction technology is not a recent invention; on the contrary, it has evolved over centuries following inspiration from materials existing in nature, such as wood, plants, mammalian bones, *etc*. Wood, for example, is a composite consisting of high-strength cellulosic fibres and a matrix of lignin. In terms of composites known today, the use of fibres/fabrics as a reinforcing phase started about 100 years ago,[1] mostly as wood products with simple and relatively cheap components. Prior to and during World War II, wood products and composites were commonly used in aeroplanes and automotives. In 1941 a car was built with wood, hemp, sisal and wheat straw, driven by hemp oil.[2] All this changed, however, with the development of high-performance fibres such as glass,

RSC Green Chemistry No. 16
Natural Polymers, Volume 1: Composites
Edited by Maya J John and Thomas Sabu
© The Royal Society of Chemistry 2012
Published by the Royal Society of Chemistry, www.rsc.org

carbon and aramid. Owing to their excellent mechanical and low-density properties, these high-performance fibres and derived composites became very popular in aerospace, marine, automotive and construction industries. Over the past decade or so there has been a renewed interest in natural fibre composites, mainly due to a big push from policy makers of many European countries to force the automotive industry to reuse and recycle materials. This has led to the use of bio-based materials in automotive manufacturing.

At present their largest usage is in the automotive industry, mainly for semi-structural applications. Semi-structural are those which require the component to support its own weight in addition to bearing light external loads such as soft knocks or impacts.[3] In automotives, natural fibre composites are used for seatbacks, headrests, front door-liners, rear door-liners, parcel shelves, boot-liners, sunroof interior shields and noise, vibration and harshness materials.[4] Flax, hemp, sisal, wool and other natural fibres are used in the Mercedes Benz E-class vehicle. Daimler-Benz uses flax and abaca in dashboards, centre armrests, seatbacks; BMW uses flax and sisal in interior door linings, kenaf and flax in package trays and door panels. Similarly, Toyota and Ford use kenaf fibres. In most cases the matrix is polypropylene (PP) or polyethylene (PE); however, poly(lactic acid) (PLA) is also becoming popular.[4] Although these fibres with high-performance resins can form composites capable of passing performance criteria for automotive exteriors, the high moisture absorption behaviour of these composites has restricted their use as exterior components. Mercedes, however, has started using abaca and flax fibre for exterior components such as front bumpers. In the construction industry, natural fibre composites (NFCs) are being used for door panels, window frames, office cabinets, marine flooring, *etc.* For structural, load-bearing applications, natural fibre composites have had limited success,[3] mainly due to their inferior mechanical properties compared to high-performance fibres such as glass, carbon and aramid.

Natural fibres from plant origins have certain advantages over glass fibres, in particular low density, low abrasive wear, availability worldwide and renewable and biodegradable nature. Their production is economical and they can be easily recycled. Physical and mechanical properties of selected natural fibres are given in Table 5.1 and compared with those of high-performance fibres commonly used for structural composites, such as glass, carbon and aramid. Although the theoretical strength of natural fibres is very high (the theoretical elastic modulus of the cellulose molecule[5] is 138 GPa), in practice the strength is very low (see Table 5.1), mainly due to many naturally occurring structural defects within the fibre and those produced during fibre processing. Their impact strength, in particular, is low compared with high-performance and other synthetic fibres.[6] Natural fibres, however, show higher elongation-at-break than glass or carbon fibres, which may be good for the performance of composites.[7] Their thermal conductivity is low (0.29–$0.32 \, \mathrm{W \, m^{-1} \, K^{-1}}$) and therefore they might make a good thermal barrier.[7] There are, however, some disadvantages of using natural fibres, such as low compatibility with hydrophobic polymer matrices, thermal sensitivity at the temperature of

Table 5.1   Physical and mechanical properties of fibres.[6–8]

Fibre	Diameter (μm)	Tensile strength (GPa)	Initial modulus (GPa)	Elongation at break (%)	Density (kg m^{-3})	Moisture absorption (%)
Cotton	16–21	0.3–0.6	5.5–12.6	2.0–10.0	1500–1600	8–25
Flax	40–620	0.3–1.5	27–80	1.2–3.2	1400–1500	7
Hemp	16–50	0.5–1.1	3–90	1.3–4.7	1400–1500	8
Jute	30–140	0.2–0.8	3–55	1.4–3.1	1300–1500	12
Ramie	40–60	0.4–0.9	44–128	2–3.8	1500	12–17
Coir	100–450	0.1–0.3	3–6	15–47	1250–1500	10
Sisal	100–300	0.5–0.9	9–28	2–3	1300–1500	11
Abaca	17–21	0.98	72	10–12	1500	–
E-Glass	3–20	2–6	50–100	2.5–3	2400–2600	–
S-Glass	3–20	3.5	87	2.5	2500	–
Carbon	5.0–6.9	1.5–7.0	150–240	1.4–1.8	1400–2000	–
Para-aramid	10–12	2–4	63–150	3.3–3.7	1400–1450	5–7

compounding processes, moisture absorption and flammability. To improve the adhesion between hydrophilic fibres and hydrophobic matrices, the fibres are usually surface treated or compatibilizers are added in the polymer matrices. Moisture absorption can also be reduced by surface treatments. Thermal stability, flammability and smoke generation properties of composites produced from these fibres, however, are also major issues these days, because, depending upon applications, they must pass some type of regulatory fire test in order to ensure public safety. This is more important if their usage is to be expanded in the marine and aerospace sectors, where fire regulations are more stringent than in the automotive sector. In this chapter the key fire issues pertaining to the thermal stability and flammability of composites and their components, the fire retardant requirements for their usage in different sections and flame retardant solutions are discussed. This chapter also complements previous book chapters on flammability and fire retardancy of high performance fibre-reinforced polymer (thermoset) composites.[8,9]

5.2   Thermal Stability and Flammability of Natural Fibres

Natural fibres from plant origins can be grouped into three categories: seed hairs (*e.g.* cotton); bast or skin fibres, collected from the phloem (the inner bark or the skin) or bast surrounding the stem of plants (ramie, jute, flax *etc.*); and leaf fibres (sisal and abaca). Out of these, jute, ramie, flax and sisal are most commonly used for composites.[10] All of these fibres consist of highly crystalline cellulose fibrils spirally wound in a matrix of amorphous hemicelluloses, lignin and pectin. Lignin and pectin act as bonding agents.

Chemically, cellulose contains glucose units linked together to form long unbranched chain with 1,4-β-glycosidic linkages. The thermal degradation of cellulose has been extensively studied in the literature[11] and is summarized here in Scheme 5.1. It proceeds through two types of reactions. At lower temperatures (200–280 °C), there is gradual degradation which includes depolymerization, dehydration and decarboxylation leading to formation of "dehydrocellulose". This "dehydrocellulose" decomposes further to various volatile products and char. At higher temperatures (280–340 °C), a rapid volatilization occurs and a non-volatile liquid intermediate product, laevoglucosan, is formed. Laevoglucosan subsequently pyrolyses to lower molecular weight and highly flammable volatiles, and leaves a charred residue. Hence slow heating favours the first type of reactions, leaving more char, whereas fast heating produces more tar and less char. This is also the basis of imparting flame retardancy to cellulosics, where condensed-phase active flame retardants promote the first type of reactions, producing more char at the expense of volatile combustible products.

Hemicellulose contains short, highly branched chains of sugars. It contains five-carbon sugars (D-xylose and L-arabinose), six-carbon sugars (D-galactose, D-glucose and D-mannose) and uronic acid.[6] Whilst cellulose is crystalline, strong and resistant to hydrolysis, hemicellulose has a random, amorphous structure with little strength. It decomposes between 200 and 260 °C but forms more non-combustible gases and less tar than cellulose,[12] mainly going through the first type of reactions shown in Scheme 5.1.

Lignin is formed by the removal of water from sugars to form aromatic structures. There are many monomers of lignin; the type depends on the source in nature. Large lignin molecules are three-dimensional and highly cross-linked.[6] Lignin starts decomposing from about 160 °C and continues to decompose until about 400 °C. At the lower temperatures, relatively weak

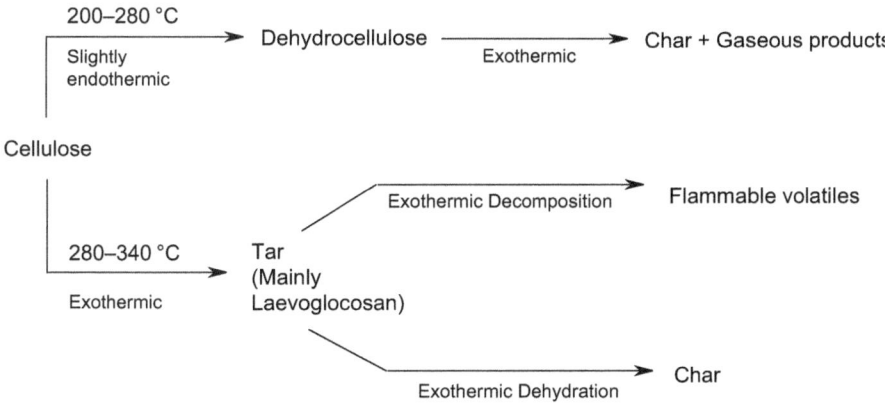

Scheme 5.1 Various stages of the pyrolysis of cellulose.

bonds break, whereas at higher temperatures, cleavage of bonds in the aromatic rings of the lignin takes place.[12] Lignin contributes more to char formation than either cellulose or hemicellulose.[12]

From the above discussion it is clear that fibres having higher cellulose content should have higher flammability than the ones with higher lignin content. The compositions of different fibres are presented in Table 5.2. Hence, coir, kenaf, sisal and jute should be less flammable than other fibres. However, their onset of decomposition temperature determines their suitability for usage with a particular polymer matrix of the thermoplastic composite.

The thermal stability of natural fibres can be studied by thermogravimetric analysis (TGA). Kozlowski *et al.*[13] have reported the TGA of two bast fibres, hemp and flax (lignin content 2–5%), and compared them with two leaf fibres, cabuya and abacca (lignin content 7–13%). All fibres started degrading at about 230 °C with mass loss curves very similar, except that the decomposition process of bast fibres was a one-step process, whereas for the two leaf fibres it was a two-step process, where the lower temperature (230–280 °C) step was attributed to hemicellulose decomposition and the higher one (280–300 °C) to lignin decomposition. They also studied their flammability by cone calorimetry at 35 kW m^{-2} heat flux. Hemp and flax fibres showed a lower peak heat release rate and mass loss rate compared with abacca and cabuya, which is contrary to the above discussion for the stability of the lignin content. They have argued that during the decomposition of lignin, relatively weak bonds break at lower temperatures, whereas cleavage of bonds in the aromatic rings takes place at higher temperatures. Hence, with a lower lignin content the degradation begins at a higher temperature, but there is a lower oxidation resistance, which would be provided by the aromatic structure of lignin. Similar observations were made by Manfredi *et al.*[14] for flax, sisal and jute fibres, with a lignin content of 2.0, 9.9 and 11.8%, respectively. These fibres in composite form using an acrylic resin showed different behaviour against fire in a cone calorimeter test. The composites containing flax and sisal caused a long duration but slow growing fire. On the other hand, jute fibre composites caused a quickly growing but short duration fire.

Table 5.2 Chemical composition of fibres.[6]

Fibre	Cellulose (%)	Hemicellulose (%)	Lignin (%)	Pectin (%)
Flax	60–81	14–19	2–3	1.8–2.3
Jute	51–72	12–20	5–13	0.2
Abaca	61–64	21	12	0.8
Sisal	43–88	10–13	4–12	0.8–2
Kenaf	36	21	18	2
Ramie	69–76	13–15	0.6–1	1.9–2
Hemp	70–78	18–22	4–5	0.9
Cotton	83–92	2–6	0.5–1	5.7
Coir	43	0.3	45	4
Wood	45–50	23	27	–

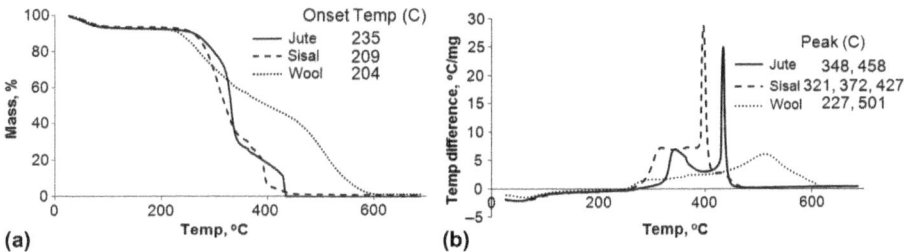

Figure 5.1 (a) TGA and (b) DTA curves of sisal, jute and wool fibres in air.

From our own work the TGA and DTA curves for jute and sisal fibres are shown here in Figure 5.1 and compared with those of wool fibres. The onset of the decomposition temperature of jute (235 °C) is much higher than sisal (209 °C) and wool (204 °C) fibres. Wool fibre, however, starts decomposing earlier and is more thermally stable in terms of mass loss than both cellulosic fibres. The DTA curves also show that exothermic peaks representing decomposition and char oxidation reactions for both cellulosic fibres are at lower temperatures and are of higher intensity compared to wool fibres, suggesting that they should be more flammable than wool fibres.

The crystallinity and orientation of fibres also affect their flammability. For cellulosic fibres, higher levels of crystallinity result in higher levels of laevoglucosan during pyrolysis and, hence, increased flammability.[15] Increasing orientation results in decreased pyrolysis and therefore decreased flammability.[16] However, most of these studies reported in the literature are in terms of TGA measurements. How much these factors affect the flammability of fibres in terms of the limiting oxygen index, or vertical flame spread, is not known.

To compare the flammability of fibres, the limiting oxygen index (LOI) test is good indicator. LOI is the minimum concentration of oxygen, expressed as volume percent, in a mixture of oxygen and nitrogen that will just support the flaming combustion of a material.[17] Usually the fibres are tested as textile structures. There is no standard method of determining the LOI of fibres. In our laboratory we have developed our own methodology for testing fibres, where the fibre is rolled into slivers about 10–12 cm long. The LOI test is undertaken in a manner similar to that used for plastic sticks. The LOI value of cotton is 18–20 vol%, and while the LOI values for other fibres are not available in the literature, these are expected to be in a similar range, with the higher lignin content being higher than cotton by 1–3 units. Hence, all these fibres are flammable, more so than protein fibres such as wool with a 25 vol% LOI value. The other flammability test for textiles is the vertical flame spread, *e.g.* BS 5438:1989,[18] which is applicable for textile structures. There is no study available in the literature where the vertical flame spread of fabrics from different natural fibres of similar area densities and weave/knits have been compared.

5.3   Flammability of Composite Matrices

Natural fibres can be used with thermoset, thermoplastic or biopolymers as long as the processing temperature of the polymer is lower than the onset of the decomposition temperature of the lignocellulosic fibre used. Natural fibres start decomposing around 200 °C, as discussed in the above section. Their use with different types of matrices reported in the literature has been summarized in Table 5.3.[10,12,19–30]

5.3.1   Thermoplastics

While the physical, thermal and flammability properties of polymers in general have been extensively reviewed in the literature, selected properties taken from different sources[31–36] are given in Table 5.4. The softening temperatures and melting temperatures of thermoplastics are used for their processing, and hence determine the suitability of their use with natural fibres. The only thermoplastic polymers processable below the decomposition temperature of

Table 5.3   Reported natural fibre–matrix combinations to produce composites.[a]

Natural fibres[b]	Thermoplastic matrices[c]	Biodegradable matrices[d]	Thermoset matrices[e]
Wood flour/ fibre	PP, PE, PVC, PS, PU	–	UP
Cellulose	PP, PE, PA66, PVC, PS	PLA	Ep
Jute	PP, PE	PLA	SBR, rubber, Ep, UP, VE, Ph, acrylic
Flax	PP, PE	PLA, PGA, PCL, PHB	Ep, VE, Ph, mela- mine, acrylic
Sisal	PP, PE, PS	–	Rubber, Ep, UP
Abaca	PP	PLA, PHBV	Ep
Kenaf	PP, PE	PLA	Ep, Ph
Ramie	PP	PLA	–
Hemp	–	–	Ep, UP, VE, Ph
Coir	–	–	Rubber, UP, Ep
Banana	–	–	UP, Ph
Bamboo	PP	–	Ep, UP
Pineapple	PE	–	UP, Ph
Sunhemp	PP	–	UP
Wheatstraw	PP	–	–
Wool	–	–	UP

[a]PP = polypropylene;   PE = polyethylene;   PS = polystyrene;   PVC = poly(vinyl chloride); PU = polyurethane;   PLA = poly(lactic acid);   PGA = poly(glycolic acid);   PCL = (poly-ε-capro-lactone), PHB = poly(hydroxybutyrate); PHBV = poly(3-hydroxybutyrate-*co*-3-hydroxyvalerate); Ep = epoxy; UP = unsaturated polyester; VE = vinyl ester; Ph = phenolic.
[b]Onset of decomposition temperature of natural fibres is in the range 200–230 °C.
[c]Taken from references 10, 12 and 19 and references cited therein.
[d]Taken from references 19–28 and references cited therein.
[e]Taken from references 10, 12, 19, 29 and 30 and references cited therein.

Table 5.4 Thermal and flammability characteristics of selected thermoplastics.[31–36]

Polymer	Glass transition temp.[a] (°C)	Melting temp.[b] (°C)	Decomp. temp. range[c] (°C)	Heat of combust.[c] (MJ kg^{-1})	LOI[d] (vol%)	Cone calorimetric data[e] at 40 kW m^{-2}		
						TTI (s)	PHRR (kW m^{-2})	THR (MJ m^{-2})
Polyethylene	(−125)–2	137–146	340–440	48.0	18	159	1408	221
Polypropylene	(−35)–1	160–220	330–410	44.0	18	86	1509	207
Poly(methyl methacrylate)	94–150	160–220	170–300	26.0	18	36	665	828
Poly(vinyl chloride)	69–98	177–212	200–300	18.0	45	73	175	24
Polyamide 6	40–100	210–220	300–350	32.0	23	65	1313	226
Poly(butylene terephthalate)	31	220–267	–	–	22	113	1313	170
Poly(ethylene terephthalate)	60–86	265–310	285–305	21.5	25	116	534	114
Polycarbonate	145–150	220–267	350–400	31.0	27	182	429	119
Polystyrene	100–109	177–277	300–400	40.0	18	97	1101	210
Poly(tetrafluoroethylene)	(−113)	327–346	510–540	4.6	95	10000	13	12
ABS	110–125	–	–	35.0	19	69	944	163
SAN	100–200	–	–	36.0	–	–	–	–

[a]Taken from references 34 and 35.
[b]Taken from references 34–36.
[c]Taken from reference 33.
[d]Taken from reference 31.
[e]Taken from reference 32.

natural fibres are polyethylene (PE, mp 137–146 °C), polypropylene (PP, mp 160–220 °C), polystyrene (PS, mp 177–277 °C) and poly(vinyl chloride) (PVC, mp 177–212 °C). As seen from Table 5.3, these are most commonly used polymers. Other polymers such as polyamide 6 (PA6, mp 210–220 °C), polyester (PET, mp 265–310 °C), polycarbonate (PC, mp 220–267 °C), *etc.*, are processable at temperatures where natural fibres start decomposing. Out of these polymers, PP is a more popular choice, usually with short fibres, where composites are produced by melt blending and subsequently compression or injection moulding.

The LOI and cone calorimetry are common methods of quantifying the flammability of polymers. LOI has been discussed in the previous section. In a cone calorimeter, samples in a horizontal orientation are exposed to an external heat flux ($25–100\,kW\,m^{-2}$) and ignited with a spark ignition, as detailed in ISO-5660.[37] Important parameters that describe the flammability behaviour of polymers include the time to self-sustained ignition (TTI), the heat release rate (HRR) and its peak value (PHRR), the total heat release (THR), the amount of smoke generated and the residual char yield. The LOI[31] and cone calorimetric results for selected thermoplastic polymers tested by Hirschler[32] at $40\,kW\,m^{-2}$ heat flux are given in Table 5.4. The results show that PE, PP and PS, commonly used for natural fibre composites, are highly flammable, their LOI values being 17–18 vol%, which are very low compared to some other polymers. The peak heat release ($1100–1510\,kW\,m^{-2}$) and total heat release ($207–220\,MJ\,m^{-2}$) of these three polymers studied at $40\,kW\,m^{-2}$ heat flux[32] are very high compared to others. However, PVC, which also can be used with natural fibres, is inherently flame retardant (LOI = 45 vol%), with very low PHRR and THR (see Table 5.4). PVC contains chlorine in the polymer backbone, which provides inherent flame retardant characteristics. PVC, however, releases HCl during burning, so has environmental issues associated with it.

5.3.2 Biodegradable Matrices

Biodegradable materials are those which are capable of decomposing into carbon dioxide, methane, water, inorganic compounds or biomass by the enzymatic action of microorganisms. Biodegradable matrices include PLA, cellulose esters, starch plastic, poly(ε-caprolactone) and aliphatic polyester-*co*-polyesters.[6] All of these biopolymers are hydrophilic polyesters and so they absorb moisture. In the case of natural fibre reinforcement, this could be even more problematic since cellulosic fibres tend to absorb rather large amounts of water when exposed to it. These biopolymers are mostly used in the packaging industry and other applications with minor strength requirements. PLA composites, however, are finding applications in other sectors as well, mainly automotive. Their use with different natural fibres in composites, as reported in the literature,[19–28] is shown in Table 5.3.

PLA is a linear aliphatic polyester derived from renewable sources such as corn.[38,39] PLA is produced from starch, extracted from plant matter (*e.g.* corn), which is converted into fermentable sugar (*e.g.* glucose) by enzymatic hydrolysis. This natural sugar then, through fermentation, is converted into lactic acid, as shown schematically in eqn (5.1):

$$\text{Starch} \xrightarrow{\text{enzyme hydrolysis} + H_2O} \text{Glucose} \xrightarrow{\text{fermentation}} \text{Lactic acid} \xrightarrow[\text{ring-opening polymerization}]{\text{polycondensation or}} \text{PLA}$$

$$(5.1)$$

PLA's properties in fibre form have been reviewed by Farrington *et al.*[39] Its mechanical properties are similar to conventional PET fibre.[39] Although the fibre is flammable, its flammability is slightly less (LOI = 26 vol%) compared to PET fibre (LOI = 22–25 vol%).[39] In matrix form the physical and mechanical properties of PLA have been reported by Van de Velde and Kiekens[20] and other researchers.[21–28] PLA's glass transition temperature is between 60 and 65 °C and melting temperature between 173 and 178 °C. Since the processing temperature is lower than the decomposition temperature of most natural fibres, this can be easily used for preparing composites. Its density, however, is higher than PP.[20]

In our recent research work, we have studied the flammability of PLA as a matrix polymer and compared it with that of PP. Nonwoven mats of PP and PLA fibres were used for nonwoven web production. These were melt pressed at 190 °C for 2.5 min, followed by quick cooling to obtain 2.5–2.7 mm thick sheets. TGA and DTA of the composites were performed on an SDT 2960 simultaneous DTA–TGA instrument from room temperature to 700 °C using 15 ± 1 mg samples heated at a constant rate of 10 °C min^{-1} in air flowing at 100 ± 5 mL min^{-1}. The results are shown in Figure 5.2.

The DTA profile of polypropylene (Figure 5.2b) shows a melting endotherm peak at 169 °C, followed by a large exotherm with a peak maximum at 382 °C representing the main decomposition step. This is followed by another very small exotherm at 423 °C, which represents the oxidation of the products formed in the first step. The TGA curve (Figure 5.2a) shows one large decomposition stage in the temperature range 197–395 °C, with 98% mass loss. Between 395 and 480 °C, 2% mass loss occurs, representing the oxidation of combustible products. No char is left at the end of the test.

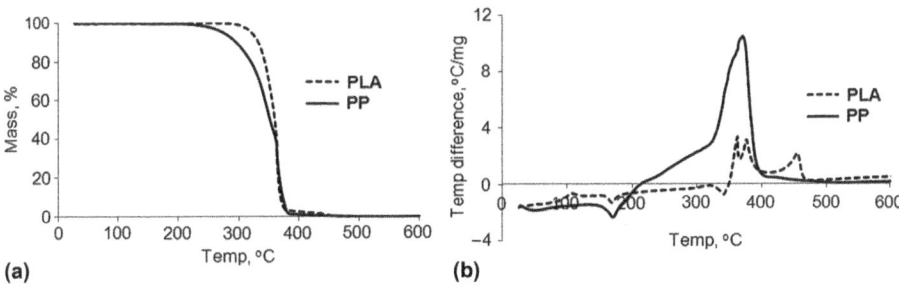

Figure 5.2   (a) TGA and (b) DTA curves of PP and PLA matrices in flowing air.

The DTA of PLA (Figure 5.2b) shows an endotherm at 58 °C, representing the glass transition temperature, an endotherm at 169 °C, representing melting, an endotherm at 342 °C followed by a double peaked exotherm (362, 376 °C), representing the main decomposition, and an exotherm at 455 °C, representing the oxidation of char. The TGA curve (Figure 5.2a) shows that the onset of decomposition occurs at 296 °C and 97% mass loss occurs up to 385 °C, followed by 3% up to 465 °C. On comparing the TGA and DTA curves of PP and PLA in Figure 5.2, PLA seems to be more thermally stable than PP.

These samples were also tested by a cone calorimeter at 35 kW m^{-2} heat flux. Mass loss, heat release rate and smoke production *versus* time curves are shown in Figure 5.3 and derived results are presented in Table 5.5. Both samples ignited at between 28 and 30 s. PP showed a PHRR of 1699 kW m^{-2} and a THR of 95.4 MJ m^{-2}. For PLA the values are much lower: PHRR = 663 kW m^{-2} and THR = 49.8 MJ m^{-2}. The effective heat of combustion of PLA (18.2 MJ kg^{-1}) is also much lower than for PP (42.9 MJ kg^{-1}). The trend for mass loss for PP and PLA samples (Figure 5.3a) is different from that observed by TGA, as here the mass loss curves are quite similar. This is due to the high heat flux, causing rapid degradation. PLA produces negligible smoke compared to PP, as shown in Figure 5.3(c). This study shows that although PLA is combustible, it poses less of a fire hazard as it releases less heat during combustion.

5.3.3 Natural Rubber

Natural rubber is an elastomer, commonly derived from latex of many plants and trees. Chemically it is polyisoprene, mainly *cis*-1,4-polyisoprene, although

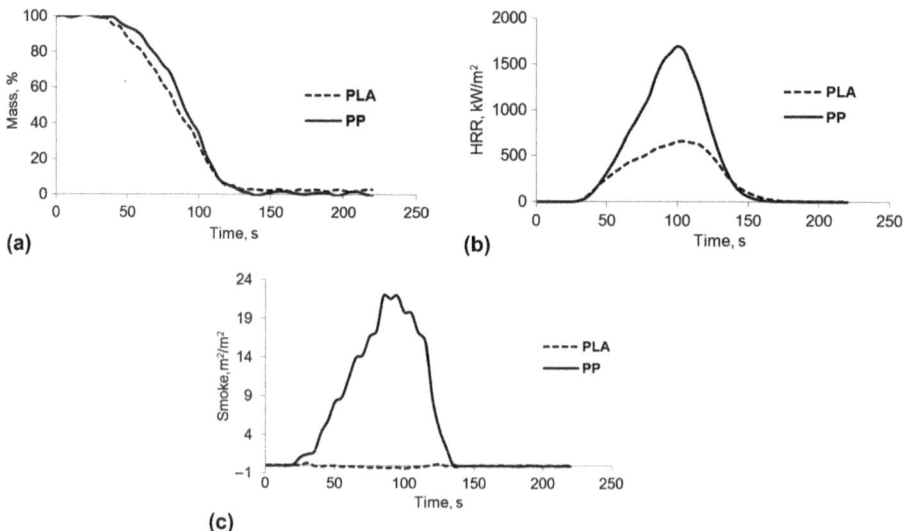

Figure 5.3   (a) Mass loss, (b) HRR and (c) smoke *vs.* time curves for PP and PLA matrices exposed to 35 kW m^{-2} heat flux.

Table 5.5 Cone calorimetric results for PP, PLA and different fibre-reinforced composites at $35\,kW\,m^{-2}$ external heat flux.

Sample	Sample thickness (mm)	TTI (s)	FO (s)	PHRR (kW m^{-2})	THR (MJ m^{-2})	Hc (MJ kg^{-1})	Smoke (m^2 m^{-2})	Char (%)
PP	2.7	30	160	1699	95.4	42.9	1302	0
PLA	2.5	28	180	663	49.8	18.2	0.7	3.0
Glass/PP	3.8	25	260	506	79.7	38.1	1308	50.9
Glass/PLA	3.6	27	280	275	46.4	16.6	0.1	43.9
Jute/PP	3.2	33	290	675	95.5	31.7	1121	3.3
Jute/PLA	2.8	38	300	393	60.7	16.1	0.9	4.4
Sisal/PP	1.3	25	140	680	43.1	30.6	532	3.2
Sisal/PLA	1.3	27	120	542	28.1	15.9	1.2	3.3
Wool/PP	3.6	30	360	632	140.9	35.1	1928	4.6

the *trans* isomer can also be isolated from some plants. Both of these forms can also be produced synthetically. Natural rubber is very flammable, with LOI = 17 vol%.[31] Rubber is generally vulcanized before use. The ingredients for vulcanization may increase or decrease the flammability of rubber, but generally increase the smoke production and toxicity.[31]

5.3.4 Thermosets

Thermoset matrices are fabricated from the respective resin, a curing agent, a catalyst or curing initiator and a solvent sometimes introduced for lowering the viscosity and improving the impregnation of reinforcements. In thermosets, solidification from the liquid phase takes place by the action of an irreversible chemical cross-linking reaction which produces a tightly bound three-dimensional network of polymer chains. The molecular units forming the network and the length and density of the cross-links of the structure will influence the mechanical and thermoplastic or non-thermoplastic and flammability properties of the material. The level of cross-linking between resin functional groups and often the degree of non-thermoplasticity is a function of the degree of cure, which usually involves application of heat and pressure. However, some resins cure at room temperature.

The most common types of thermoset matrices for composites are unsaturated polyester, vinyl esters and phenolics. All of these matrices used with different natural fibres and reported in literature are summarized in Table 5.3. Unsaturated polyester resins are prepolymer mixtures containing unsaturated groups and styrene, with the latter serving as both a diluent and cross-linking agent during the radical polymerization process. Vinyl ester resins are similar in structure to unsaturated polyester resins, differing only in the location of the primary vinyl reactive groups, which are positioned at the ends of the monomeric chains. Both unsaturated polyester and vinyl esters are room temperature curing resins, often post-cured at 80 °C, and hence are suitable for use with natural fibres. These, however, are very flammable with LOI values of 20–23 vol%,[8,9,40] and burn with heavy smoke and soot.

Epoxies refer to a wide variety of cross-linked polymer chains based on polymer monomeric units containing an epoxide group. Epoxy resins are typically formed from the reaction between a diepoxide with a primary diamine. Reactants with higher reactive functionalities result in a highly cross-linked, stiff and tough epoxy network suitable for aerospace applications.[9] These, however, are cured at higher temperatures than those with lower functionalities. The latter are room temperature curing resins. The flammability also differs: the ones with high functionalities are highly cross-linked and less flammable than the low functionality ones. In our recent work we have tested two resins with different functionalities;[41] their flammability behaviour is shown in Table 5.6. Epoxy resins generally have superior functionalities, which result in better mechanical properties compared to unsaturated polyester resins or vinyl esters, but their use is limited due to high cost which can be several times higher than the latter.

Phenolics are inherently flame retardant resins, but they are brittle and do not possess high mechanical properties. We have discussed the flammability of these resins in detail in previous reviews,[8,9,40] where it was shown that the flammability of these resin can be ranked as:

Phenolic < polyimide < bismaleimide < epoxy < polyester and vinyl ester

It must be emphasized that this is a generic discussion on these resins. Since thermosets contain different components (resin, curing agent and sometimes hardener), the choice of components for each resin type will affect the flammability properties of the cured resin, as shown for the two different epoxy resins in Table 5.6. Hence, the trend is very generic. In most cases the flammability of low-temperature curing epoxies is similar to unsaturated polyester or vinyl ester resins. Similarly, some phenolics are better than others.

Table 5.6 Flammability characteristics of selected thermoset matrices.[41]

Resin	Formulation	Curing conditions	LOI (vol%)	Cone calorimetric data at $50\,kW\,m^{-2}$		
				TTI (s)	PHRR (kW m^{-2})	THR (MJ m^{-2})
Epoxy-1	Tetraglycidyl-4,4'-diaminodiphenyl-methane, cured with 4,4'-diaminodiphenyl sulfone	180 °C for 3 h	27.8	24	823	56
Epoxy-2	Butane-1,4-diol diglycidyl ether, cured with cycloaliphatic amines	22 °C for 24 h, post-cured at 80 °C for 1 h	21.5	34	1861	77

5.4 Flammability of Composites

5.4.1 Performance Requirements Depending Upon Application Areas

For the usage of natural fibre composites in the transport and construction industries, it is essential to assess the fire risk associated with the application. All products have to conform to certain specified regulations for particular applications. A comprehensive list of international standards and codes used for fire-safety design of fibre-reinforced polymer composites for different sectors is beyond the scope of this chapter. The reader is referred elsewhere to such reviews and book chapters;[33,42,43] only selected relevant ones are discussed here briefly.

For the automotive sector, in the USA the Federal Motor Vehicle Safety Standard 302 (FMVSS 302)[44] regulates the flammability of materials used in the interiors of cars and trucks. The test is run on samples mounted horizontally and ignited with a 3-mm diameter, 20-mm long Bunsen burner flame. To pass this test the horizontal burn rate of the sample should not exceed 102 mm min^{-1}. The FMVSS 302 test is virtually an international standard, as it has been harmonized with many equivalent designations.[42] Other examples include ISO 3795, BS AU 169 (UK), ST 18-502 (France), DIN 75200 (Germany), JIS D 1201 (Japan) and SAE J369 (automotive industry). This test is not as stringent as in other sectors.

For the rail industry the rate of fire spread and toxicity of the fumes produced during their combustion are of vital importance. In the US the Federal Railroad Administration (FRA) has a standard, Federal Register 49:192 (1984), which gives requirements for all materials used in rail carriages. In Europe there is not a common standard across the different countries. Each European country has its own regulations, *e.g.* DIN 5510 (Germany), BS6853 (UK) and NF F16-101 (France). Most of these include flame spread, smoke development and toxicity. Other tests include the well-established BS 5852:2006 for upholstered furniture composites and/or complete pieces of furniture, such as seats in rail compartments when they are subjected to a smouldering cigarette, or to a flaming ignition source of thermal output ranging from a burning match to that approximating to the burning of four double sheets of full-size newspaper.[45] On the other hand, smoke production may be determined using ISO 5659-2:2006 based on the well-known NBS smoke chamber method[46] and is similar to ASTM E662.

For marine applications, all commercial passenger and cargo ships in European countries have to comply with the fire performance requirements contained in the International Convention for the Safety of Life at Sea (SOLAS)[47] as IMO/HSC Code (Code of Safety for High Speed Craft of the International Maritime Organization). The fire tests to be carried out and the acceptance criteria are defined in the IMO/FTP code (International Code for Application of Fire Test Procedures), which have been mandatory since 1998.[48] This code allows for use of non-conventional shipbuilding materials, defined as "fire restricting", materials which have low flame spread characteristics, limited rate of heat release and smoke emissions. Furthermore, for areas of moderate and major fire hazard (*e.g.*, machinery spaces, storerooms), the materials used should be "fire resisting", *i.e.* they should prevent the fire and smoke propagation to adjacent compartments during a defined period of time (60 min for high hazard and 30 min for low hazard areas). Composites used for load-bearing structures should be able to maintain their load-bearing capacity within the specified period of time (30 or 60 min). For fire-restricting material characterization, specified tests include the room corner test, ISO 9705, and the cone calorimeter test, ISO-5660.

For aerospace applications the polymer-based composites used for the interior design of aircraft must be self-extinguishing and exhibit low flame, smoke and toxicity (FST) characteristics; they are required to comply with industry and international regulations such as FAR (the US Federal Aviation Regulations), which govern the requirements for materials used for such applications.[49] The most important test is the Heat Release Rate Test (FAR 25.853, Part IV, Appendix F), according to which when the sample is tested at $35 \, kW \, m^{-2}$ external heat in a Ohio State University calorimeter, both the peak heat release rate and total heat release rates measured over 2 min should not exceed $65 \, kW \, m^{-2}$. For detailed descriptions of all other fire tests, the reader is referred to the FAR regulations.

5.4.2 Key Parameters Affecting the Flammability of Composites

The physical, mechanical and flammability properties of composites not only depend upon the matrix and fibre type, but also on the reinforcement type. The latter also determines the manufacturing process. If short fibres of only a few millimetres in length or particles such as cellulosic wood flour are used as the reinforcing element, then for thermoplastic composites, melt mixing in a continuous extrusion process is often used. Both polymer and fibres are introduced into the extruder as a means of optimizing the wetting and mixing needed to obtain good fibre dispersion. With this technique the fibre content achieved is usually less than 40% by weight of fibres owing to processing issues. The fibre length is reduced further under the processing conditions, resulting in fibres being oriented in complex and often non-optimal patterns.[50] Since short fibres function mainly as fillers, the flammability of the composite produced should be different from the one where reinforcement is in fabric form. For composites with woven fabric reinforcement, the polymer as a cast sheet or extruded film is introduced between layers of reinforcement fabrics. The assembly is passed through a series of rollers that melt the polymer and enables impregnation. After a chilling roll, the impregnated material is collected as a sheet. Commingling of reinforcing fibre and matrix in a fibrous form is another common method of composite fabrication. On heating under pressure, the matrix fibres melt and a consolidated structure is obtained.

For thermoset composites, short fibres are mixed with resin before casting into moulds of the desired shape. In textile form, woven/knitted fabrics may be impregnated with resins using techniques such as hand lay-up, resin transfer moulding, *etc.*, and then stacked to produce samples of desired thicknesses.

Although the effect of different reinforcement types on the flammability of the composites is not documented in the literature, it is expected that the flammability of the resulting composites would be different.

Various other factors affecting the flammability of composites in general have been reviewed previously.[8,9,40] The composite thickness of a structure has a significant effect on its flammability. Thermally and physically thin samples ignite early, show large peak heat release values, burn for a shorter time and produce less total heat release compared to thermally and physically thick samples which burn slowly but for a longer time.[51,52] "Thermally thick" means the heat wave penetration depth is less than the physical depth, whereas in "thermally thin" the composite has the same temperature through this limiting thickness. The transition from thermally thin to thermally thick is not a constant since it depends on the material thermal properties, including fibre and resin thermal conductivities. For a given composite of defined thickness, the condition depends on the intensity of the fire or, more correctly, the incident heat flux. While many large-scale fire tests involve heat sources or "simulated fires" having constant and defined fluxes, in real fires, heat fluxes may vary. For example, a domestic room filled with burning furniture at the point of flashover presents a heat flux of about $50 \, kW \, m^{-2}$ to the containing wall and door surfaces; larger building fires present fluxes as high as $100 \, kW \, m^{-2}$ and hydrocarbon fuel "pool fires" may exceed $150 \, kW \, m^{-2}$. The heat flux dependence on burning behaviour can be examined using calorimetric techniques such as the cone calorimeter.[8,9,40]

The overall burning behaviour of a composite will be the sum of its component fibres and thermoplastic polymer/resin plus any positive (synergistic) or negative (antagonistic) interactive effects.

5.4.3 Flammability Data

Despite widespread interest in natural fibre composites and their increasing usage in different sectors, there is little published literature on their fire performance. This is mainly due to the fact that so far they have only been widely used in the automotive industry and fire safety regulatory requirements for automotives are not very stringent. All available literature is from the research prospective, where mainly cone calorimetry has been used to assess their flammability. Most of the work reported in the open literature and some recently carried out in our laboratories is reviewed here in the following sections.

5.4.3.1 *Thermoplastic Composites Reinforced with Short Fibres*

A number of studies have been conducted on the burning behaviour of PP composites reinforced with short fibres.[53–56] In general, the presence of fibres reduces the time-to-ignition, peak (PHRR) and total (THR) heat release rates, but increases the total burning time or time-to-extinguishment in cone calorimetric tests. PP shows one peak heat release rate, whereas in natural composites two peaks are observed.[53–56] The first major peak represents the surface ignition and burning, and is usually reduced in intensity compared to pure PP owing to the charring behaviour of lignocellulosic fibres. The second peak represents the char-surface rupture and eventual burn-out of the whole sample. Borysiak et al.[53] have reported the flammability behaviour of PP/wood composite samples, prepared by co-extruding PP and Scotch pine wood particles (50 wt%) in a single screw extruder and then press moulding them. The flammability of the composites was tested in a cone calorimeter at 35 kW m^{-2} external heat flux and compared with PP plaques of similar thicknesses. The time-to-ignition (TTI) of the composite decreased from 47 s for pure PP to 31 s, the PHRR from 1407 to 810 kW m^{-2}, the THR from 120 to 92 MJ m^{-2} and the average specific extinction area from 514 to $393 \text{ m}^2 \text{kg}^{-1}$. This indicates that natural fibres reduce the TTI of composites owing to their easily ignitable characteristics. However, once ignited, they are not as flammable as hydrocarbons, and hence reduce the overall flammability of the composites. Schartel et al.[36] and Le Bras et al.[54] have studied flax/PP (40:60 wt%) composites containing short fibres and reported that with fibres the PHRR decreased to 640 kW m^{-2} in composites compared to 1800 kW m^{-2} in PP,[54] when tested at 50 kW m^{-2} external heat flux.

Helwig et al.[56] prepared PP/flax composites with long flax fibres (150–300 mm), where fibres were placed between PP sheets and then melt pressed. A number of composites with fibre contents of 12.5, 20, 30 and 40% were prepared and tested with a cone calorimeter at 35 kW m^{-2} heat flux. The results showed that the burning behaviour of PP was significantly reduced in the presence of fibres. The composites ignited earlier than the PP sample, but peak and total heat release values were decreased, deceasing further with increasing fibre content. Important results are shown in Table 5.7.[56]

5.4.3.2 Thermoplastic Composites Reinforced with Textile Structures

In our recent research work we have prepared a number of composite samples using PP and PLA as matrices and jute, sisal, wool and glass fabrics as reinforcements. The reinforcement was in a woven fabric form. The area densities of the reinforcing fabrics varied according to the fibre type used and are indicated here: $\text{wool} = 172 \text{ g m}^{-2}$, $\text{jute} = 174 \text{ g m}^{-2}$, $\text{sisal} = 62 \text{ g m}^{-2}$ and $\text{glass} = 280 \text{ g m}^{-2}$. An assembly of each type of nonwoven mat of thermoplastic fibres and eight layers of a woven reinforcing fabric was melt pressed in a

Table 5.7 Cone calorimetric results for flax/PP composites at 35 k W m^{-2} heat flux.[56]

Fibre content in PP/flax composite	TTI (s)	PHRR[a] (kW m^{-2})	THR (MJ m^{-2})	EHC (MJ kg^{-1})	Average MLR (g s^{-1} m^{-2})
0	62	1200	156	41.6	17.9
12.5	45	480, 750	161	39.2	13.7
20	45	470, 560	165	37.2	11.7
30	39	440, 380	125	34.2	9.8
40	35	290, 380	109	33.1	9.2

[a]PHRR values are approximate, calculated from graphs reported by Helwig and Paukszt.[56]

mould at 190 °C for 2.5 minutes, followed by quick cooling to obtain a composite sample sheet. The fibre/polymer ratio for each composite was kept as 40:60 by weight. The details of the composite fabrication technique are presented in our forthcoming publication.[57] Since all of the fabrics were of different area densities, the samples prepared were of different thicknesses, as shown in Table 5.5. However, for each fabric type the PP and PLA composites were of similar thicknesses. All of these samples were tested for flammability in a cone calorimeter at 35 kW m^{-2} external heat flux. Since all samples are of different thicknesses and the thickness of a sample affects its burning behaviour, samples of similar thicknesses (produced from fabrics of similar area densities) are grouped together in Table 5.5. In Figure 5.4, HRR, mass loss and smoke production *versus* time curves for all samples of each type of matrix are shown.

All fibres had a minimal effect on the TTI of PP or PLA matrices. All types of reinforcement reduced the PHRR of PP from 1699 kW m^{-2} to 506–680 kW m^{-2},

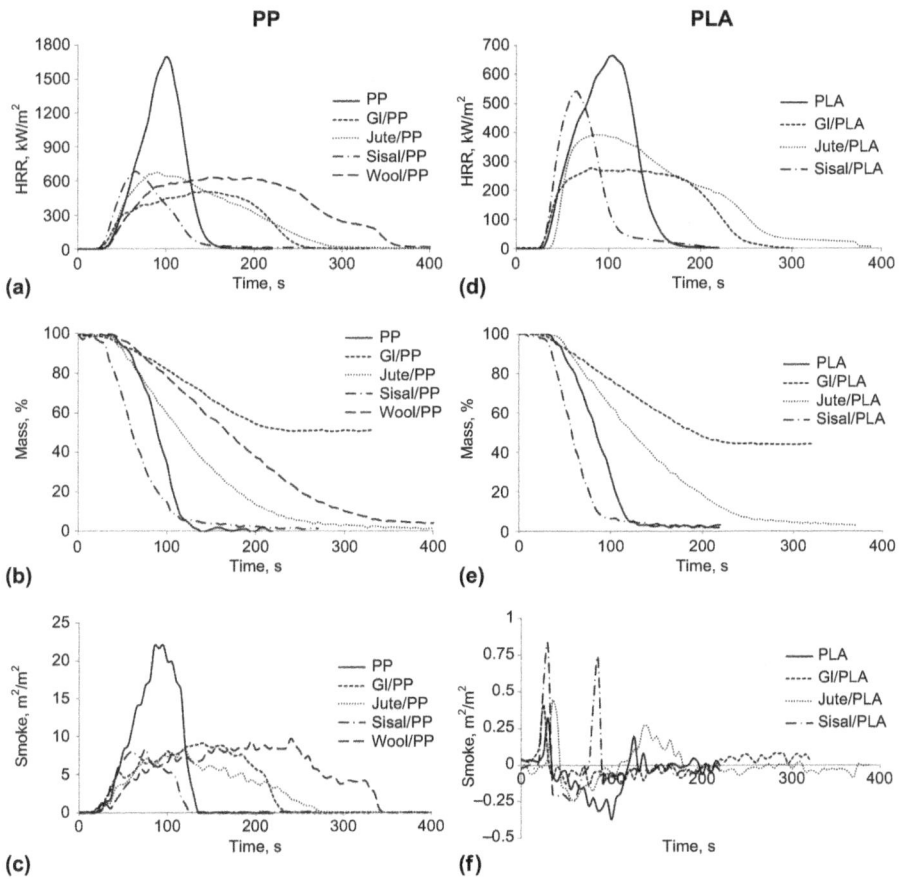

Figure 5.4 (a, d) HRR, (b, e) mass loss and (c, f) smoke *vs.* time curves for glass, jute, sisal and wool fibre-reinforced PP (a–c) and PLA (d–f) composites exposed to 35 kW m^{-2} heat flux.

which is a 60–70% reduction. From all these fibres, glass displayed the greatest reduction, followed by wool and then jute and sisal. This trend is as expected from the flammability of these fibres. However, the THR increased for all natural fibres. The low value of the THR shown by the sisal/PP composite is due to its lesser thickness than other samples (Table 5.5). As discussed earlier in Section 4.2, thermally and physically thin samples ignite early, burn for a shorter time and produce less total heat release compared to thermally and physically thick samples which burn slowly but for a longer time.[51,52] Natural fibre-reinforced composites burn slowly but for a longer time, producing more total heat and smoke. This effect is more pronounced in wool/PP composites.

In the case of PLA composites, glass fibres helped in reducing the PHRR by 58% in comparison to pure PLA, jute by 40% and sisal by 18%. This reduction is much less than that seen for PP composites. The THR for the jute/PLA sample is increased, whereas smoke production in all samples is minimal.

On comparing the results for PP and PLA composites, the difference between any one type of fibre-containing composite was similar to that seen for the two matrices in Figure 5.3. However, owing to different sample thicknesses, the results cannot be directly compared. In a subsequent publication,[56] these results will be analysed in more detail.

5.4.3.3   Thermoset Composites

The only work reported on the flammability of natural fibre/reinforced thermoset composites is by Manfredi *et al.*[14] Composite samples tested include jute, flax, sisal and glass fibre reinforced unsaturated polyester and modified acrylic (modar) resins. These samples were prepared by Rodriguez[29] by placing fabric or mats in a mould, impregnating with resin by a vacuum infusion technique and curing at 60 °C for 2 h and post curing at 110 °C for 3 h. Jute/unsaturated polyester composites showed the best results on flexural and tensile strengths and the lowest in impact energy, because of the strong interphase developed. Flax composites showed higher impact energy than the other natural fibre composites, due to the existence of the effective energy dissipation mechanisms, such as pull-out and axial splitting of the fibres.

Manfredi *et al.*[14] studied the flammability by cone calorimeter at $35\,kW\,m^{-2}$ external heat flux and, from the data reported, the TTI and THR values are given in Table 5.8. The TTI depends upon the ease of ignition of the material on the surface, *i.e.* resin-rich surfaces show low TTI and high peak heat releases, whereas the THR depends upon the amount of flammable material in the composite and the duration of burning. From the results in Table 5.8 it can be seen that there is no real trend of results which can clearly show the effect of different reinforcing fibres or matrices. To assess the fire risk, the total heat evolved was plotted against the peak of the heat release rate divided by the TTI.[29] Among the composites with an acrylic matrix, the one reinforced with sisal fibre showed the highest fire risk and in global terms the worst fire resistance. Jute fibre composites showed a quick growing but short duration fire and, conversely, flax fibre composite developed a long duration but slow

Table 5.8 Mechanical and cone calorimetric results at $35\,kW\,m^{-2}$ heat flux for natural fibre reinforced polyester and acrylic composites.[14,29]

Sample[a]	Reinforcement type	Mechanical properties				Cone results at $35\,kW\,m^{-2}$			
		Flexural modulus (GPa)	Tensile modulus (GPa)	Impact energy (kJ m⁻²)	Water absorp. (%)	TTI (s)	PHRR[b] (kW m⁻²)	THR (MJ m⁻²)	Av. SEA (m² kg⁻¹)
Jute/acrylic	Woven jute fabric	4.6	6.9	8.8	3.0	72	520, 910	74.2	225.5
Jute/polyester	Woven jute fabric	6.6	8.0	10.6	2.7	51	395, 515	77.6	736.1
Sisal/acrylic	Nonwoven mat	2.5	3.4	12.7	3.2	52	310, 470	101.4	432.9
Sisal/polyester	Nonwoven mat	3.9	5.3	12.2	2.7	–	–	–	–
Flax/acrylic	Nonwoven mat	5.9	6.3	15.0	2.8	110	580	104.9	460.7
Flax/polyester	Nonwoven mat	4.8	6.3	13.2	–	–	–	–	–
Glass/acrylic	Nonwoven mat	11.3	13.3	98.7	0.36	62	340, 310	54.4	491.6
Glass/polyester	Nonwoven mat	11.4	14.9	106.5	0.4	56	275	32.3	709.5

[a]All composites had 30% fibre by volume; sample thickness = 3–4 mm.
[b]PHRR values are approximate, calculated from graphs reported by Manfredi et al.[14]

growing fire. Glass fibre composites showed more fire resistance than the natural fibre composites. Comparing the different matrices, it was observed that the fire risk was similar between the composites with unsaturated polyester and acrylic matrices. The jute fibre composites released lower quantities of smoke than the other composites (see Table 5.8). Comparing the different matrices, unsaturated polyester composites produced more smoke than the acrylic composites.

5.5 Flame Retardant Strategies for Components/ Composites

Since composites are composed of two or more components, any one or both of these can be rendered flame retardant prior to making a composite structure. Alternatively, the finished composite structure can be surface treated with protective coatings. These are discussed here as three generic methods of rendering the composites flame retardant:

5.5.1 Flame Retardant Matrices: Use of Reactive or Additive Flame Retardants

A matrix can be flame retarded by adding flame retardant chemicals into it or by chemical modification of the polymer backbone by introducing flame retardant elements into it. Alternatively, an inherently flame retardant matrix can be used, e.g. PVC for thermoplastics and phenolics for thermosets. While additives can be used for both thermoplastics and thermosets, chemical modification of the resin is only feasible in thermosets due to a vast choice of resins, curing agents and hardeners. For example, in unsaturated polyesters the use of halogenated resins or replacement of the curing agent from styrene to bromostyrene, introduction of halogen elements such as chlorine into the epoxy backbone in the diglycidyl ether of bisphenol C (DGEBC), fluorine in the diglycidyl ether of bisphenol F (DGEBF) or bromine in tetrabromobisphenol A (TBBA). For a detailed discussion on flame retardancy of thermosets, the reader is referred to previous reviews.[8,9]

The use of additives is the most popular method of flame retarding thermoplastic composites containing short fibres, where flame retardant (FR) chemicals can be added into the polymer and fibre prior to melt compounding. For long fibre composites, FR chemicals can be introduced into the polymer prior to forming sheets. The choice of flame retardant depends upon the type and processing temperature of the polymer used in the composite. There are a number of literature reviews available on flame retardants for different polymer types.[58-63] Most commonly used flame retardants (as reported in literature) for thermoplastic polymers include aluminium hydroxide, magnesium hydroxide, expandable graphite, ammonium polyphosphate, intumescent systems and nanoclays. Metal hydroxides such as aluminium hydroxide and magnesium hydroxide act as flame retardants by undergoing endothermic reactions at elevated temperatures, reducing the polymer temperature and releasing water vapour, which effectively dilutes the volatile species emanating from the polymer degradation. However, metal hydroxides are effective only at very high percent loadings (typically > 50 wt%), which might have an adverse effect on the mechanical properties of the polymer matrix. Nitrogen- and phosphorus-containing additives such as ammonium polyphosphate, intumescent chemicals, *etc.*, on the other hand, chemically interfere with the polymer decomposition and promote char formation. Chemically reactive type flame retardants are required at comparatively lower levels (20–30 wt%). Some specific examples of the use of these flame retardants in natural fibre composites are discussed here.

Hapuarachchi *et al.*[64] have reported the use of ATH (alumina trihydrate or aluminium hydroxide) (40 wt%) to flame retarded hemp/unsaturated polyester composites. They showed that ATH increased the TTI from 54 s in the control to 78 s and reduced the PHRR from 361 to 176 kW m^{-2}, when subjected to 50 kW m^{-2} external heat flux. Sain *et al.*[65] have used magnesium hydroxide on sawdust and rice husk-filled polypropylene composites. Replacement of 25% of the natural filler with magnesium hydroxide reduced the horizontal burning rate (ASTM D 635) by around 50%. No synergetic effect was observed when magnesium hydroxide was used in combination with boric acid and zinc borate.

Schartel *et al.*[36] have compared the effectiveness of ammonium polyphosphate and expandable graphite as flame retardants for flax/PP (40:60 by wt%) composites containing short fibres, prepared by melt blending all components together. Thermogravimetry in a nitrogen atmosphere was used for studying the pyrolysis behaviour, whereas flammability was studied by LOI, UL94, the glow wire test and cone calorimetry at different heat fluxes ranging from 30 to 70 kW m^{-2}. Selected results extracted from their paper are presented in Table 5.9. Both flame retardants promoted char formation, as observed by thermogravimetry and cone results. Ammonium polyphosphate and other (mono- and diammonium) phosphates on heating decompose to phosphoric acid, which esterifies the hydroxyl groups of the cellulosic structure of natural fibres. The resulting cellulose ester increases the char amount of the decomposition.[66] Polyphosphoric acid does not chemically react with decomposing PP as PP undergoes decomposition *via* random chain scission involving formation of some unsaturated end groups, *e.g.* $-C(CH_3)=CH_2$, which on oxidation yields further scission products and has no tendency to cross-link and form char.[67] Expandable graphite undergoes endothermic decomposition and expansion up to 300 times its initial volume under the impact of heat. The expanded charred structure creates a heat insulating layer for the underlying substrate, hence imparting flame retardancy. All flax/PP composites showed two peaks for the heat release rate in the cone calorimeter: the first one at the beginning of burning and the other one at the end, when the char cracks on the thermal feedback of the sample holder. The double peak behaviour is due to the lignocellulosic content of flax fibres, similar to the behaviour known for wood.[68] In flame retarded samples the first peak is reduced significantly due to char formation, which reduces the burning propensity of the composite. The

Table 5.9 Flammability results for flax/PP composites containing APP and expandable graphite.[36]

| | | | | Cone results[a] at (35) and 50 kW m^{-2} | | | |
Sample and mass composition	TGA residue at 490 °C	LOI (vol%)	UL94[c]	PHRR (kW m^{-2})	THR (MJ m^{-2})	Critical HRR[b] (kW m^{-2})	Char residue at flame-out (%)
PP/flax (70:30)	13.9	21	HB	(410), 520	(130), 135	167	10.6
PP/flax/APP (45:30:25)	27.4	26	HB	(210), 270	(95), 95	131	29.3
PP/flax/expandable graphite (45:30:25)	29.4	30	V1	(130), 160	(85), 95	35	>29.6

[a]The values are approximate, extracted from a figure in Schartel et al.[36]
[b]Calculated by extrapolating the steady heat release rate obtained from different heat fluxes in the cone calorimeter to an external heat flux of zero.
[c]HB = the sample was tested in a horizontal position and found to burn at a rate less than a specified maximum.
V1 = the sample was tested in a vertical position and self-extinguished within a specified time after the ignition source was removed.

values extracted at 35 and 50 kW m^{-2} external heat fluxes in Table 5.9 show that expandable graphite is more effective in reducing both peak and total heat release values compared to APP (ammonium polyphosphate). Critical heat flux values, obtained by extrapolating the steady heat release rate obtained from different heat fluxes in the cone calorimeter to an external heat flux of zero, also indicate the efficacy of expandable graphite compared to APP. This is further supported by LOI and UL94 results, where only expandable graphite-containing samples could achieve a V1 rating, whereas both control and APP-containing samples failed the test.

In a similar study, Le Bras et al.[54] studied the effect of APP on the pentaerythritol/melamine system. Flax/PP (40:60 by wt%) composites were prepared by extruding PP and short flax fibres (20 mm mean length). On adding APP to the sample as PP/flax/APP (31:46:23 wt%), the TTI, PHRR and THR values decreased. Even with a decrease in PHRR from 640 to 300 kW m^{-2} the sample could not pass the UL94 test due to a flaming time higher than 30 s and burning drops. However, when APP was replaced by an intumescent system [flax/PP/(APP/pentaerythritol/melamine), 31:46:23 wt%], the sample showed the formation of a stable intumescent char with no polymer dripping and instantaneous complete extinction after the second flame exposure, obtaining V0 rating. It shows that char-forming fibres need an intumescent composite structure maintaining coherence of the protective shield and thermal barrier properties to give a high level of flame resistance to the material. However, in the cone calorimetry test the intumescent system did not show any further improvement in PHRR (270 kW m^{-2}) and THR values.

Matko et al.[55] studied the efficiency of diammonium phosphate and ammonium polyphosphate in polypropylene, polyurethane and starch-based biocomposites. Natural fibre reinforcement was in the form of wood flake (pinewood as sawdust, assorted of 1.2 mm) or corn shell (3–12 mm size). APP was added to the polymer mix during polymer processing, whereas diammonium phosphate was used to treat the lignocellulosic fillers by mixing two components in 1:1 ratio in aqueous solution and then drying the material under an infrared lamp. Polypropylene biocomposites were prepared by compounding different components in a Brabender mixer and then compression moulding into sheets. Polyurethane biocomposites were prepared by mixing the wood flake/corn shell fillers and flame retardant into the mixture of polyol and an isocyanate at room temperature, followed by pressing into sheets. Starch samples contained corn starch and glycerol, as plasticizing agent, in a 1:1 mass ratio and required amount of flame retardant and compounding was carried out in a Brabender compounder. Composition and flammability results for these samples are summarized in Table 5.10. As can be seen from the results in the table for wood flake/PP composites, the sample containing 10% wood flakes and 10% APP had a LOI of 22 vol%, which is higher than that of the pure PP (LOI = 18 vol%); the value, however, hardly changed between 10 and 40% wood flake content. On increasing the fibre content up to 50% the LOI reached 30 vol% and achieved the V0 rates of flame retardancy. Hence, the combination of wood flake reinforced PP composite with APP shows flame retarded character only at high concentration of reinforcing component. The 20% APP containing samples were slightly better, with LOI reaching 24 vol% with 10% wood flakes and 32 vol% with 50% wood flakes.

Table 5.10 Flammability results of wood flake and corn shell containing polypropylene (PP) and polyurethane (PU) composites and their flame retarded counterparts.[55]

Composite	Flame retardant[a]	LOI (vol%)	UL94[b]
PP + wood flake (10–40%)	APP (10%)	22–25	HB
PP + wood flake (10–40%)	APP (20%)	24–27	HB
PP + wood flake (50%)	APP (10%)	30	V0
PP + wood flake (50%)	APP (20%)	32	V0
PU (80%) + wood flake (20%)	–	23	HB
PU (60%) + wood flake (20%)	APP (20%)	31	V0
PU (60%) + wood flake (20%)	DAP (20%)	30	V0
PU (80%) + corn shell (20%)	–	20	HB
PU (60%) + corn shell (20%)	APP (20%)	27	V0
PU (60%) + corn shell (20%)	DAP (20%)	29	V0
Corn starch + glycerol (1:1)	–	23	NR
Corn starch + glycerol (1:1)	APP (5%)	28	HB
Corn starch + glycerol (1:1)	APP (10%)	33	V0
Corn starch + glycerol (1:1)	APP (20%)	39	V0
Corn starch + glycerol (1:1)	APP (30%)	60	V0

[a]DAP = diammonium phosphate.
[b]HB = the sample tested in a horizontal position and burned at a rate less than a specified maximum; NR = No rating; V0 = the sample was tested in a vertical position and self-extinguished after the ignition source was removed.

For PU composites, both samples containing wood flakes or corn shell had low LOI values, *i.e.* 23 and 20 vol%, respectively. However, adding 20% of APP increases them to a great extent (to 31 and 27 vol%, respectively). A V0 rate was achieved with all the flame retarded systems. The LOI value of plasticized thermoplastic starch is quite low; the introduction of a small amount of APP (10%) increased the flame retardancy up to the self-extinguishing V0 rating. On increasing the concentration of APP, a sample with LOI of 60 vol% could be obtained. The outstanding flame retardancy results of starch-based thermoplastic biopolymer systems, which outperform the other studied biocomposites considerably, can be explained by the polyol character of the polymer matrix and the presence of polyol plasticizer. Thus the flame retardancy of such biopolymer systems can be solved quite easily by introducing phosphorus charring components. From this study it can be observed that that both the polysaccharides and polyurethane act as a charring agent and their content in the biocomposites correlates well with the achievable flame retardancy level. Plasticized starch thermoplastic, a fully degradable biopolymer, can be flame retarded efficiently by the introduction of as low as 10% APP and a LOI = 60 vol% value can be achieved with 30% flame retardant additive. Generally, it can be concluded that although bio-based polymers seem to be more expensive than the traditional mass polymers, their flame retardancy can be achieved in relatively simple and inexpensive ways.

Since the 1990s, polymer nanocomposites obtained by the addition of nanoparticles such as organoclays, nanosilica and carbon-based nanotubes into polymers have attracted considerable interest due to their ability to improve the physical and mechanical properties of polymeric materials while also enhancing their thermal stability and flammability. Although the use of nanoclays is vastly reported for bulk polymers[69] such as polypropylene,[66,67,70] polystyrene,[69] unsaturated polyester[71] and epoxy resins,[41] no specific examples for natural fibre-reinforced composites could be found in the literature.

5.5.2 Treatment of Natural Fibres with Flame Retardants

For cellulosic fibres a number of phosphorus- and nitrogen-based flame retardants (phosphates, phosphoramides, *etc.*) are available. Many reviews have been published on the flame retardancy of natural fibres/textiles.[66,72–75] Most of the flame retardants used for textile applications can also be used on the fibres/fabrics prior to composite formation, provided that the decomposition temperature of the flame retardant is higher than the composite's processing temperature. Candidate flame retardants are ammonium phosphates such as mono- or diammonium phosphates, ammonium bromide in combination with ammonium phosphates to provide some vapour-phase flame retardant action, borax and boric acid, ammonium sulfamate and sulfates. Durable flame retardant finishes used for cotton fabrics can also be applied. Examples include *N*-methylol dialkyl phosphonopropionamides, from which a well-known product, Pyrovatex CP (Ciba), is derived and tetrakis(hydroxymethyl)phosphonium derivatives [Proban CC (Rhodia, previously

Albright & Wilson)].[66] Most of these flame retardants function in the condensed phase and promote char formation by converting the organic fibre structure into carbonaceous char and hence reduce volatile fuel formation.[11] These condensed-phase active flame retardants are very effective in natural fibres owing to the polymer backbone having reactive side-groups, which on removal lead to unsaturated carbon bond formations and eventually a carbonaceous char following elimination of most of the non-carbon atoms present.[11] Vapour-phase active halogenated flame retardants are also very effective on cellulosis, which function in the vapour phase by interfering with the flame chemistry through free radical generation. The detailed mechanisms of action of these flame retardants on textiles are similar to the ones in bulk polymers, for which readers are referred to other reviews.[58–60]

Natural fibres/fabrics are generally treated with flame retardant finishes in the finished fabric form. Finishes are usually applied by a pad-dry method where the fabric is passed through the chemical formulation (mostly in aqueous form), passed through rollers to squeeze out the excess and then dried in an oven at 120 °C. This gives a non-durable finish. To obtain a semi-durable or durable finish, the fabric is passed through another oven set at a higher temperature (usually 160 °C), where a curing stage allows a degree of interaction between the finish and the fibre. These treatments can also be applied to nonwoven webs of fibres or short/long fibres by spraying the flame retardant finish and then drying the fibres prior to composite formation.

Intumescent chemicals can also be used for treating the fibres/fabrics. Intumescent coatings have been around for nearly 70 years, normally used as surface treatments for structural materials and metals to protect them from heat and fire. On heating, they form a foamed char, which thermally insulates the underlying structure. In textiles, although they can be used a surface coating on the fabric, they are usually applied in-between different fabric layers, which are then needle-bonded to consolidate the whole structure.[66] In our work on flexible composites[76] we applied an aqueous formulation of intumescent chemicals and a binder on a nonwoven web using a brushing technique. After drying, a second nonwoven web was attached to it and the whole assembly was then needle punched into a coherent single web.[76] In a later study the intumescent formulation was applied by padding on the nonwoven web.[77] We have demonstrated that intumescent chemicals can chemically interact with cellulosic fibres and this results in high levels of flame and heat resistance.[76,78] We also have shown that on introduction of these intumescent/flame retardant fibre systems to thermoset (epoxy, polyester and phenolic) resins, there is a physical and chemical interaction of the three components leading to enhanced char formation.[79] Glass fibre reinforced rigid composite structures containing these intumescent/flame retardant cellulosic fibres have shown superior flame retardant properties without any detrimental effect on their mechanical properties.[51,77,79,80] The intumescent/FR fibre combinations may be introduced either as a pulverized additive to the resin or as an additional textile fabric layer to the composite structure.[77]

With regards to specific examples in natural fibre composites, this methodology of flame retarding composites has not much been exploited. The only example available is by Misra *et al.*,[30] where tin(II) chloride treatment was used for brominated coir fibre in an epoxy-based composite. The coir fibre was treated with saturated bromine water for increased electrical properties and then with tin(II) chloride solution to increase the fire resistance. The fibre was ground to a powder and then mixed with epoxy resin to make the composite. They found that a 5% tin-based halogen compound reduced the smoke density by 25% and increased the LOI value from 36 vol% in a control to 39 vol% in the FR composite.

5.5.3   Surface Treatments of Composites

The most efficient way to protect materials against fire without modifying their intrinsic properties (*e.g.*, mechanical properties) is the use of fire retardant coatings.[81] In the case of glass/carbon fibre-reinforced thermoset composites, thermally insulative surface coatings can protect the resin-rich surface from ignition. This means of protecting flammable materials is called "passive fire proofing", as it serves to decrease heat transfer from the fire to the structure. Most of the effective coatings are ceramic- or intumescent-based. Intumescence is defined as the swelling of a substance when exposed to heat, typically forming a multi-cellular, carbonaceous or ceramic layer, which acts as a thermal barrier that effectively protects the substrate against a rapid increase of temperature, thereby maintaining its structural integrity.[82] The alternative is to use ceramic fabrics, ceramic coatings, hybrids of ceramic and intumescent coatings, silicone foams or a phenolic skin, all of which showed good performance for glass-fibre reinforced thermoset composites.[83] Use of mineral and ceramic claddings is quite popular for naval applications to fireproof conventional composite structures. These barriers function as insulators and reflect the radiant heat back towards the heat source, which delays the heat-up rate and reduces the overall temperature on the reverse side of the substrate. One commercial example of this sort of product is the Tecnofire range of ceramic webs produced by Technical Fibres in the UK; these are available with a number of different inorganic fibres, including glass and rock wool either with or without an associated exfoliated graphite present. They are designed to be compatible with whatever resin is used in composite production. We have shown their effectiveness for glass fibre-reinforced polyester composites.[84,85] 3 M Nextel 312 woven fabrics from alumina/boria/silica fibres are also used as composite fire barriers. Phenolic foams are also very effective fire barriers and particularly used for bulkhead structures in military ships.

There is no example of usage of the above-mentioned surface barrier coatings/ceramic claddings for natural fibre composites available in the literature, either for thermoset or thermoplastic composites. While surface treatment can be easily used for thermoset composites, the problem with thermoplastic composites could be that on nonpolar matrix surfaces the polar intumescent chemicals or inorganic/ceramic coatings do not stick. Use of compatiblizers, however, can solve this problem.

5.6 Summary

Natural fibre composites are becoming popular in the automotive industry owing to their renewability, biodegradability and cost. The use of natural fibres in a composite reduces the flammability of the composite compared to the pure polymer. However, for these composites to be used in other sectors, such as marine, aerospace or construction, they will have to pass commercial fire tests, such as UL94, flame spread, *etc.* For this, some sort of flame retardant treatment is required. Flame retardants can be added into the thermoplastic polymer or the natural fibre textile reinforcement can be flame retarded using conventional textile finishing processes. The composite can also be surface treated with flame retardant chemicals such as intumescent paints.

References

1. J. W. S. Hearle, in *Proceedings of the 7th European Conference on Composite Materials, London, 1996*, Woodhead, Cambridge, 1996, vol. 2, p. 377.
2. A. K. Bledzki, A. Jaszkiewicz, M. Murr, V. E. Sperber, R. Lutzendorf and T. Reussann, in *Properties and Performance of Natural-Fibre Composites*, ed. K. L. Picering, Woodhead, Cambridge, 2008, ch. 4.
3. M. P. Staiger and N. Tucker, in *Properties and Performance of Natural-Fibre Composites*, ed. K. L. Picering, Woodhead, Cambridge, 2008, ch. 8.
4. M. S. Huda, L. T. Drzal, D. Ray, A. K. Mohanty and M. Misra, in *Properties and Performance of Natural-Fibre Composites*, ed. K. L. Picering, Woodhead, Cambridge, 2008, ch. 7.
5. T. Nishono and K. Takano, *J. Polym. Sci., Part B*, 1995, **33**, 1647–1651.
6. J. Biagiotti, D. Puglia and J. M. Kenny, *J. Nat. Fibres*, 2004, **1**, 37–68.
7. R. Kozlowski and M. W. Przybylak, in *Proceedings of the 4th International Conference of the Textile Research Division, Cairo, Egypt, April 2007*, pp. 272–283.
8. B. K. Kandola and A. R. Horrocks, in *Fire Retardant Materials*, ed. A. R. Horrocks and D. Price, Woodhead, Cambridge, 2001, pp. 182–203.
9. B. K. Kandola and E. Kandare, in *Advances in Fire Retardant Materials*, ed. A. R. Horrocks and D. Price, Woodhead, Cambridge, 2008, ch. 15.
10. N. Saheb and J. B. Jog, *Adv. Polym. Technol.*, 1999, **18**, 351–363.
11. B. K. Kandola, A. R. Horrocks, D. Price and G. V. Coleman, *J. Macromol. Sci., Rev. Macromol. Chem. Phys.*, 1996, **C36**, 721–794.
12. S. Chapple and R. Anandjiwala, *J. Thermoplas. Compos. Mater.*, 2010, **23**, 871–893.

13. R. Kozlowski and M. W. Przybylak, *Polym. Adv. Technol.*, 2008, **19**, 446–453.
14. L. B. Manfredi, E. S. Rodríguez, M. Wladyka-Przybylak and A. Vázquez, *Polym. Degrad. Stab.*, 2006, **91**, 255–261.
15. M. Lewin, *Polym. Degrad. Stab.*, 2005, **88**, 13–19.
16. M. Lewin, and A. Basch, in *Flame-Retardant Polymeric Materials*, ed. M. Lewin, S. M. Atlas and E. M. Pearce, Plenum, New York, vol. 2, pp. 1–41.
17. ASTM D2863-00, Standard method for measuring the minimum oxygen concentration to support candle-like combustion of plastics (oxygen index).
18. BS 5438:1989, British Standard Methods of test for flammability of textile fabrics when subjected to a small igniting flame applied to the face or bottom edge of vertically oriented specimens.
19. D. Puglia, J. Biagiotti and J. M. Kenny, *J. Nat. Fibres*, 2004, **1**, 23–65.
20. K. Van de Velde and P. Kiekens, *Polym. Testing*, 2002, **21**, 433–442.
21. N. Graupner, A. S. Herrmann and J. Müssig, *Composites, Part A*, 2009, **40**, 810–821.
22. A. K. Bledzki, A. Jaszkiewicz and D. Scherzer, *Composites, Part A*, 2009, **40**, 404–412.
23. S. Ochi, *Mech. Mater.*, 2008, **40**, 446–452.
24. B. H. Lee, H. S. Kim, S. Lee, H. J. Kim and J. R. Dorgan, *Compos. Sci. Technol.*, 2009, **69**, 2573–2579.
25. M. S. Huda, L. T. Drzal, A. K. Mohanty and M. Misra, *Compos. Sci. Technol.*, 2008, **68**, 424–432.
26. T. Yu, J. Ren, S. Li, H. Yuan and Y. Li, *Composites, Part A*, 2010, **41**, 499–505.
27. K. Oksman, M. Skrifvars and J.-F. Selin, *Compos. Sci. Technol.*, 2003, **63**, 1317–1324.
28. D. Plackett, T. L. Andersen, W. B. Pedersen and L. Nielsen, *Compos. Sci. Technol.*, 2003, **63**, 1287–1296.
29. E. Rodríguez, R. Petrucci, D. Puglia, J. M. Kenny and A. Vázquez, *J. Compos. Mater.*, 2005, **39**, 265–282.
30. R. K. Misra, S. Kumar, K. Sandeep and A. Misra, *J. Thermoplas. Compos. Mater.*, 2008, **21**, 71–101.
31. P. Joseph and J. Ebdon, in *Fire Retardant Materials*, ed. A. R. Horrocks and D. Price, Woodhead, Cambridge, 2000, pp. 220–263.
32. M. Hirschler, in *Heat Release in Fires*, ed. V. Babrauskas and S. Grayson, Elsevier, New York, 1992, pp. 375–422.
33. *Plastics Flammability Handbook: Principles, Regulations, Testing and Approval*, ed. J. Troitzsch, Hanser, Cincinnati, 3rd edn., 2004.
34. J. Brandrup, E. H. Immergut, E. A. Grulke, A. Abe and D. R. Bloch, *Polymer Handbook*, Wiley, New York, 4th edn., 1999.
35. C. A. Harper, *Handbook of Plastics, Elastomers & Composites*, McGraw-Hill, New York, 4th edn., 2002.

36. B. Schartel, U. Braun, U. Schwarz and S. Reinemann, *Polymer*, 2003, **44**, 6241–6250.
37. ISO 5660-1: 1993, *Fire Tests on Building Materials and Structures – Part 15: Method for Measuring the Rate of Heat Release of Products*.
38. H. Tsuji and Y. Ikada, *J. Appl. Polym. Sci.*, 1998, **67**, 405–415.
39. D. W. Farrington, J. Lunt, S. Davies and R. S. Blackburn, in *Biodegradable and Sustainable Fibres*, ed. R. S. Blackburn, Woodhead, Cambridge, 2005, ch. 6.
40. A. R. Horrocks and B. K. Kandola, in *Design and Manufacture of Textile Composites*, ed. A. C. Long, Woodhead, Cambridge, 2005, pp. 330–363.
41. C. Katsoulis, E. Kandare and B. K. Kandola, *Polym. Degrad. Stab.*, 2011, 96, 529–540.
42. (a) J. Troitzsch, in *Advances in Fire Retardant Materials*, ed. A. R. Horrocks and D. Price, Woodhead, Cambridge, 2008, p. 291; (b) M. M. Hirschler, in *Advances in Fire Retardant Materials*, ed. A. R. Horrocks and D. Price, Woodhead, Cambridge, 2008, p. 443; (c) U. Sorathia, in *Advances in Fire Retardant Materials*, ed. A. R. Horrocks and D. Price, Woodhead, Cambridge, 2008, p. 527; (d) R. E. Lyon, in *Advances in Fire Retardant Materials*, ed. A. R. Horrocks and D. Price, Woodhead, Cambridge, 2008, p. 573.
43. A. P. Mouritz and A. G. Gibson, *Fire Properties of Polymer Composite Materials*, Springer, Dordrecht, 2006, p. 313.
44. FMVSS 302: *Motor Vehicle Safety Standard No. 302, Flammability of Materials – Passenger Cars, Multipurpose Passenger Vehicles, Trucks and Buses*, National Highway Traffic Safety Administration, Washington.
45. BS 5852: 2006: *Methods of Test for Assessment of the Ignitability of Upholstered Seating by Smouldering and Flaming Ignition Sources*; http://www.bsi-global.com/en/.
46. ISO 5659-2: *Plastics – Smoke Generation – Part 2: Determination of Optical Density by a Single-Chamber Test*; http://www.iso.org/iso/home.htm.
47. SOLAS, *Consolidated Edition, 2004; Consolidated Text of the International Convention for the Safety of Life at Sea, 1974; and its Protocol of 1978: Articles, Annexes and Certificates*.
48. Fire Test Procedure (FTP) Code, *International Code for Application of Fire Test Procedures, Resolution MSC.61 (67)*, International Maritime Organization, London, 1998.
49. Federal Aviation Regulation (FAR), *Airworthiness Standards*, Department of Transportation, Federal Aviation Administration, FAA specification JAR 25.853, Part IV, Appendix F.
50. J. L. Thomason and M. A. Vlug, *Composites, Part A*, 1997, **28**, 277–288.
51. B. K. Kandola, A. R. Horrocks, P. Myler and D. Blair, *Composites, Part A*, 2002, **33**, 805–817.
52. A. P. Mouritz and A. G. Gibson, *Fire Properties of Polymer Composite Materials*, Springer, Dordrecht, 2006.

53. S. Borysiak, D. Paukszta and M. Helwig, *Polym. Degrad. Stab.*, 2006, **91**, 3339–3343.
54. M. Le Bras, S. Duquesne, M. Fois, M. Grisel and F. Poutch, *Polym. Degrad. Stab.*, 2005, **88**, 80–84.
55. Sz. Matkó, A. Toldy, S. Keszei, P. Anna, Gy. Bertalan and Gy. Marosi, *Polym. Degrad. Stab.*, 2005, **88**, 138–145.
56. M. Helwig and D. Paukszt, *Mol. Cryst. Liq. Cryst.*, 2000, **354**, 373–380.
57. B. K. Kandola, B. Mottershead, S. C. Anand and S. I. Mistik, *Composites, Part A*, to be submitted.
58. *Fire Retardant Materials*, ed. A. R. Horrocks and D. Price, Woodhead, Cambridge, 2001.
59. *Advances in Fire Retardant Materials*, ed. A. R. Horrocks and D. Price, Woodhead, Cambridge, 2008.
60. *Fire Retardancy of Polymeric Materials*, ed. A. B. Morgan and C. A. Wilkie, CRC Press, London, 2nd edn., 2009.
61. S. V. Levchik and E. D. Weil, *J. Fire Sci.*, 2006, **24**, 345–364.
62. S.-Y. Lu and I. Hamerton, *Prog. Polym. Sci.*, 2002, **27**, 1661–1712.
63. E. D. Weil and S. V. Levchik, *J. Fire Sci.*, 2008, **26**, 5–43.
64. T. D. Hapuarachchi, G. Ren, M. Fan, P. J. Hogg and T. Peijs, *Appl. Compos. Mater.*, 2007, **14**, 251–264.
65. M. Sain, S. H. Park, F. Suhara and S. Law, *Polym. Degrad. Stab.*, 2004, **83**, 363–367.
66. B. K. Kandola, in *Fire Retardancy of Polymeric Materials*, ed. A. B. Morgan and C. A. Wilkie, CRC Press, New York, 2010, pp. 725–761.
67. B. K. Kandola, A. Yenilmez, A. R. Horrocks and G. Smart, in *Fire and Polymers*, ed. A. Morgan, G. L. Nelson and C. A. Wilkie, *ACS Symp. Ser. 1013*, American Chemical Society, Washington, 2009, pp. 47–69.
68. W. J. Parker and H. C. Tran, in *Heat Release in Fires*, ed. V. Babrauskas and S. J. Grayson, Elsevier, Barking, UK, 1992, vol. 4, pp. 331–372.
69. *Flame Retardant Polymer Nanocomposites*, ed. A. B. Morgan and C. A. Wilkie, Wiley-VCH, Weinheim, 2007.
70. B. B. Marosfoi, S. Garas, B. Bodzay, F. Zubonyai and G. Marosi, *Polym. Adv. Technol.*, 2008, **19**, 693–700.
71. S. Nazare, B. K. Kandola and A. R. Horrocks, *Polym. Adv. Technol.*, 2006, **17**, 294–303.
72. A. R. Horrocks, in *Fire Retardant Materials*, ed. A. R. Horrocks and D. Price, Woodhead, Cambridge, 2001, pp. 128–181.
73. P. Bajaj, in *Handbook of Technical Textiles*, ed. A. R. Horrocks and S. C. Anand, Woodhead, Cambridge, 2000, pp. 223–263.
74. S. Bourbigot, in *Advances in Fire Retardant Materials*, ed. A. R. Horrocks and D. Price, Woodhead, Cambridge, 2008, pp. 9–40.
75. E. D. Weil and S. Levchik, *J. Fire Sci.*, 2008, **26**, 243–281.
76. B. K. Kandola and A. R. Horrocks, *Text. Res. J.*, 1999, **69**, 374.
77. B. K. Kandola, A. R. Horrocks, P. Myler and D. Blair, *Composites, Part A*, 2003, **34**, 863.
78. B. K. Kandola and A. R. Horrocks, *Polym. Degrad. Stab.*, 1996, **54**, 289.

79. B. K. Kandola, A. R. Horrocks, P. Myler and D. Blair, in *Fire and Polymers*, ed. G. L. Nelson and C. A. Wilkie, *ACS Symp. Ser. 797*, American Chemical Society, Washington, 2001, p. 344.
80. B. K. Kandola, P. Myler, A. R. Horrocks and M. El-Hadidi, *Fire Safety J.*, 2008, **43**, 11–23.
81. X. Wang, E. Han and W. Ke, *Surf. Coat. Technol.*, 2006, **201**, 1528–1535.
82. B. K. Kandola, E. Kandare, P. Myler, C. Chukwudolue and W. Bhatti, in *Proceedings of the Fire and Materials 2009 Conference, San Francisco*, 2009, pp. 33–45.
83. U. Sorathia, C. M. Rollhauser and W. A. Hughes, *Fire Mater.*, 1992, **16**, 119–125.
84. E. Kandare, C. Chukwudolue and B. K. Kandola, *Fire Mater.*, 2010, **34**, 21–38.
85. E. Kandare, A. K. Chukwunonso and B. K. Kandola, *Fire Mater.*, 2011, **35**, 143–155.

CHAPTER 6

Natural Fibre Composites: Automotive Applications

S. C. R. FURTADO,[a] A. J. SILVA,[a] C. ALVES,*[b]
LUÍS REIS,[a] MANUEL FREITAS[a] AND PAULO FERRÃO[a]

[a] Instituto Superior Técnico, Avenida Rovisco Pais 1049-001, Lisbon, Portugal; [b] Universidade Federal do Rio Grande do Norte, Campus Universitário - Lagoa Nova, Av. Sen. Salgado Filho, 3000, 59078-970, Natal, RN, Brazil
*Email: cralves@dcdesign.com.br

6.1 Introduction

Composite materials have been used by humans for some millennia. Presently, the most common polymer matrix composites, able to carry significant loads, use carbon, aramid (aromatic polyamides), or glass fibres as reinforcements. Significant contributions can be found in the literature regarding their mechanical behavior in static and dynamic conditions, their ageing behavior and, more recently, their environmental impact.

Industry in general is looking for "greener" materials (from renewable sources) to use in its products, both as a way of reducing the environmental burden and as a way of appealing to a growing environmentally conscious market. The automobile industry in particular is constantly looking for products that are lighter, more eco-friendly, and are still adequate for mass production at low cost.

A tentative shift from steel to light alloys and composite materials in the later part of the last century was immediately followed by a reaction from the steel industry with the introduction of high-strength steels and other special alloys that could also reduce the weight of products. The major focus of the steel

RSC Green Chemistry No. 16
Natural Polymers, Volume 1: Composites
Edited by Maya J John and Thomas Sabu
© The Royal Society of Chemistry 2012
Published by the Royal Society of Chemistry, www.rsc.org

industry is the recyclability of steel, compared with the recyclability of other metal alloys or composite materials.

One advantage of using natural fibres such as kenaf, hemp, flax, jute, and ramie, in substitution of aramid or glass as fibre-reinforced composites, is their renewable nature and inherent biodegradability, rendering the issue of reuse or recyclability meaningless. Using natural fibres in automotive structures (especially in enclosures) could be an advantage over other non-recyclable reinforced polymers, provided that their mechanical behavior is fully understood and comparable with more traditional solutions.

Auto companies are seeking materials that combine different properties, like sound abatement capability and reduced weight. It is estimate that ~75% of a vehicle's energy consumption is directly related to factors associated with the vehicle's weight.[1,2] To improve the fuel economy, manufacturers have focused on reducing automobile weight, primarily through manufacturing methods that use raw materials offering the strength of steel without its heft.[1] Composites reinforced by vegetable fibres, like jute, flax, hemp, or ramie, can be a good choice. It was shown that they can be a realistic alternative, for example, to glass-reinforced composites, because they can deliver the same performance for lower weight, can be 25–30% stronger for the same weight, and can exhibit favorable non-brittle fracture on impact, which is an important requirement in the passenger compartment.[2] According to BMW, it is possible to manufacture bio-based composites that are as much as 40% lighter than equivalent injection-molded plastic parts.[3]

Environmental legislation in different countries is becoming much more stringent, leading to an increased use of natural fibre-reinforced plastics in automotive construction and pushing the automakers to evaluate the environmental impact of a vehicle's entire lifecycle, from raw materials to manufacturing to disposal.[4] A regulation example is the European Directive 2000/53/EC, in which vehicles have to be partially decomposable or recyclable (95%) by 2015.[3,5] Japan also requires that 95% of a vehicle be recovered (which includes incineration of some components) by 2015.[3] In the USA, 10–11 million vehicles reach the end of their useful lives each year. A network of salvage and shredder facilities process about 96% of these old cars; about 25% of the vehicles weight, including plastics, fibres, foams, glass, and rubber, remains waste. A car made mostly of heated, treated, and molded bio-fibre would simply be buried at the end of its lifetime and consumed naturally by bacteria,[2] leading to less necessity of landfill.

The use of natural composites in the automotive industry started with Henry Ford. Around 1942, he began experimenting, initially using compressed soybeans to produce composite plastic-like components (Figure 6.1). During that period the petroleum-based chemicals were very cheap, and therefore soy-based plastic could not find economic importance. With the new environmental regulations and depletion and uncertainty of petroleum sources, scientists and manufacturers have revived their interest to derive a new generation of composite materials from renewable sources like soybean-based plastic or natural fibre.[6] Many major automakers have now a real interest in

Figure 6.1   Henry Ford performing a durability test with a fire axe on a prototype car made of plastic derived from soybeans.[6]

Table 6.1   Some fibre prices; vegetable fibre prices as of 2008;[10] E-glass fibre price for 2011.[11]

Fibre	Flax	Hemp	Jute	Abaca	Sisal	Coir	E-glass
Price (US $/kg)	0.747	0.754	0.332	1.155	0.825	0.314	2

global sustainability. DaimlerChrysler identifies bio-based materials as one of the two key parts of its plan to create a global sustainability network.[7]

Over the last few years, a number of researchers have been involved in investigating the exploitation of natural fibres as load bearing constituents in composite materials.[8] In Europe, automotive companies have developed components made of vegetable composites, motivated mainly by regulations that have played an important role toward sustainable mobility. In the USA, about 1.5 million vehicles are already using vegetable fibres as reinforcement for thermoplastic and thermosetting polymers. In Brazil, some automotive initiatives are concerned with the selection of "greener" materials (from renewable sources). Several studies point out that automotive textile usage has grown in terms of quantity, quality, and product variety worldwide.[5]

The use of such materials in composites has increased owing to their relative cheapness (Table 6.1), the fact that surface treatments avoid the use of organic solvents, and that they can compete well in terms of strength per weight of material.[2,8] They also contribute to a reduction in fossil fuel consumption and greenhouse gas (GHG) emissions of vehicles by consuming about 45% less energy in their incineration, which complies with environmental directives and the Kyoto Protocol.[5] NPSP Composieten have introduced bio-composites in external and internal body panels for trains and train furniture for the Dutch Railways, which resulted in a lower train weight (1000 kg per train), saving 5 MWh of electricity per train per year, which led to less pollution of about 0.5 metric tons of CO_2 a year.[9]

Automakers now see strong promise in natural fibre composites.[2,5] Targeting the environmental requirements of vehicles warrants a shift of vehicle design from oil-derived polymers and synthetic reinforcements to natural materials.

Undoubtedly, vegetable or "green" composites are playing a major role in the product design process, presenting many environmental advantages compared to oil-based composites, and offering a possibility to developing countries to use their own natural resources in a wide range of industries, as well as resulting in social aspects such as new jobs in impoverished local communities.[5] Currently, at least 315 000 metric tons of bio-composites reinforced by vegetable fibres are already being used in the European industry, mainly in the automotive and construction sectors. By 2020, this quantity could be more than doubled.[12]

6.2 Composite Materials: Definition and Classification

Composites are one of the oldest materials used by humans, and there have been significant developments due to the booming of the aviation industry in the mid-20th century plus the emergence of the aerospace industry. Since then, many investments have been used to develop and produce polymeric resins and glass fibres, which motivated their use widely, with glass fibres becoming the main composite reinforcement worldwide.[13] Several industrial applications require properties that are not found in neat materials; thus composite materials can be feasible to meet industrial requirements presenting useful mechanical properties through system synergy, in which an applied load stress is supported by the matrix and reinforcement. In this sense, it is clear that the interface reinforcement/matrix plays an important role in composites, influencing their final properties. Besides the mechanical advantages of composite materials, they also give a great advantage for design processes, allowing designers to assume a role beyond merely "material picker".[14] Therefore, the success of a product made of composite materials will depend on the best designer choice related to the matrix/reinforcement combination, besides other aspects such as cost and production.[15]

According to the American Society for Testing and Materials (ASTM), composite materials are the combination of two or more different materials to form a new and useful engineering material, differing from the neat materials that originated them. They can be produced by combining metals, ceramics, or polymers,[16] and their constituent elements should be distinguished on a macroscopic scale. Most composites have better mechanical behavior than these phases or even new properties that emerged from their combination.[17,18] According to Jones,[17] some of the main improved properties are strength, hardness, fatigue and corrosion resistance, weight, and thermal conductivity.

Generally, composites are made mainly of two different materials, well known as matrix and reinforcement. The matrix of a composite performs essential functions such as to protect the reinforcement from the environment, ensure an efficient distribution of an applied load stress to the reinforcement, and to provide a stable shape of the composite. On the other hand, the reinforcement performs the structural function, increasing one or more properties of the composite.

6.2.1 Matrices

Matrices can be distinguished as metallic, ceramic, or polymeric. The polymeric ones are widely used to produce components for various industrial markets, since they present lower costs than other matrices. They also present low density, which implies lightweight composites. Polymeric matrices are divided into two categories: thermoplastic and thermoset. Thermoplastic matrices are composed of large molecules bound to each other by van der Waals and hydrogen bonds. When heated, these bonds are temporarily broken, allowing molecular mobility; after cooling the molecules assume new positions, reshaping the material. Usually, thermoplastic polymers have higher impact strength and toughness than thermosetting polymers. They are easily processed by injection or thermoforming and can be recycled, which is an important behavior related to their environmental performance. Nowadays, they represent about 35% of the global composite market and the thermoplastic polymers most used are the following: polypropylene, polyamide, polycarbonate and PEEK (polyether-ether-ketone).[19] Thermosetting matrices are composed of three-dimensional molecules, forming rigid and permanent structures that do not allow their reprocessing and/or reshaping. Generally, they are processed by adding accelerator and catalyst into the resin to start the polymerization, forming a three-dimensional structure in a process called "curing". Worldwide, the most used thermoset matrix to manufacture industrial composites is the unsaturated polyester, mainly due to its low cost, good properties, and low viscosity, which implies better wettability and impregnation of the reinforcement. These matrices can also be divided into three groups: orthophthalic for general industrial usage, isophthalic that presents good thermal behavior, and bisphenol used in aggressive environments.[20] Table 6.2 shows some thermoplastic and thermoset resins as well as some characteristics.

6.2.2 Reinforcements: Fibres and Fillers

In composite materials, reinforcements are the components which improve their structural properties as a whole and support most load stress, with a percentage also influencing the final cost of the composites. More than one type of reinforcement can be used in a composite to obtain the intended properties, forming a combined composite (Figure 6.2f). Most reinforcements have large length/diameter ratios (fibre form), which are widely accepted as

Table 6.2 Mechanical properties of thermoplastic and thermoset resins.[a]

Resin		Density (g m^{-3})	Young's modulus (MPa)	Tensile strength (MPa)
Thermoplastic	Nylon	1.1	1.3–3.5	55–90
	PEEK	1.3–1.35	3.5–4.4	100
	PPS	1.3–1.4	3.4	80
	Polyester	1.3–1.4	2.1–2.8	55–60
	PC	1.2	2.1–3.5	55–70
	PTFE	2.1–2.3	–	10–35
Thermoset	Epoxy	1.2–1.4	2.5–5.0	50–110
	Phenolic	1.2–1.4	2.7–4.1	35–60
	Polyester	1.1–1.4	1.6–4.1	35–95

[a]Adapted from Mazumdar.[21]

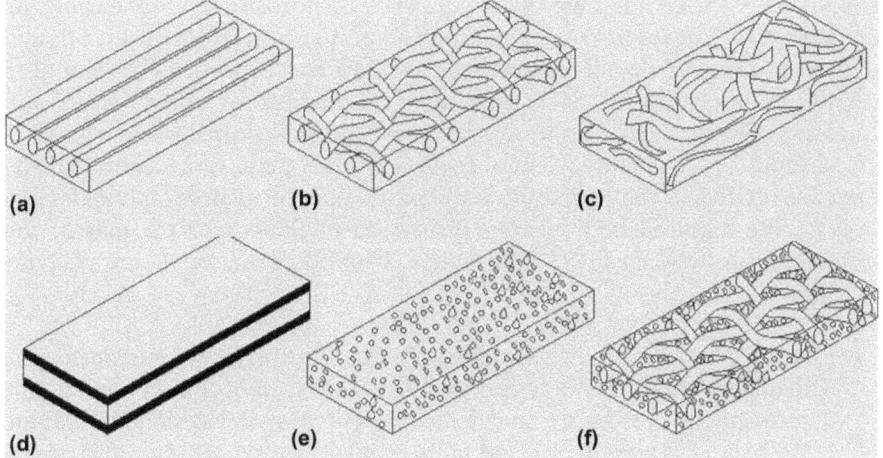

Figure 6.2 Composite classification.

reinforcements for structural and non-structural industrial applications. In polymeric matrix composites the most used fibres are carbon, aramid, boron, and glass. Among them, glass fibre is widely used to produce industrial components, mainly due to its great cost/benefit ratio. Different types of glass fibres can be produced, based on chemical combinations to meet specific requirements from different markets. Currently, two types of glass fibres represent about 90% of the global fibre consumption:[13]

- E-type. This represents more than 90% of total glass fibre usage in the industrial market owing to its good mechanical and electrical properties and moisture resistance.
- S-type. This is widely used in the construction sector owing to its higher mechanical properties (about 20–30%) than E-type glass fibre, besides its

better thermal and fatigue resistance. However, its high cost restricts its industrial usage.

Fillers are another type of reinforcement, which are particles dispersed into the matrix to achieve the intended composite properties such as cost, weight, electrical conductivity, mechanical strength, color, and to decrease contraction.[20,22] Usually, fillers are obtained from natural deposition, the most widely used being calcium carbonate, silica, quartz, graphite, and a large number of metal powders.[23]

Nowadays, calcium carbonate ($CaCO_3$) is the mineral most used as filler in several industrial sectors, such as paper, ceramic, paints, pharmaceuticals, and plastics. It is also used in environmental applications, *e.g.* water treatments and to neutralize acid rain. There are many types of $CaCO_3$ filler, depending on their origin and their particle treatment, resulting in different composite properties.[22,23] Some filler characteristics such as porosity and dimensions are very important because they define the system interface, in which the strength increases as the particle dimensions decrease. Moreover, the dispersion of fillers into the matrix must be as homogenous as possible to avoid tensile concentrations, which decrease the composite mechanical properties.

According to Nezbedová[24] and Eylisson,[25] the main constraints to filler usage are related to the properties and compounds of their components, filler porosity, the required homogenous dispersion of fillers into the matrix, and the interaction between fillers and matrix. Composite materials are classified as:[17,26]

- Fibrous composite materials: composed of fibres (reinforcement material with a large length/diameter ratio) in the matrix. In fibrous composites the reinforcement can be arranged in different directions: uniaxial (Figure 6.2a), biaxial (Figure 6.2b), and multi-axial (Figure 6.2c). They can also be divided into short fibre webs (chopped fibre random fleece) and endless fibres webs (continuous fibre random mat).[20]
- Laminated composite materials: composed of layers from at least two different materials, which are linked to form a laminate (Figure 6.2d).
- Particle composite materials: composed of macroscopic particles (fillers) of one or more materials, dispersed into matrix (Figure 6.2e).
- Combined composite materials: combination of fibrous and particle composite materials described above (Figure 6.2f).

Composite materials are also classified based on their matrices and/or reinforcement origins, which can be metallic, ceramic, or polymeric.[18]

6.3 Natural Fibres

Natural fibres are subdivided based on their origins, coming from plants, animals, or minerals. All plant (vegetable) fibres are composed of cellulose, while animal fibres consist of proteins (hair, silk, and wool). Vegetable fibres

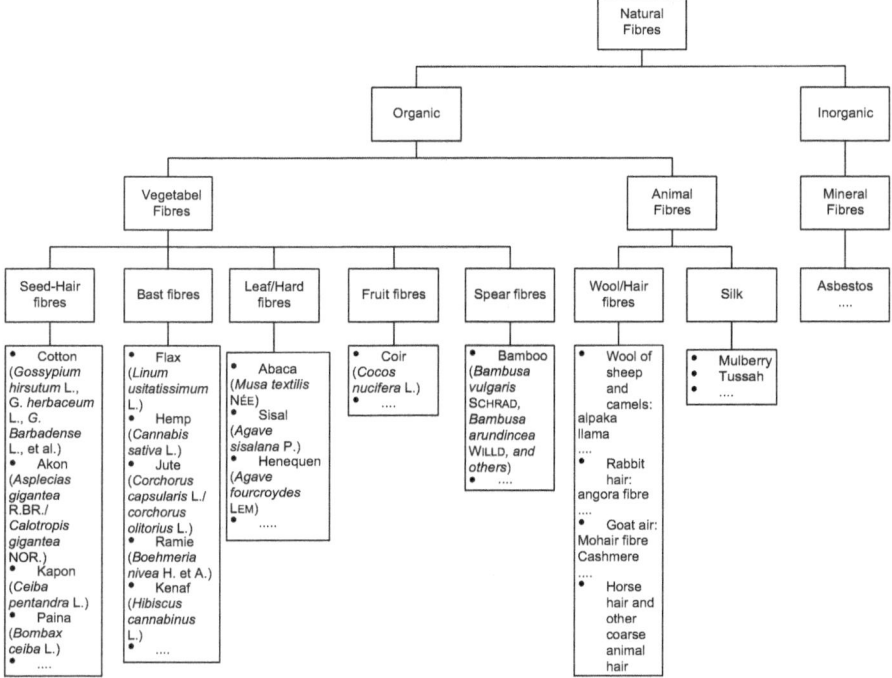

Figure 6.3 Classification of natural fibres.[27]

include bast (or stem or soft sclerenchyma) fibres, leaf or hard fibres, and seed fibres.[8] Figure 6.3 presents a classification of natural fibres.[27]

Plant/vegetable fibres are the natural fibres most utilized; because of this, the expression "natural fibres" is widely used to designate plant/vegetable fibres.

Vegetable fibres can be considered as naturally occurring composites.[28] They are cell walls that occur in stem and leaf parts and comprise cellulose, hemi-celluloses, lignins, and aromatics, waxes and other lipids, ash, and water-soluble compounds (Figure 6.4).[29] The main components of vegetable fibres are cellulose (α-cellulose), hemicellulose, lignin, pectin, and waxes.[8]

The properties of each constituent contribute to the overall properties of the fibre. The reinforcing efficiency of vegetable fibre is related to the nature of cellulose and its crystallinity. A high cellulose content and low microfibril angle are desirable properties in a fibre to be used as reinforcement in polymer composites. The cellulose fibrils are aligned along the length of the fibre, which render maximum tensile and flexural strengths, in addition to providing rigidity.[28] Hemicellulose is responsible for the biodegradation, moisture absorption, and thermal degradation of the fibre as it shows least resistance, whereas lignin is thermally stable but is responsible for the UV degradation.[31]

The percentage composition of each of these components varies for different fibres. Generally, the fibres contain 60–80% cellulose, 5–20% lignin, and up to 20% moisture (Table 6.3).[31]

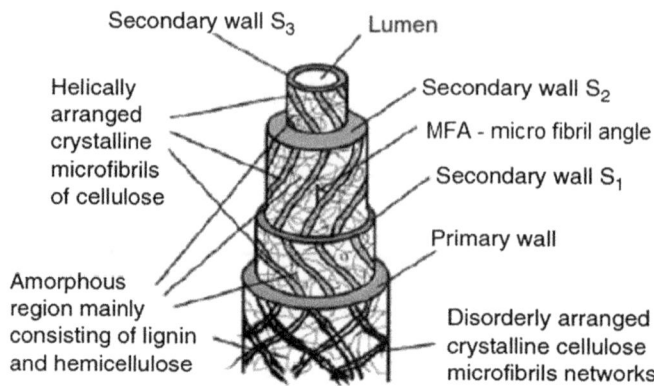

Figure 6.4   Structure of a bio-fibre.[30]

Table 6.3   Mean chemical composition of some vegetable fibres in (% of dry mass).[32]

	Flax	*Hemp*	*Jute*	*Abaca*	*Coir*	*Sisal*
Cellulose	71.2	78.3	71.5	70.2	35.6	73.1
Hemicelluloses	18.5	5.4	13.3	21.7	15.4	13.3
Pectin	2.0	2.5	0.2	0.6	5.1	0.9
Lignin	2.2	2.9	13.1	5.6	32.7	11.0
Extractives	4.3	–	1.2	1.6	3.0	1.3
Fat and wax	1.6	–	0.6	0.2	–	0.3

6.4   Vegetable Fibres: Advantages and Disadvantages

Interest in vegetable fibres is growing owing to their better properties. They possess many advantages and disadvantages compared to synthetic fibres like glass or carbon fibres. At this time, for example, glass fibre-reinforced plastics have been proven to meet the structural and durability demands of automobile interior and exterior parts. However, glass-reinforced plastics exhibit short-comings, such as their relatively high fibre density (approximately 40% higher than vegetable fibres), machining difficulties, poor recycling properties, and the potential health hazards posed by particulates.[3] The lignocellulosic fibres, besides their environmental advantage over synthetic ones, offer the capacity to buckle rather than break during processing and fabrication. In addition, cellulose possesses a flattened oval cross-section that enhances stress transfer by presenting an effectively higher aspect ratio.[8] The advantages and disadvantages are detailed next, for each property or characteristic of interest.

6.4.1   Production

Vegetable fibres need less energy in their production, resulting in lower pollutant emissions. Glass fibre production requires 5–10 times more non-renewable energy than vegetable fibre production.[33]

However, vegetable fibres cannot be made to grow anywhere in the world, whereas a plant for glass or carbon fibre production can be installed nearly anywhere, so the cost of transportation can play a significant role in the overall cost of production, and care needs to be taken in calculating the cost of production of each type of fibre. This difficulty can also lead to a higher environmental impact. It was shown that, for a buggy bonnet LCA (life cycle assessment) in Brazil, in the jute fibre-reinforced composite production phase, about 52% of the environmental impact is caused by transport fuel consumption.[34]

6.4.2 Environment/Health

The most interesting aspect about vegetable fibres is their positive environmental impact. Vegetable fibres are a renewable resource, with production requiring little energy. They are carbon dioxide neutral, *i.e.* they do not return excess carbon dioxide into the atmosphere when they are composted or combusted;[8] they also provide carbon dioxide sequestration and are biodegradable.[28] They also present safer handling and working conditions compared to synthetic reinforcements.[8] It has been shown that despite the higher energy consumption to treat the jute fibres for a buggy bonnet production, their lighter weight characteristic ensures better environmental performance compared to glass fibre.[35] The use phase of vehicles is the most pollutant phase. For similar recycling efficiency, Silva[5] has shown that the energy required to reach their respective efficiency is about 76% higher for glass composites than for jute composites.

6.4.3 Mechanical Properties

The growing interest in lignocellulosic fibres is mainly due to their economical production, with few requirements for equipment and low specific weight compared to glass fibre composites (Table 6.4). For example, using bast fibres in the automotive industry leads to weight savings of between 10 and 30% and corresponding cost savings.[8] It has been shown that vegetable fibres perform better than the glass fibre composites for all fibre architectures and better than aluminum or steel.[9] A practical translation of this parameter is that a panel with the same weight and surface area can be more than 10% stiffer in vegetable fibre composites compared to glass.

Dent resistance is also a major factor when selecting a material for automobile enclosures. It is proportional to the square of the strength of the material and to the fourth power of the panel thickness and inversely proportional to the panel bending stiffness. Panels made of vegetable fibre composites, being inherently lighter, will lead to an increase in dent resistance when compared to glass fibre composite panels with the same weight.

One of the disadvantages is the high moisture absorption of vegetable fibres, leading to swelling and the presence of voids at the interface, which results in poor mechanical properties and reduces dimensional stability of composites.[8,37] It was shown in jute fibre/unsaturated polyester composites that, owing to

Table 6.4 Mechanical properties of fibres.[5,11,36]

Fibre	Density, ρ (g cm^{-3})	Young's modulus, E (GPa)	Specific elastic modulus, E/ρ	Tensile strength, σ (GPa)	Specific tensile strength, σ/ρ	Elongation at break (%)
Cotton	1.51	12	7.95	0.4	0.26	3–10
Flax	1.4	60–80	42.86–57.14	0.8–1.5	0.57–1.07	1.2–1.6
Hemp	1.48	70	47.30	0.55–0.9	0.37–0.61	1.6
Jute	1.45	10–32	6.89–22.07	0.45–0.55	0.31–0.38	1.1–1.5
Abaca	1.5	31.1–33.7	20.73–22.47	0.43–0.81	0.29–0.54	2.9
Coir	1.33	4–6	3.01–4.51	0.14–0.15	0.11	15–40
Sisal	1.45	26–32	17.93–22.07	0.58–0.61	0.40–0.42	3–7
E-glass	2.6	73	28.07	1.8–2.7	0.69–1.04	2.5

water absorption in a weathering test, the flexural modulus can decrease by about 30%.[38] Another drawback of vegetable fibres is the quality variation (even between individual plants in the same cultivation), depending on growth conditions, processing, and other reasons, leading to non-uniformity and variation of the dimensions and of their mechanical properties.[8,39] Therefore, quality control must be very rigorous for accurate results. Vegetable fibres have problems at the interfacial adhesion between the fibre and polymer matrix, which determines the composite physical properties, owing to the polar and hydrophilic nature of lignocellulosic fibres and the non-polar characteristics of most organic polymers. Because of this, it is usually necessary to make compatible or couple the blend.[3,6,37]

6.4.4 Physical and Chemical Properties

Vegetable fibres possess high electrical resistance. They have a hollow cellular structure, providing excellent sound absorption efficiency.[40] Vegetable fibres are more shatter resistant and have better energy management characteristics than glass fibre in their respective composites.[2] Vegetable fibres in the manufacture of fire resistant upholstery have already been produced, where non-woven flax (LinFR) is used to act as a fire barrier that reduces the vulnerability of filling material to the development and spread of fire.[8]

An important drawback of vegetable fibres is their low microbial resistance and susceptibility to rotting, which is a restriction to the successful exploitation of bio-fibres for durable composites. These properties pose serious problems during shipping, storage, and composite processing.[8]

6.4.5 Processing

Vegetable fibres are non-abrasive to mixing and molding equipment, which can contribute to a significant cost reduction.[8] In automotive parts, compared to glass fibre composites the vegetable fibre composites reduce the mass of the component, lowering the energy needed for production by 80%.[2,5]

The major disadvantage of vegetable fibre reinforced composites is their inherent polar and hydrophilic nature of the lignocellulosic fibres and the non-polar characteristics of most thermoplastics, resulting in compounding difficulties and leading to non-uniform dispersion of fibres within the matrix, which impairs the efficiency of the composite. Another problem is that the processing temperature of composites is restricted to 200 °C, as vegetable fibres undergo degradation at higher temperatures.[31] The result of prolonged high-temperature exposure may be discoloration, volatile release, poor interfacial adhesion, or embrittlement of cellulose components.[28,30]

It is quite clear that the advantages outweigh the disadvantages and most of the shortcomings have remedial measures in the form of chemical treatment.[8] However, several major technical considerations must be addressed before the engineering, scientific, and commercial communities gain the confidence to enable wide-scale acceptance of vegetable fibres, particularly in exterior parts where a class A surface finish is required. Challenges include the homogenization of the fibre's properties and a full understanding of the degree of polymerization and crystallization, adhesion between the fibre and matrix, moisture repellence, and flame retardant properties.[3] Care has to be taken so as not to erode the environmental advantage of vegetable fibres with chemical treatments.

6.5 Applications in the Automotive Industry

Regardless of the need for low variability in mechanical properties,[41] the use of vegetable fibre reinforced composites has extended to almost all fields. In the past decades, vegetable fibres composites with thermoplastic and thermoset matrices have been embraced by car manufacturers and suppliers for door panels, seat backs, headliners, package trays, dashboards, and interior parts.[3,5,8] The following are some examples of principal car component applications for various vegetable fibre types:[3,8,16,42]

- Flax, sisal and hemp are processed into door cladding, seatback linings and floor panels (Figure 6.5). Flax fibres are also used in car disk brakes to replace asbestos fibres.
- Coconut fibre is used to make seat bottoms, back cushions, and head restraints.
- Cotton is used to provide sound proofing; it was embedded in phenolic resin and used in the body of the East German Trabant car, the first production car manufactured with natural fibres.
- Abaca is used in underfloor body panels.
- Wood fibres are used in a large number of applications in decks, docks, window frames, and molded panel components.
- Kenaf is used in door inner panels (Figure 6.6).

Figure 6.5   The underbody of a DaimlerChrysler A-class compression molded flax/polypropylene.[3]

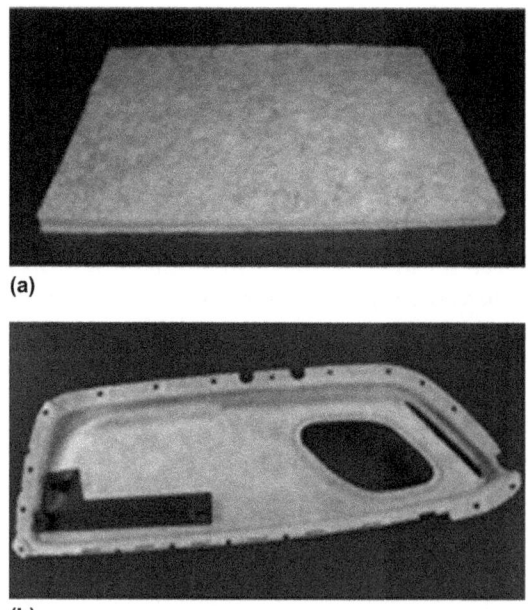

(a)

(b)

Figure 6.6   (a) A kenaf fibre mat; (b) a door inner panel comprising 50% kenaf mat and 50% polypropylene.[3]

Recently, three-layer particleboards were produced from a mixture of sunflower stalks and poplar wood at certain ratios utilizing urea/formaldehyde (UF) adhesives. Results show that all the panels provide properties required by the normal standards for general purpose-use particleboards.[8]

All the major car manufacturers now use bio-based composites in various applications. Some of these applications are presented next:[2,3,7,8,40,43,44]

- DaimlerChrysler is the biggest proponent, with up to 50 components in its European vehicles being produced from bio-based materials. They have been exploring the idea of replacing glass fibres with natural fibres in automotive components since 1991. A subsidiary of the company, Mercedes-Benz, pioneered this concept with the "Beleem project" based in Sao Paolo, Brazil. In this case, coconut fibres were used in the commercial vehicles over a nine-year period. This initiative translates into new jobs in coconut production, allowing the local population to improve their life conditions. Mercedes also used jute-based door panels in its E-class vehicles in 1996. In September 2000, DaimlerChrysler began using vegetable fibres for their vehicle production. Johnson Controls has started production of door-trim panels from vegetable fibre and polypropylene for DaimlerChrysler. DaimlerChrysler has now increased its research and development in flax-reinforced polyester composites for exterior applications. In Figures 6.7 and 6.8 are presented some examples of the application of natural fibres in Mercedes cars.
- In 2000 Audi launched the A2 midrange car in which door trim panels were made of polyurethane reinforced with mixed flax/sisal needle felt.
- Honda embarked on using vegetable fibre materials, such as wood-fibre parts in the floor area of the Pilot sport utility vehicle (SUV), a decision that was driven by engineering considerations as well as corporate philosophy.
- BMW Group incorporates a considerable amount of renewable raw materials into its vehicles, including 10 000 tonnes of vegetable fibres in 2004.
- General Motors kenaf and flax mixture has gone into package trays and door panel inserts for Saturn L300s and European-market Opel Vectras, while wood fibre is being used in seatbacks for the Cadillac Ville and in the cargo area floor of the GMC Envoy and Chevrolet TrailBlazer.
- Ford mounts Goodyear tires that are made with corn on its fuel-sipping Fiestas in Europe. Goodyear has found its corn-infused tires have lower rolling resistance than traditional tires, providing better fuel economy. The sliding door inserts for the Ford Freestar are made with wood fibre. Ford made the automotive industry's first application of wheat straw-reinforced plastic for the third-row storage bins of the 2010 Ford Flex. In its Start Concept, sisal fibre is used to form the interior panels. Figure 6.9 present a front-end grill opening reinforcement used on Ford cross members.
- Toyota has shown interest in using kenaf to make Lexus package shelves, and incorporated it into the body structure of Toyota's i-foot and i-unit concept vehicle. Over the course of the last five years, Toyota has evaluated numerous materials made from renewable resources to assess their performance, appearance, safety, and mass production capability. They have introduced environmentally preferable parts in a number of their

vehicles, and currently they are investigating new materials for fabrics and carpets as well as additional applications of PP/PLA-based and natural fibre-based materials in North American vehicles.

An overview of some automotive usage of vegetable fibres is presented in Table 6.5.

The recent reports are pointing out the importance of vegetable fibres and bio-composites in the automotive industry, where many of their advantages, like the above, are presented. The vegetable fibres incorporation in automotive components can also play an important role towards legislations accomplishment. Using materials, like vegetable fibres, into automotive parts, that can be simple buried at the end of their life, represent an aid for accomplishment of legislations, like the end of life vehicle (ELV) directive in Europe, which states

(a)

(b)

Figure 6.7   (a) Under floor protection trim of Mercedes A-class made from abaca (*Musa textilis*) fibre reinforced composites. (b) Newest Mercedes S-class automotive components made from different vegetable fibre reinforced composites.[8]

Figure 6.8   Mercedes E-class components. By using flax/sisal thermoset in the door panels, a 20% weight saving was achieved.[40]

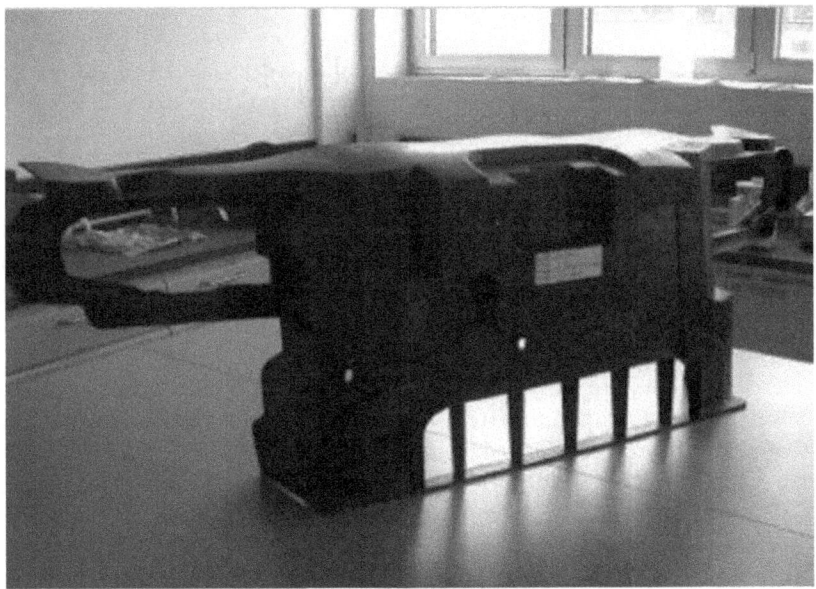

Figure 6.9   A front-end grill opening reinforcement for the Ford cross member.[3]

that by 2015, that vehicles must be constructed of 95% partially decomposable or recyclable materials. Therefore, the end of life process becomes easier, with lower costs and less energy involved. These aspects will definitely lead to an increased use of vegetable fibres.

Table 6.5 Vegetable fibres usage in the automotive industry.[45]

Automotive company	Vehicle	Part
Audi	A2, A3, A4, A6, A8, Avant, Roadstar, Coupe	Seat back, side and back door panels, boot lining, hat rack, spare tire lining
BMW	3, 5, and 7 series and others	Door panels, headliner panel, boot lining, seat back
Daimler Benz	A, C, E, S class	Door panels, windshield/dashboard, business table, pillar cover panel
Fiat	Punto, Brava, Marea, Alfa Romeo 146 and 156	–
Ford	Mondeo CD 162 and Focus	Door panels, B-pillar, boot liner
Mitsubishi	Space star, Colt	Door panels, instrumental panels
Opel	Astra, Vectra and Zafira	Headliner panel, door panels, pillar cover panel, instrument panel
Peugeot	New model 406	–
Renault	Clio	–
Rover	Rover 2000 and others	Insulation, rear storage shelf/panel
Volkswagen	Golf A4, Passat Variant, and Bora	Door panel, seat back, boot lid finish panel, boot liner
Volvo	C70 and V70	–

6.6 Future Perspectives

Currently, there is on-going global research into the use of vegetable fibre composites, and automakers are producing prototypes that provide a hint into the future of manufacturing:[3,46,47]

- The U.S. Agricultural Research Service has been developing industrial and commercial uses for a wide variety of agricultural products, including waste items, and groups such as the Soybean Checkoff and the National Corn Growers Association that focus on researching and promoting new markets for member's crops are supporting research efforts into new applications for their feed sources.
- Tier 1 suppliers are actively involved in producing prototype parts. Thus, Visteon has developed a system for making flax-based instrument panels; Composites Products has developed a process to produce door panels from flax; Findlay Industries, which makes the cargo area floors for GM and Honda SUVs and the package shelves for Saturn and Opel, also manufactures headliners for Mack Trucks that are made with a hemp, flax, kenaf, and sisal mixture; and soy-resin body panels have been developed that are currently used on John Deere Tractors.

As can be seen in Table 6.6, the consumption of vegetable composite is expected to increase greatly by 2020, including in the automotive industry.

By maintaining the optimization of the vegetable fibre cloth production for composites and improving the adhesion between fibre and resin, it is clear that the performance of vegetable fibre composites will be improved even further.[9]

Table 6.6   Estimated consumption evolution from 2010 to 2020.[12]

Vegetable composites	*Estimated quantities in the EU, 2010*	*Estimated quantities in the EU, 2020*
Compression molding Flax, hemp, jute, kenaf, sisal, abaca, coir (>95% automotive, 5% cases and others)	40 000 t	120 000 t
Cotton fibre (automotive, mainly trucks)	100 000 t	100 000 t
Wood fibre (mainly automotive)	50 000 t	150 000 t
Extrusion and injection molding Wood/plastic composite (construction, furniture, automotive, consumer goods)	120 000 t	120 000 t
Flax, hemp, jute, kenaf, sisal, cork (construction, furniture, automotive, consumer goods)	5000 t	5000 t
Vegetable composites in total	315 000 t	830 000 t
Composites (glass, carbon, and natural fibres reinforced plastics) in total	2.4 Mt	3,0 Mt
Vegetal-based share	*ca.* 13%	*ca.* 28%

In the end, the path will be in the direction of an important question: will a 100% green composite for automotive components be possible? By a 100% green composite, we mean a composite where both fibres and matrix are from renewable sources. Therefore, to complete the set, a bio-based thermoset or thermoplastic is needed.

Bio-resin is a term used to describe a resin or a resin formulation derived from a biological source.[48] A bio-resin can be a thermoplastic or a thermoset biodegradable plastic.[39] The most widely applied renewable resources include plant oils, polysaccharides (mainly cellulose and starch), and proteins.[39,49]

In the last century, the cultivation of oilseed crops, primarily for food use, led to the investigation of vegetable oils as precursors for resin systems.[48] Examples of vegetable oils used are from rapeseed, soybean, castor, pine, *etc.*[39,50]

Another example of a bio-resin is poly(lactic acid) (PLA), which is a bio-based thermoplastic synthesized from lactic acid monomer. The lactic acid monomer is produced *via* fermentation from corn starch. Over recent years, owing to its good mechanical properties, much attention has been given to potential applications for PLA as a replacement for petroleum-based products.[51]

The major barriers for widespread acceptance of biopolymers, as substitutes for traditional non-biodegradable polymers, are performance limitations and the high cost. For example, PLA is sensitive to hydrolytic degradation under melt processing conditions in the presence of small amounts of moisture; thus, the hygroscopic nature of the natural fibres has a negative effect on the adhesion mechanism as well as the biodegrability of bio-based composites.[52]

Figure 6.10   Bio-based electric scooter, developed by NPSP Composieten and Qwic, where the entire metal frame has been replaced by a natural fibre-reinforced monocoque structure.[9]

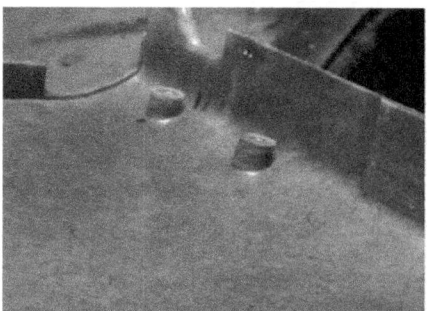 

Figure 6.11   Homogeneous components in industrial tests with hemp/PTP fibres, showing a high surface quality.[53]

The high cost of biopolymers compared to traditional plastics is not due to the raw material costs for biopolymer synthesis; rather, it is mainly attributed to the low volume of production. Another challenge for the development of biodegradable polymers lies in the fact that such biopolymers should be stable during storage or usage and then degrade once disposed of after their intended lifetime.[2] With the rapid advancements in fully bio-based polymers that may be processed with vegetable fibres, it is only a matter of time before fully bio-based composites are prevalent within the automotive industry.[3] Some examples can be seen in Figures 6.10 and 6.11, and the components that were produced illustrate clearly that lightweight and high surface quality construction can be achieved by the use of natural composites.[53]

6.7 Conclusions

It was seen that the use of vegetable fibres in the automotive sector is real, and the tendency is to increase their adoption. This result is mainly due to even more stringent environmental legislations that result from the depletion and uncertainty of petroleum sources, and because natural fibres have better environmental performance than other traditional ones such as fibreglass. They are from renewable sources and deliver the same performance for a lower weight.

Research on natural fibre composites will continue. Increased awareness of environmental issues will only foster the tendency to push the boundaries even further. The major automakers have already started using components produced by vegetal fibres in their products, and new developments are in progress.

In the direction of more sustainability, 100% "green" materials are being developed. Composite materials that are 100% bio-based are being researched, where a bio-fibre and a biopolymer are used. However, the performance limitations and the high cost are major drawbacks to the widespread use of biopolymers. Despite this, good results are already achieved, such as a fully bio-based composite with high surface quality and excellent burning characteristics. It can be said that "plant-based cars are the wave of the future".

References

1. Anon., *Composite Materials Promise Increased Fuel Efficiency*, The Budd Company, Troy, MI, 2001, pp. 26–33.
2. A. K. Mohanty, M. Misra and L. T. Drzal, *J. Polym. Environ.*, 2002, **10**, 19–26.
3. J. Holbery and D. Houston, *JOM*, 2006, **58**, 80–86.
4. E. Ghassemieh, in *New Trends and Developments in Automotive Industry*, ed. M. Chiaberge, In Tech Europe, Rijeka, Croatia, 2011, pp. 365–394.
5. C. A. Silva, *Sustainable Design of Automotive Components Through Jute Fiber Composites: An Integrated Approach*, PhD thesis, Instituto Superior Técnico, Lisbon, 2010.
6. A. K. M. Lawrence, T. Drzal and M. Misra, presented at the Automotive Composites Conference, Troy, MI, 2001.
7. Anon., *Special Report: Cars Made of Plants?*, 2009; http://www.edmunds.com/fuel-economy/special-report-cars-made-of-plants.html.
8. M. J. John and S. Thomas, *Carbohydr., Polym.*, 2008, **71**, 343–364.
9. J. Breuer and W. Böttger, *JEC Compos.*, 2011, no. 66, 42–44.
10. S. Piotrowski and M. Carus, in *Industrial Applications of Natural Fibres: Structure, Properties and Technical Applications*, ed. *J.* Mussing, Wiley, New York, 2010, pp. 73–86.
11. G. Koronis, A. J. Silva and M. Fontul, presented at the 16th International Conference on Composite Structures, Porto, 2011.
12. J. Hobson and M. Carus, *Biowerkstoff-Report*, 2011, no. 8, 24–25.
13. N. P. Giacomini, *Compósitos Reforçados com Fibras Naturais para a Indústria Automobilística*, Master Thesis, Universidad de São Paulo, Brazil, 2003 (in Portuguese).

14. C. A. Silva, *Design Sustentável: a Importância das Fibras de Juta, Sisal e Coco, no Planejamento Deprodutos e Éticas Sustentáveis*, Master's thesis, Universidade Estadual Paulista, Bauru, São Paulo, Brazil, 2006 (in Portuguese).
15. M. Baxter, *Product Design: Practical Methods for the Systematic Development of New Products*, Thornes, Kingston upon Thames, UK, 2nd edn., 1999.
16. S. C. Furtado, *Cálculo Estrutural Numérico e Experimental da Carenagem de um Veículo*, Master's thesis, Instituto Superior Técnico, Lisbon, 2009 (in Portuguese).
17. R. M. Jones, *Mechanics of Composite Materials*, Taylor & Francis, Philadelphia, 2nd edn., 1999.
18. J. N. Reddy, *Mechanics of Laminated Composite Plates and Shells: Theory and Analysis*, CRC Press, Boca Raton, 2nd edn., 2004.
19. N. M. A. Lopes, *Análise e Caracterização da Degradação de Compósitos Reforçados com Fibra de Jutae Fibra de Vidro em Ambiente Controlado (Temperatura & Humidade)*, Master's thesis, Instituto Superior Técnico, Lisbon, 2009 (in Portuguese).
20. M. F. S. F. de Moura, A. B. de Morais and A. B. Magalhães, *Materiais Compósitos: Materiais, Fabrico e Comportamento Mecânico*, Publindústria, Porto, 2005 (in Portuguese).
21. S. Mazumdar, *Composites Manufacturing: Materials, Product, and Process Engineering*, Florida, CRC Press, Boca Raton, 2002.
22. A. Y. Goldman and C. J. Copsey, *Mater. Res. Innov.*, 2000, **3**, 302–307.
23. J. A. L. Caetano, *Aplicação de Cargas Vegetais de Cortiça em Compósitos como Substituição de cargas Minerais no Âmbito do Eco Design*, Master's thesis, Instituto SuperiorTécnico, Lisbon, 2009 (in Portuguese).
24. E. Nezbedová, *Chem. Listy*, 2007, **101**, 28–29.
25. E. A. Santos, *Avaliação Mecânica e Micorestrutural de Compósitos de Matriz de Poliéster com Adição de Cargas Minerais e Resíduos Industriais*, Master Thesis, Universidade Federal do Rio Grande do Norte, Rio Grande do Norte, Brazil, 2007 (in Portuguese).
26. D. Hull, *An Introduction to Composite Material*, Cambridge University Press, Cambridge, 2nd edn., 1996.
27. J. Mussig and T. Slootmaker, in *Industrial Applications of Natural Fibres: Structure, Properties and Technical Applications*, ed. J. Mussig, Wiley, New York, 2010, pp. 44–48.
28. G. I. Williams and R. P. Wool, *Appl. Compos. Mater.*, 2000, **7**, 421–432.
29. D. E. Akin, in *Industrial Applications of Natural Fibres: Structure, Properties and Technical Applications*, ed. J. Mussig, Wiley, New York, 2010, pp. 13–22.
30. A. Dufresne, in *Monomers, Polymers and Composites from Renewable Resources*, ed. M. N. Belgacem and A. Gandini, Elsevier, Amsterdam, 2008, pp. 401–418.
31. D. N. Saheb and J. P. Jog, *Adv. Polym. Technol.*, 1999, **18**, 351–363.
32. C. Jayasekara and N. Amarasinghe, in *Industrial Applications of Natural Fibres: Structure, Properties and Technical Applications*, ed. J. Mussig, Wiley, New York, 2010, pp. 197–217.

33. S. V. Joshi, L. T. Dzral, A. K. Mohanty and S. Arora, *Composites, Part A*, 2004, **35**, 371–376.
34. C. Alves, P. M. C. Ferrão, A. J. Silva, L. G. Reis, M. Freitas, L. B. Rodrigues and D. E. Alves, *J. Cleaner Prod.*, 2011, **18**, 313–327.
35. C. Alves, L. G. Reis, P. M. C. Ferrão and M. Freitas, in *New Trends and Developments in Automotive Industry*, ed. M. Chiaberge, InTech Europe, Rijeka, Croatia, 2011, pp. 223–254.
36. P. Wambua, J. Ivens and I. Verpoest, *Compos. Sci. Technol.*, 2003, **63**, 1259–1264.
37. P. Vidal Alonso, A. Tielas Macia and D. Garcia Murias, *JEC Compos.*, 2011, no. 66, 39–41.
38. B. N. Dash, A. K. Rana, H. K. Mishra, S. K. Nayak and S. S. Tripathy, *J. Appl. Polym. Sci.*, 2000, **78**, 1671–1679.
39. L. Laine and L. Rozite, *Eco-efficient Composite Materials: State of the Art*, EU, Lille, 2010.
40. B. Suddell, presented at the Joint Meeting of the 32nd Session of the Inter-governmental Group on Hard Fibres and the 34th Session of the Inter-governmental Group on Jute, Kenaf and Allied Fibres, Salvador, Brazil, 2003.
41. G. Davies, *Materials for Automobile Bodies*, Butterworth-Heinemann, Oxford, 2003.
42. S. Horn, H. J. Bader and K. Buchholz, *Plastics from Renewable Raw Materials and Biologically Degradable Plastics from Fossil Raw Materials*; http://www.rsc.org/education/teachers/Resources/green/docs/plastics.pdf.
43. Anon. Ford is making greener vehicles through increased use of renewable and recyclable materials; http://blog.ford.com/article_display.cfm?article_id=32474.
44. Toyota, *North America Environmental Report: Challenge, Commitment, Progress*, 2011; http://www.toyota.com/about/environmentreport2011/splash.html.
45. O. Faruk, *Cars from Jute and Other Bio-Fibers*; http://biggani.com/files_of_biggani/mashiur/interview/omar_faruk.pdf.
46. G. Marsh, *Next Step for Automotive Materials*, 2003; http://www.materialstoday.com/pdfs_6_4/marsh.pdf.
47. L. M. Sherma, *Plast. Technol.*, 1999, **45**, 62–68.
48. P. A. Fowler, V. V. Tverezovskiy, R. M. Elias, C. G. Chappell, C. S. Fitchett, N. G. Laughton and J. F. Seefeld, presented at COST Action E49, Nantes, France, 2009.
49. J. M. Raquez, M. Deléglise, M. F. Lacrampe and P. Krawczak, *Prog. Polym. Sci.*, 2010, **35**, 487–509.
50. Anon. *Clean, Green Technology Makes Sustainable Resins from Vegetable Oil*, Oakdene Hollins, Aylesbury, UK, 2006.
51. A. H. Harris and E. C. Lee, *Injection Molded Polylactide (PLA) Composites for Automotive Applications*, Ford Motor Company, 2006.
52. T. Mukherjee and N. Kao, *J. Polym. Environ.*, 2011, **19**, 714–725.
53. J. Müssig, M. Schmehl, H. B. von Buttlar, U. Shonfeld and K. Arndt, *Ind. Crops Prod.*, 2006, **24** 132–145.

CHAPTER 7

Water Vapour Sorption of Natural Fibres

C. A. S. HILL

Forest Products Research Institute, Joint Research Institute for Civil and
Environmental Engineering, Edinburgh Napier University, Merchiston
Campus, Colinton Road, Edinburgh, EH10 5DT, UK
Email: c.hill@napier.ac.uk

7.1   Introduction

The mechanical properties of natural fibres are strongly influenced by their
moisture content and it is important to understand how and why water can
affect their behaviour. Although the water sorption properties of natural fibres
are dependent upon the presence of hydroxyl groups within the cell wall
polymers, this is not the only factor determining this behaviour. A critical
consideration is the cell wall composition and in particular the structure and
composition of the interfibrillar matrix of the cell wall. The cell wall of natural
fibres can be viewed as a composite structure, with the reinforcing elements
comprising the crystalline cellulose of the microfibrils embedded within a
matrix that is variously composed of hemicelluloses, lignin, pectin and other
substances. Water molecules sorbed within the cell wall do not penetrate the
microfibrils, but are solely located within the cell wall matrix. Water acts as a
plasticizer for the cell wall matrix macromolecules, which consequently affects
the mechanical behaviour of the fibre. Above a threshold moisture content,
natural fibres become susceptible to microbiological degradation. The presence
of water in natural fibres can also lead to deleterious properties arising from the

RSC Green Chemistry No. 16
Natural Polymers, Volume 1: Composites
Edited by Maya J John and Thomas Sabu
© The Royal Society of Chemistry 2012
Published by the Royal Society of Chemistry, www.rsc.org

presence of defects in the composite formed during processing. Shrinking and swelling of fibres in response to changes in cell wall moisture content can result in the decoupling of fibre reinforcement from the enveloping matrix, leading to loss of performance. However, the hygroscopic properties of natural fibres can be exploited in textile or insulation applications in the built environment. It is clear that water sorption must be understood and if necessary controlled. In this chapter the water sorption behaviour of natural fibres is discussed and some methods for reducing the hygroscopic behaviour are evaluated.

7.2 The Sorption Isotherm

When a natural fibre is exposed to a constant relative humidity (RH) at a fixed temperature, it will gradually attain a stable moisture content (MC). This occurs when an equilibrium condition is realised and for this reason it is referred to as the equilibrium moisture content (EMC). The equilibrium state is dynamic and this point is reached when the rate at which water molecules are entering the cell wall is equal to the rate at which they exit the cell wall. Determining the EMC at a constant temperature over a range of RH values produces a sorption isotherm. The sorption isotherm is characterized by having a sigmoidal form and is classified as being International Union of Pure and Applied Chemistry (IUPAC) Type 2.[1] Apart from the sigmoidal shape, the sorption isotherm of natural fibres also displays the property of hysteresis, where the EMC values attained during adsorption are lower than those associated with desorption at a given RH, as shown in Figure 7.1.

The sorption isotherm therefore consists of two boundary curves (adsorption and desorption) and a scanning curve where desorption is initiated at a specific RH and the curve passes through the region bounded by these two curves.[2]

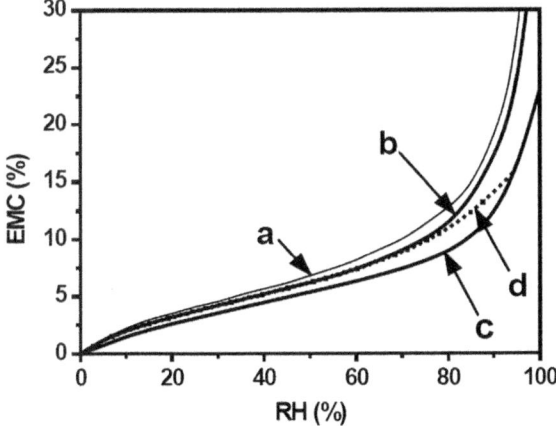

Figure 7.1 Example isotherm showing (a) first desorption from saturation, (b) subsequent desorption boundary curve, (c) adsorption boundary curve and (d) scanning curve.

In principle, every point within the space delineated by the boundary curves is accessible by starting a scanning curve at an appropriate point on the adsorption or desorption boundary curves. These properties have been known for over a century, yet there is still not a consensus as to why the sigmoidal isotherm is observed, nor is there a definitive explanation for hysteresis. The isotherm is generally reproducible, but this property can change if the fibre is subjected to prolonged drying at temperatures in excess of 100 °C, which can lead to degradation of the cell wall polymers and also irreversible hydrogen bond formation in the matrix polymeric network. It is also quite commonly observed that when a never-dried fibre is taken through the initial desorption stage, the path of this first desorption branch of the isotherm is not reproduced in any subsequent desorption experiments. Two explanations can be posited for this behaviour. Firstly, it cannot be guaranteed that all of the water present in the fibre when drying down from a saturated state is actually located within the cell wall; secondly, it is possible that there is some irreversible hydrogen bond formation (hornification) occurring in the initial drying process, irrespective of whether this takes place at elevated or ambient temperatures. In practice, the determination of sorption isotherms is undertaken over a RH range from zero to 95%, since it becomes very difficult to make accurate measurements much above this upper RH limit. There is also an increasing likelihood of a contribution from capillary condensation to the sorption process at RH values exceeding 90%.[3]

There is considerable variation in the sorption isotherms determined for different fibre types.[3] The reasons for this are complex, but are undoubtedly related to the cell wall composition and microfibril angle (MFA), although the importance of hydroxyl group availability is not clear. The importance of the cell wall composition is related to the mechanical response of the cell wall to the presence of sorbed moisture and the MFA will also influence the mechanical properties of the cell wall. For example, a high MFA will allow for considerable linear extension of the fibre when moisture is adsorbed onto the internal surface. Figure 7.2 shows comparative isotherms for flax and coir, where it can be seen that the fibre with the highest lignin content (coir) also exhibits the highest levels of water sorption. Coir also exhibits a high MFA (of the order of 40°) compared with flax (around 3°).

Given that lignin has a much lower hydroxyl (OH) to carbon ratio compared with polysaccharides, this result at first appears to be counter-intuitive. However, although the presence of hydroxyl groups undoubtedly is a major factor in making the fibres hygroscopic, the ability and extent of swelling of the cell wall to accommodate the sorbed water molecules is probably the most significant factor in determining the EMC over most of the hygroscopic range. Thus a mechanical model for describing water sorption would appear to be more appropriate.

Although sorption isotherms are not determined at RH values much in excess of 95%, it is quite common practice to extrapolate the adsorption isotherm line through the data points and project onto the EMC axis at 100% in order to determine a "fibre saturation point" (FSP).[4] The term "fibre saturation point" is used to represent the moisture content of the fibre in a theoretical

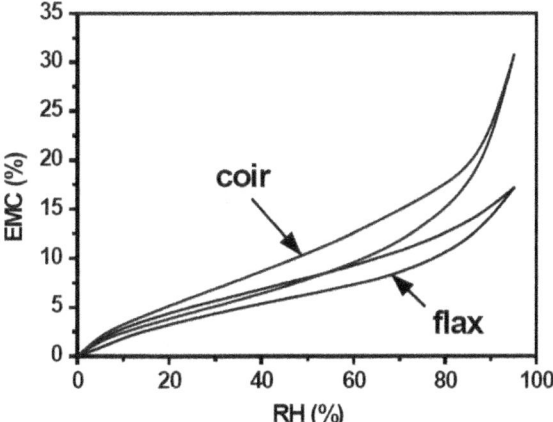

Figure 7.2 Isotherms for coir and flax.

condition where the cell wall moisture content is at a maximum, but there is no water present in the macrovoids of the fibre. In the case of wood, it is established that the FSP is a point where many of the physical properties (*e.g.* swelling, modulus of elasticity, modulus of rupture, impact toughness) no longer change and this is often found to be in the region of 30% MC for wood and many plant fibres.[2,5] However, in practice it becomes very difficult to measure the changes in properties in the region 30–40% MC and there is some dispute as to where the "true" FSP may lie. Methods which produce total saturation of the cell wall with water indicate that the FSP may lie nearer to 40% MC.[3] The use of the projection of adsorption isotherms to evaluate a FSP has been criticized on a number of grounds.[6] One objection relates to the fact that the sorption isotherm is changing rapidly at this point and small errors in curve fitting lead to large errors in the projected FSP (p-FSP) value. It is also not possible to determine isotherms with any degree of accuracy much above 95% RH and it is then necessary to resort to methods such as tension plates. In addition, it has been found that there is some variation in the isotherm at the higher end of the hygroscopic range below 95% RH even with identical fibre samples, as is shown for flax in Figure 7.3.

Despite these criticisms, the projection method is still commonly used to report FSP. This has been done for a range of natural fibres, where it can be seen that there is considerable variability in the p-FSP (Table 7.1).[3] It is immediately obvious from these data that cotton has a much lower p-FSP compared to all the other fibres in this table.

Although caution must be exercised when interpreting the p-FSP values for reasons noted above, it is possibly instructive to compare the results of Table 7.1 with the polymeric composition of these fibres (Table 7.2).[3]

Cotton, with a very high cellulose content, has the lowest p-FSP, which could perhaps indicate that it is the lack of accessible OH groups that is determining the EMC. It is striking that coir, which has the highest lignin content of the

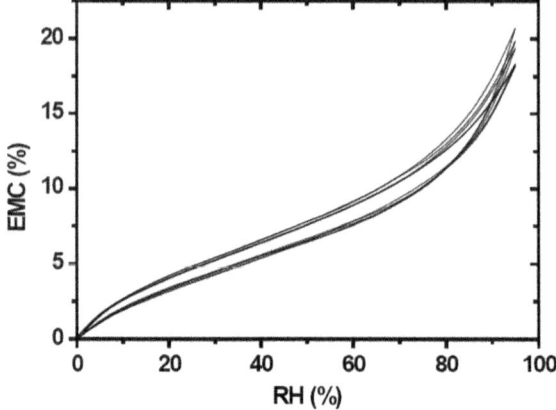

Figure 7.3   Reproducibility of the sorption isotherm of flax.

Table 7.1   The projected fibre saturation point of different natural fibres.

Fibre	p-FSP (%)
Jute	27.0
Coir	34.0
Flax	25.6
Hemp	30.4
Cotton	17.0

Table 7.2   The polymeric composition (wt%) of the natural fibres listed in Table 7.1, where "polysaccharide" refers to all cell wall polysaccharides other than cellulose.

Fibre	Cellulose	Polysaccharide	Lignin
Jute	72	14	13
Coir	43	5	45
Flax	81	16	3
Hemp	74	19	4
Cotton	94	6	0

fibres in this study, nonetheless also has the highest p-FSP. This suggests that invoking accessible OH content as the only variable affecting EMC is unlikely to be correct.

7.3   Water Sorption Models

A large number of models exist describing the sorption process; these can be classified as localized models, which can be broadly split into layering or cluster models,[7,8] and dissolution models.[9] In localized models, the water molecules are considered to bind to specific sites in the polymeric matrix. The layering model requires the initial formation of monolayers on the cell wall internal surface, with the subsequent creation of multiple layers as the sorption process proceeds. Indeed, there is a significant component of the literature which reports on the monolayer content, often determined by the application of Brunauer–Emmett–Teller (BET) isotherm analysis.[10] The application of such a model can only strictly be made in the case of an inert substrate, not a swelling cell wall (elastic gel) which will present an evolving surface as the adsorption process takes place. Nonetheless, BET analyses are commonly made with water vapour sorption isotherms and BET surface areas are reported from these analyses. A modification to the BET model has led to the development of the Guggenheim–Anderson–de Boer model and its variants.[11] In dissolution models a macroscopic approach is adopted where there is mixing between macromolecules and water molecules; examples include the Hailwood and Horrobin (HH) model, which was originally developed to explain the sorption isotherm of cotton,[12] and the Flory–Huggins model.[13] The HH model is based upon consideration of the thermodynamics of the system where a sorbent is in equilibrium with the atmospheric moisture. The analysis considers the equilibrium existing between the solid dry polymer of the sorbent, hydrated polymer and "dissolved" water. The hydrated polymer component is water that is strongly bound to the sorption sites of the polymeric constituents of the substrate, whereas the dissolved water is water that is not so strongly bound but is still located within the cell wall nanopores. These two types of water can be loosely interpreted as monolayer and polylayer water analogously to the BET model, although the concept of a monolayer existing with the complex, dynamic geometry of the cell wall internal surface is unrealistic. An HH analysis of the sorption isotherms of jute is shown in Figure 7.4. The sum of the two types of cell wall water produces a sigmoidal curve that very closely fits the data, except at the upper end of the

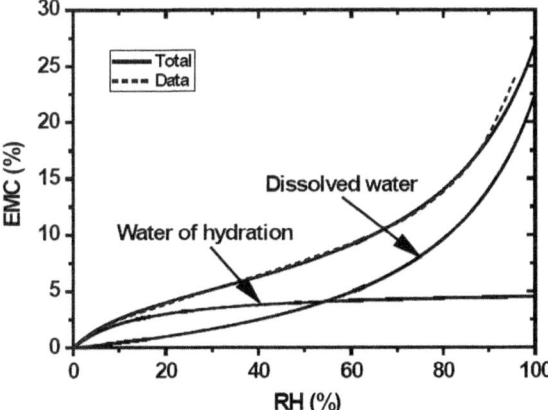

Figure 7.4 Hailwood–Horrobin model of the sorption isotherm showing the contributions from dissolved water and water of hydration.

hygroscopic range where it is possible that there is an additional contribution from capillary condensation. It is important to note that the HH model cannot be used to analyze the desorption isotherm curve, since this is composed of a boundary and a scanning curve.

7.4 Hysteresis

A number of explanations have been given in the plant fibre and wood science literature to explain hysteresis, but these have been unsatisfactory for a variety of reasons.[3] Early explanations relied upon the assumption that hysteresis merely represented the failure to achieve "true" equilibrium in the experiment and that if a sufficient amount of time was allowed then the equilibrium state would be achieved. It has been shown in numerous studies that this is not the case and that hysteresis is a true phenomenon and not an experimental artefact. Other explanations have considered the nature of the adsorption process occurring on the surfaces of the cell wall nanopores, resulting in a stepwise filling which continues until the nanopore is filled with liquid water. Desorption then occurs from the surface meniscus of the water in the nanopore, giving rise to hysteresis. However, the concept of the existence of a meniscus in liquid water in these cell wall nanopores, which are of the order of 2–4 nm in diameter, is problematical, making such an explanation highly suspect. It has been argued that hysteresis arises due to the formation of irreversible hydrogen bonds during the first desorption cycle, which can certainly be used to explain why the first compared to subsequent desorption boundary curves do not follow the same path, but cannot explain why adsorption and desorption boundary curves follow different paths on subsequent sorption cycles. Another model relies upon consideration of the geometry of the nanopores in the cell wall, where the diameter of the throat of such a capillary is smaller than the interior, akin to an ink bottle. Other explanations invoke differences in contact angle between the adsorption and desorption processes. During the wetting cycle the contact angle of the water with the internal surface of the cell wall is different compared with that for desorption, where the water is in contact with an already wet surface. The problem with such an explanation is that the use of concepts related to liquid water in such small capillaries is unlikely to have any physical meaning; furthermore, during the adsorption process, the water is entering from a vapour state and not as a liquid. It is however possible that such an explanation may have some application in the upper part of the hygroscopic range (above 70% RH), where capillary condensation becomes more important. It has been suggested that hysteresis is caused by the presence of permanent gases in the cell wall, leading to incomplete wetting, but it has been shown that hysteresis is still observed even if sorption isotherms are determined in a vacuum. With this model under conditions of adsorption the sorption sites are thought to be partially masked by the presence of gas molecules. It has also been argued that the presence of permanent gases in such samples only contributes towards hysteresis if such materials are non-swelling.[3]

When water vapour enters the cell wall, it causes swelling of the material. This is because the water is occupying space between the microfibrils, thereby forcing them apart (water cannot penetrate the microfibrils). This process of expansion is resisted primarily by the interfibrillar matrix polymers; as water is removed from the cell wall, the matrix collapses to its previous configuration, *i.e.* the nanopores are not permanent but transient. The cell wall of plant fibres and many other natural materials (*e.g.* wood) can be considered to be a swelling gel.[14] The sorption behaviour of natural fibres has been reported upon recently, where a model for sorption hysteresis was presented based upon the micromechanical behaviour of the cell wall matrix during the adsorption or desorption process.[3] This model was originally developed to describe sorption in glass polymers[15] and subsequently applied to sorption hysteresis effects with humic soils.[16] Essentially, the process of adsorption results in the creation and expansion of nanopores within the cell wall matrix, whereas desorption leads to the collapse of these voids in the cell wall matrix. This process is inelastic on the time scale of molecular diffusion and as a consequence the adsorption and desorption processes take place in a material that is in different states. This phenomenon is observed in glassy polymers below the glass transition temperature (T_g).[3,17] The glassy state is defined as being one in which the molecular scale nanopores are embedded in a matrix that is unable to fully relax to a thermodynamic equilibrium state due to the stiffness of the matrix macromolecules. It is therefore logical to consider that there is a link between the matrix mechanical properties and the hysteresis effect. It has been found that the extent of hysteresis (*i.e.* the area bounded by the adsorption and desorption curves) is related to the cell wall composition of fibres and that fibres with higher lignin contents display greater hysteresis, which is consistent with the model.[3] This is illustrated in Figure 7.5, showing a comparison of the hysteresis between the adsorption and desorption isotherms (obtained by subtracting the

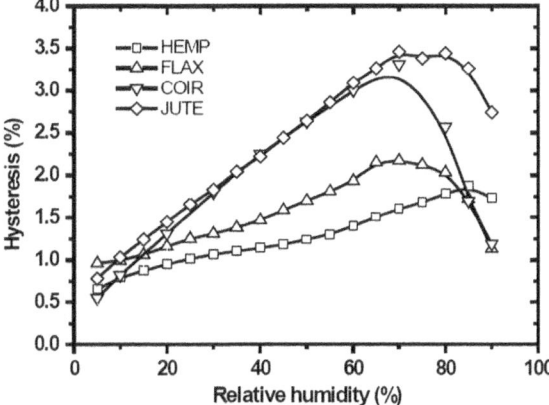

Figure 7.5 Differences in hysteresis (calculated by subtracting adsorption from desorption moisture contents).

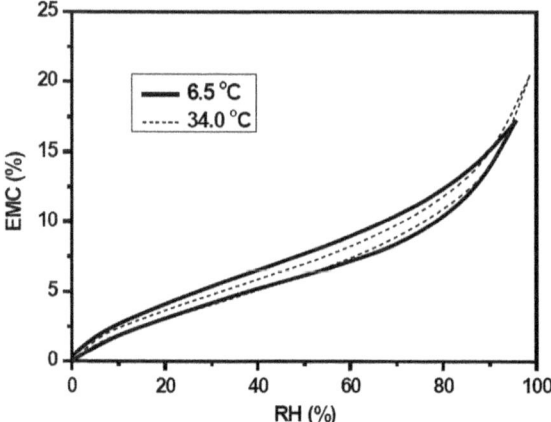

Figure 7.6   Effect of temperature on the isotherm of flax.

EMC of adsorption from the EMC of desorption) for flax, hemp, coir and jute. The fibres with a higher lignin content display a greater amount of hysteresis between the adsorption and desorption boundary curves.

According to the model, the presence of a cross-linked matrix polymer (such as lignin) would be expected to result in a higher level of hysteresis, but the model also predicts that as the isotherm temperature increases, the extent of hysteresis should decrease and become zero at the T_g of the matrix. Figure 7.6 shows the effect of temperature upon the sorption isotherm of flax. As predicted by the model, the hysteresis decreases as the isotherm temperature is raised. This property has been previously noted with Sitka spruce[18] and cotton.[19]

7.5   Water Adsorption and Cell Wall Swelling

As noted in the previous section, the adsorption of moisture into the cell wall of a fibre leads to swelling of the material. The diffusion of water molecules into the cell wall matrix that surrounds the microfibrils requires the breaking of hydrogen bonding networks, with the creation of nanopores.[3,20] It is thought that the initial water molecules entering the cell wall are relatively tightly bound to the primary sorption sites (OH groups), but as the adsorption process continues, the water molecules that diffuse into the cell wall nanopores are less constrained in their translational freedom.[21] As the cell wall swells, this creates elastic strain in the matrix. The sorbed water molecules therefore perform work upon the cell wall and during the reverse process this strain energy is released as water molecules leave the cell wall. However, the cell wall deformation process is not perfectly elastic since hysteresis occurs in the sorption isotherm. This is further discussed in Section 7.7.

7.6 Sorption and Heat of Wetting

Water molecules in the vapour state have considerably more translational energy than they do in the liquid state and this energy is released as heat when fibres adsorb water. This is the same as the latent heat of vaporization of water, but there is an additional amount of heat (heat of wetting) that is released because of the additional constraint imposed on the water molecules by being confined within the cell wall nanopores. There are two ways of measuring the heat of wetting: either directly by using calorimetry, or indirectly by applying the Clausius–Clapeyron equation.[3] Although widely adopted within the scientific literature, the application of a thermodynamic analysis to a non-reversible system is unsound and indeed this is one criticism of the HH model, which incidentally is unable to explain hysteresis. Below the glass transition temperature of the matrix, the sorption phenomenon is not path independent.

7.7 Kinetics of Water Sorption

Many models describing the kinetics of water sorption processes within natural fibres have unconvincingly invoked Fick's Law.[22] One of the confounding factors when attempting to model sorption kinetic behaviour has been the difficulty of obtaining sufficiently accurate data sets, a requirement for reliable curve-fitting. However, with the commercial development of dynamic vapour sorption (DVS) equipment over the past decade, it has now become a routine matter to obtain extremely accurate experimental kinetic data. Using a DVS apparatus, it was clearly demonstrated by Kohler *et al.*[23] that the sorption kinetic behaviour with flax fibres is actually very accurately described by what is termed a parallel exponential kinetics (PEK) model, which has the mathematical form shown in eqn (7.1):

$$MC = MC_0 + MC_1[1 - \exp(-t/t_1)] + MC_2[1 - \exp(-t/t_2)] \qquad (7.1)$$

where MC is the moisture content at time t of exposure of the sample to a constant RH and MC_0 is the moisture content of the sample at time zero. The sorption kinetic curve is composed of two exponential terms which represent a fast and slow process having characteristic times of t_1 and t_2, respectively. The terms MC_1 and MC_2 are the moisture contents at infinite time associated with the fast and slow processes, respectively. The PEK model is not just applicable to flax fibres,[23,24] but also to a wide range of plant fibre types,[25] regenerated cellulose fibres[26–28] and microcrystalline cellulose,[29] as well as unmodified and modified wood,[30,31] but has also been found to be applicable to the drying behaviour of foodstuffs and the swelling properties of cereal grains.[32–34] Although not yet proven, it is likely that the PEK model can be applied to all swelling gel materials, with the proviso that the sample size (or more correctly the volume to surface ratio) is small, although what the limiting ratio is has not been determined. An example sorption kinetics curve is shown in Figure 7.7 for flax and jute fibre, with the PEK fit to the data given and the fast and slow adsorption curves also reproduced.

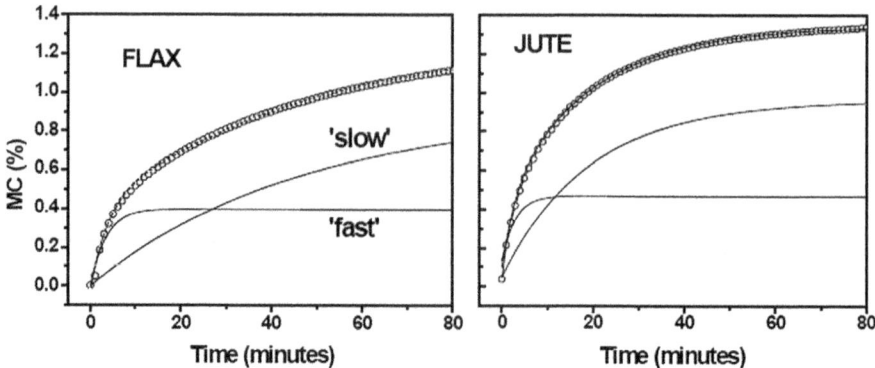

Figure 7.7   Sorption kinetic curves of flax and jute deconvoluted into fast and slow components according to the parallel exponential kinetics model.

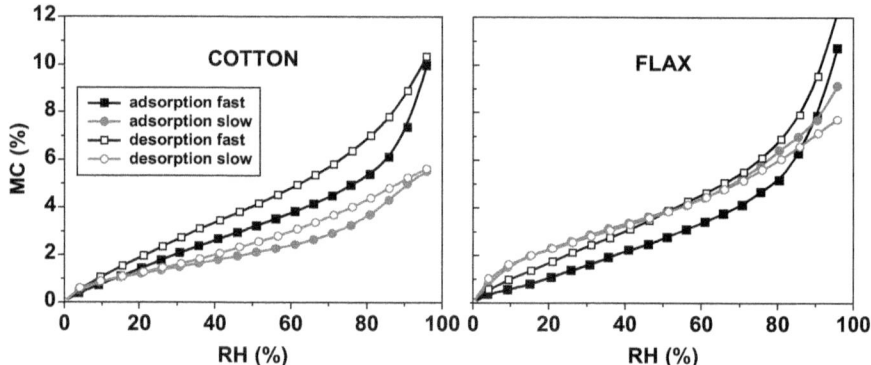

Figure 7.8   Pseudo-isotherms for the fast and slow processes shown for cotton and flax.

It is not known what these two processes represent, but they have been previously attributed to the existence of "fast" and "slow" sorption sites in the cell wall.[23] The fast sites have been attributed to the sorption of water molecules associated with the external surface and amorphous regions of the fibre and the slow sites to indirect sorption on the inner surface and crystallites.[26] By cumulatively adding the MC_1 or MC_2 values for adsorption or the desorption cycles, it is possible to construct pseudo-isotherms associated with the fast and slow processes and consequently determine what the contribution of the two processes is to the hysteresis effect. This is shown for cotton and flax in Figure 7.8.

It has been noted by workers that the adsorption pseudo-isotherms resemble the isotherms attributable to monolayer (Langmuir) and poly-layer adsorption.[23] However, it has been shown that such an interpretation of the fast and slow processes to "fast" and "slow" sorption sites is not appropriate.[35,36]

An alternative model for interpreting the sorption kinetics is presented in Section 7.8. It is important to note when comparing different fibre samples that the sample weights should be closely comparable ($\pm 10\%$) and that the same RH steps should be used in the isotherms. This is because the time constants are dependent not only upon morphological and compositional differences between fibres, but also upon sample weight and RH step size.

7.8 Water Sorption and Mechanical Behaviour

It is well known that the modulus of elasticity and strength of wood[37] and paper[38] decrease as the moisture content is raised up to the fibre saturation point, whereas the toughness of the material increases with the MC. However, reliable data for the influence of moisture on isolated plant fibre mechanical properties are not abundant. The reasons for this can be attributed to:

- The inherently high variation in fibre properties, plus the effect of fibre damage on mechanical properties.[39]
- Variation due to test fibre length or testing speed.[40]
- Difficulties in obtaining accurate fibre cross-sections (the reason why the textile industry uses linear density), plus the change in cross-sectional area as the MC of the fibre varies.
- Small sample sets.
- Poorly reported data (RH and temperature not reported, or not controlled).
- Conditioning regime not reported (conditioning down from a high to a lower specific RH will not give the same EMC as when conditioning from a low to a higher specific RH).
- Not testing using the same RH and temperature at which the fibres were originally conditioned.

It is well known that plant fibres exhibit a decrease in modulus as the RH of the testing environment increases, although the change in ultimate tensile strength does not display such a simple relationship.[2] Fibres containing a high content of amorphous polysaccharide material would be expected to exhibit reductions in tensile strength and modulus from consideration of the properties of the isolated material.[41] Such systems can be considered to act as viscoelastic gels, which will display creep behaviour when subjected to long-term loading effects. The mechanical behaviour of such systems can be modelled as a combination of elastic and plastic deformations, as is well known when examining the mechanical behaviour of wood.[42] Since water is able to plasticize the cell wall matrix substances of plant fibres, it also follows that the static and dynamic mechanical properties of the fibres will be modified accordingly.[2] The literature appears to be silent on the effect of duration of load upon natural fibre reinforced composites and the effect of variations in atmospheric RH upon this behaviour. However, there has been some work remarking upon the non-linear stress–strain behaviour of unidirectional flax reinforced composites,

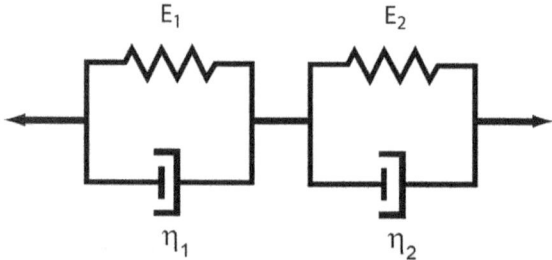

Figure 7.9 Two Kelvin–Voigt elements in series.

behaviour attributed to the annealing of microcompressive deformations in the fibres.[43] This is clearly an area where research needs to be directed if natural fibre reinforced composites are to be used in structural applications.[44]

While the mechanical behaviour of plant fibres is modified by the presence of cell wall moisture, it is also probable that the sorption behaviour is determined by the mechanical properties. It has recently been argued that the sorption kinetics of these gel-like materials is rate limited by the rate of swelling of the sorbent and that the kinetic process is in fact controlled by the matrix polymeric relaxation processes.[45,46] Based on this premise, the PEK model has been further interpreted in terms of two Kelvin–Voigt (KV) viscoelastic elements acting in series (Figure 7.9) with E_1 and E_2 being the moduli associated with the fast and slow processes, respectively, and η_1 and η_2 being the equivalent matrix viscosities.

The two elements represent the fast and slow kinetics processes. The kinetic response of a KV element when subjected to an instantaneous stress increase (σ_0) is as shown in eqn (7.2):[47]

$$\varepsilon = (\sigma_0/E)[1 - \exp(-t/\varphi)] \qquad (7.2)$$

where ε is the strain at time t, E is the elastic modulus and φ is a time constant which is defined as the ratio η/E, with η the viscosity. In the case of a plant fibre subjected to a change in RH, there is a change in the swelling pressure (Π, equivalent to σ_0) exerted within the cell wall when the atmospheric water vapour pressure is raised from an initial value p_i to a final value p_f given by eqn (7.3):[48]

$$\Pi = -(\rho/M)RT \ln(p_i/p_f) \qquad (7.3)$$

where ρ is the density and M is the molecular weight of water, R is the gas constant and T is the isotherm temperature in kelvin. In the model described herein, the strain of the system is assumed to be equivalent to the volume change of the cell wall as a result of water vapour adsorption or desorption. This volume change is further assumed to be linearly related to the change in the mass fraction of the water present in the cell wall.

The adsorbed water vapour molecules exert a pressure within the cell wall leading to dimensional change, which is equivalent to the extension of the

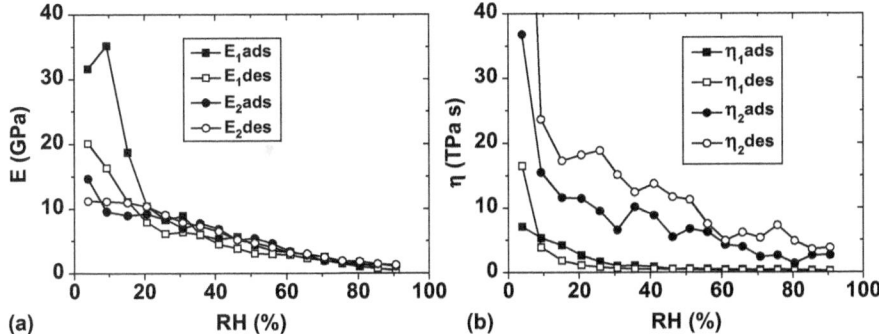

Figure 7.10 Moduli and viscosities for the cell wall calculated from the PEK parameters and applying the KV model.

spring in the KV model. This expansion/extension represents work and results in an increase in the free energy of the system. Expansion will continue until the free energy of the system is equal to the free energy of the water vapour molecules in the atmosphere. The spring modulus (E_1, E_2) therefore defines the water content of the system at infinite time (MC_1, MC_2), as shown in eqn (7.4). The fact that there is a relationship between the EMC of a fibre and modulus was established over 50 years ago.[49]

$$MC_1 = \sigma_0/E_1; MC_2 = \sigma_0/E_2 \qquad (7.4)$$

The rate at which water molecules are adsorbed or desorbed by the system is a function of the viscosity of the dashpot in the model. This viscosity is in turn related to the micro-Brownian motion of the cell wall macromolecular network. The more rapidly the matrix is able to deform, the faster the rate of water ingress or egress into or out of the cell wall. The rate of local deformation is related to the energy barrier associated with the local relaxation process and whether there is sufficient free volume to allow the relaxation process to take place. In glassy solids below the glass transition temperature (T_g) there is insufficient free volume to allow a local relaxation to take place without the cooperative motion of adjacent relaxors (a relaxor is defined as the smallest molecular segment of relaxation in each polymeric unit). This gives rise to the concept of cooperative domains within the matrix.[50–52] As the glass transition temperature is approached, the domain size decreases until T_g is reached. At this point the domain contains only one relaxor and there is sufficient free volume to allow for relaxation without the cooperation of neighbours. When this occurs, matrix relaxation becomes instantaneous and the behaviour is perfectly elastic. Hysteresis is no longer observed.

Results from the KV interpretation of the PEK data for x and y are given in Figure 7.10. By applying the sorption kinetic parameters to the KV model representing the viscoelastic response of the cell wall matrix, it is possible to derive values for the matrix modulus and viscosity. This has been done for the matrix modulus associated with the fast process (E_1) and the slow process (E_2)

and these data are presented in Figure 7.10. Values for the fast and slow moduli are of the order of 10–30 GPa at the low cell wall moisture contents associated with the bottom end of the hygroscopic range. These moduli decrease to very low values at the higher RH range, as would be predicted for a matrix that is being plasticized by the presence of sorbed water.

These modulus values at low RH are comparable with tensile modulus properties reported in the literature for these fibres.[53] Given that the expansion and contraction of the cell wall takes place within the intermicrofibrillar matrix, it seems reasonable to assume that the modulus values obtained are dominated by those associated with the matrix polymers, e.g. pectin, hemicellulose and lignin. Unfortunately, there is little in the way of reliable direct measurement of the mechanical properties of these substances in the scientific literature.

Salmén[41] has discussed the subject of cell wall polymer elastic constants at some length. Although the elastic modulus values for isolated lignin and hemicellulose have been determined, there is no reason to suppose that this is a true measure of their behaviour within the cell wall. Assuming that these reported values are representative, it is apparent that the modulus values calculated using the KV model in this study for the dry cell wall are considerably higher at low cell wall MC. This indicates that the moduli also have a contribution from the deformation of the cell wall microfibrils. Although Salmén quotes axial microfibrillar modulus values of 134 GPa, the off-axis modulus is given as 27.2 GPa, but is not affected by moisture.[41] Thus the dry modulus values calculated using the KV model in this study appear to be of the order expected, given contributions both from the matrix and microfibrils, but the microfibrillar contribution decreases at higher cell wall MC. The explanation for this lies in consideration of the geometry of the microfibrils within the cell wall. For example, it has been shown with jute fibres that water adsorption and desorption result in a rotation of the fibre about the long axis as well as an increase in cell wall volume.[54] This indicates that the distortion of the microfibrils occurring as a result of cell wall volume changes does not involve a simple off-axis bending motion. It may be that verification of the applicability of the KV interpretation of the sorption kinetics could be obtained by nano-indentation methods. The author of this chapter is not aware of any such studies with plant fibres, but there have been investigations of the properties of the wood cell wall. Nano-indentation methods have given values for the modulus of the order of 13–21 GPa for the cell wall of spruce (moisture content not stated),[55–57] comparable with the low MC modulus values reported herein.

Differences in behaviour between fibres are also found when the matrix viscosities are examined for the fast process (η_1) and the slow process (η_2). For the viscosities associated with the slow process, the η_2 values are higher for desorption, although there are differences in the behaviour between the fibres studied. Since the fast and slow processes have not been assigned to specific physical properties, it is not at this stage possible to comment upon the significance of such differences. What is clear is that the viscoelastic interpretation does seem to provide an insight into the sorption process and that there are important variations in behaviour that require further study.

It has previously been argued that an appropriate model describing sorption hysteresis in plant fibres and wood is related to the matrix response during adsorption/desorption.[3] As discussed earlier, the model describes the creation of nanopores in the matrix during the adsorption step and the collapse of these nanopores during desorption.[15–17] However, below the T_g of the matrix, the matrix is unable to respond instantaneously to the ingress or egress of water molecules because of the structural rigidity of the matrix. This results in an increased nanopore volume in the matrix during the desorption compared to the adsorption process, leading to an increased affinity for the water molecules. The timescale of the nanopore formation/annihilation processes are molecular (*i.e.* of the order of 10^{-10} s). By invoking the concept of cooperative relaxation in polymer systems below T_g, Matsuoka[50] was able to show that relaxation times of the order of minutes can be observed as the size of the cooperative domain increases (*i.e.* as the temperature decreases). This suggests that there should be a link between the hysteresis effect and the relaxation times observed in the sorption process.

7.9 Methods to Reduce Water Sorption

As noted previously, thermal modification of plant fibres by prolonged heating at temperatures above 160 °C reduces the hygroscopicity of fibres, but is also responsible for degrading the mechanical properties.[58] A commercial thermal modification process (Duralin) was developed for flax fibres over a decade ago, but this is no longer in commercial production. Thermal modification removes hygroscopic polysaccharide material in the cell wall and can also result in crosslinking promoted by the formation of furfuryl alcohol produced from degraded pentoses.[59] Other methods of reducing water vapour sorption require the chemical modification of the cell wall macromolecules, or impregnation of the cell wall with monomers, with subsequent polymerization in order to prevent leaching in service. A comparison of the water vapour sorption isotherms of flax and Duralin is shown in Figure 7.11.

Acetylation has been studied extensively as a means of improving the dimensional stabilization of wood and its potential for protecting natural fibres from degradation and the effects of water. The method used in commercial wood modification is to react the substrate with acetic anhydride at temperatures in excess of 120 °C. This leads to esterification of the cell wall polymeric OH groups and generates acetic acid as a by-product which must then be removed from the modified material (Figure 7.12).

Apart from removal of the OH groups by reaction, the bonded acetyl groups also occupy space within the cell wall, resulting in a permanent swelling (bulking). Removal of the OH groups by substitution with the less polar acetyl groups affects wettability, but the reduction in EMC is due to the bulking effect.[60] In studies of wood modified with a variety of anhydride reagents, it was shown that the reduction in EMC was related only to the weight gain due to bonded reagent, but was not related to the number of OH groups substituted.[60]

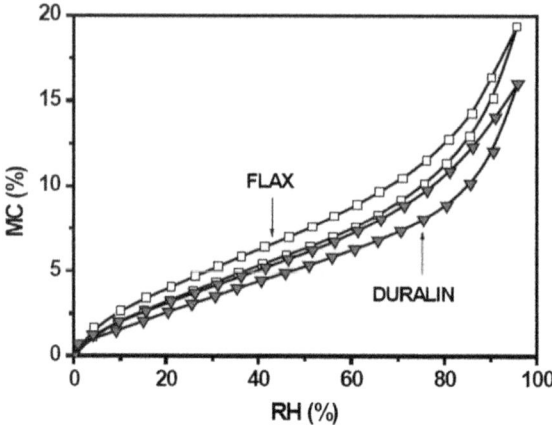

Figure 7.11   Sorption isotherms of flax and Duralin.

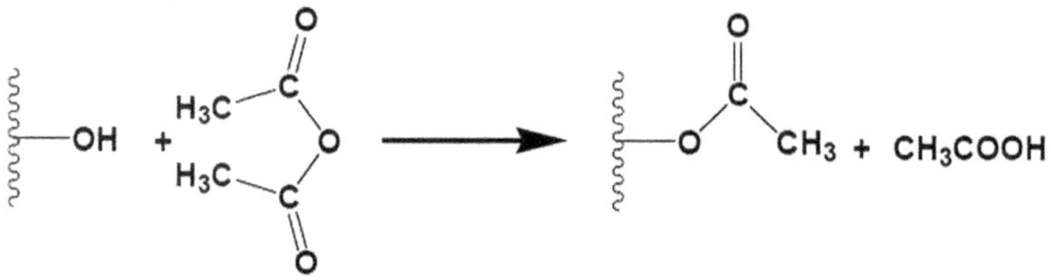

Figure 7.12   The acetylation reaction with acetic anhydride.

It has been shown that significant improvements in performance are found when acetylated fibres are subjected to biological degradation.[61] The improved resistance to biodegradation is again found to be independent of the extent of OH substitution and appears to be due to the reduction of the cell wall MC caused by the filling of internal space in the cell wall due to the presence of bonded reagent.[62] It is known that fungal attack of the wood cell wall requires the MC to be in excess of 20% and it appears that acetylation is able to provide protection by ensuring that the cell wall MC is below this threshold.

   It has also been found that improvements in resistance to the degradation of composites reinforced with silane-treated fibres can be achieved.[63] Silanes have long been used for sizing of glass fibres and they can also be used as a means of promoting adhesion between fibre and matrix, depending on the chemistry of the silane. There has been considerable interest in the use of organo-functionalized alkoxysilanes (Figure 7.13) in order to improve the bonding between natural fibres and matrix in composite materials,[64] but under certain conditions it is also possible to modify the interior of the cell wall and hence reduce water sorption. However, it is necessary to carefully control the modification conditions in order to optimize cell wall penetration by the reagent. Polymerization of the silane can occur once the alkoxy groups are removed by hydrolysis, but

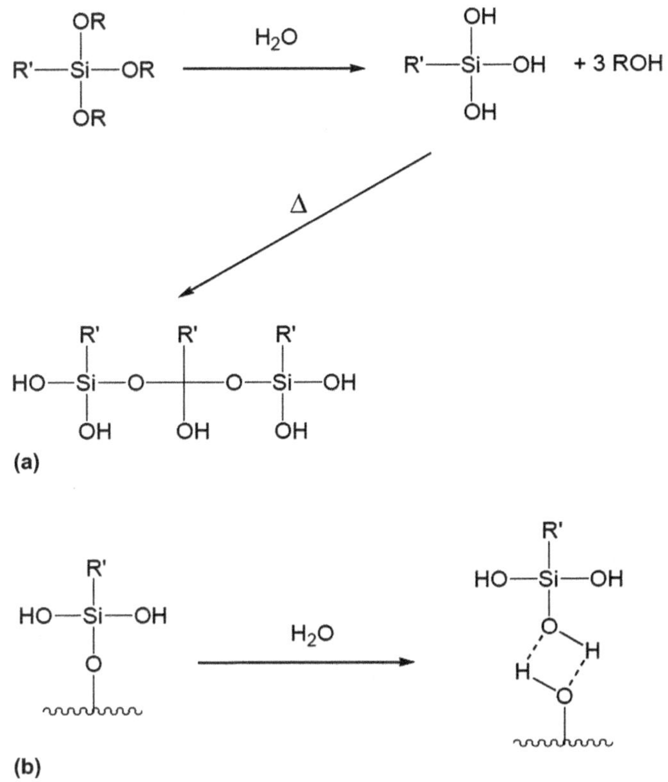

Figure 7.13 (a) Reaction of an alkoxy organosilane with water and subsequent polymerization, and (b) reaction of an organosilane with the cell wall polymers and subsequent hydrolysis.

such reactions are undesirable, at least until the silane monomer has penetrated the cell wall. Premature oligomerization of the silanes prevents penetration of the cell wall and restricts reactions to the surface of the fibre.[65]

7.10 Conclusions

The water sorption behaviour of natural fibres is an important property that influences their mechanical behaviour. It is proposed herein that there is a link between the mechanical properties of the fibre cell wall, the sorption kinetics and the property of hysteresis. The sorption kinetic behaviour can be very accurately described using the parallel exponential kinetics model, whereas Fickian behaviour is not observed. It is proposed that the sorption kinetics behaviour is described by consideration of the viscoelastic relaxation properties of the cell wall polymers. In addition, the cell wall polymer relaxation behaviour can be invoked as a means of explaining hysteresis. This then suggests a link between mechanical properties and water sorption behaviour that should prove a promising line of research for the future.

Acknowledgement

The author acknowledges the support of the Scottish Funding Council for funding of the Joint Research Institute in Civil and Environmental Engineering, which is part of the Edinburgh Research Partnership in Engineering and Mathematics.

References

1. S. J. Gregg and K. S. W. Sing, *Adsorption, Surface Area and Porosity*, Academic Press, London, 2nd edn., 1982.
2. W. E. Morton and J. W. S. Hearle, *Physical Properties of Textile Fibres*, CRC Press, Boca Raton, 4th edn., 2008.
3. C. A. S. Hill, A. Norton and G. Newman, *J. Appl. Polym. Sci.*, 2009, **112**, 1524.
4. C. Skaar, *Water in Wood*, Syracruse University Press, New York, 1972.
5. J. F. Siau, *Transport Properties in Wood*, Springer, Berlin, 1984.
6. A. J. Stamm, *J. Wood Sci.*, 1971, **4**, 114.
7. A. Venkateswaram, *Chem. Rev.*, 1970, **70**, 619.
8. W. Simpson, *Wood Fiber Sci.*, 1980, **12**, 98.
9. G. K. van der Wel and O. C. G. Adan, *Prog. Org. Coat.*, 1999, **37**, 1.
10. S. Brunauer, P. H. Emmett and E. Teller, *J. Am. Chem. Soc.*, 1938, **60**, 309.
11. J. Chirife, O. Timmermann, H. A. Inglesias and R. Boquet, *J. Food Eng.*, 1992, **15**, 75.
12. A. J. Hailwood and S. Horrobin, *Trans. Faraday Soc.*, 1946, **42B**, 84.
13. P. J. Flory, *Principles of Polymer Chemistry*, Cornell University Press, New York, 1953.
14. W. W. Barkas, *Trans. Faraday Soc.*, 1942, **38**, 194.
15. J. S. Vrentas and C. M. Vrentas, *Macromolecules*, 1996, **29**, 4391.
16. Y. Lu and J. J. Pignatello, *J. Environ. Qual.*, 2005, **34**, 1072.
17. Y. Lu and J. J. Pignatello, *Environ. Sci. Technol.*, 2002, **36**, 4553.
18. C. A. S. Hill, A. Norton and G. Newman, *Holzforschung*, 2010, **64**, 469.
19. A. R. Urquhart and N. Eckershall, *J. Text. Res. Inst.*, 1929, **20**, T125.
20. A. Stamboulis, C. A. Baillie, S. K. Garkhail, H. G. N. Van Melick and T. Peijs, *Appl. Compos. Mater.*, 2000, **7**, 273.
21. J. C. F. Walker, *Primary Wood Processing: Principles and Practice*, Springer, Dordrecht, 2nd edn., 2006.
22. M. S. Sreekala and S. Thomas, *Compos. Sci. Technol.*, 2003, **63**, 861.
23. R. Kohler, R. Dück, B. Ausperger and R. Alex, *Compos. Interfaces*, 2003, **10**, 255.
24. C. A. S. Hill, A. Norton and G. Newman, *J. Appl. Polym. Sci.*, 2010, **116**, 2166.

25. Y. Xie, C. A. S. Hill, Z. Jalaludin, S. F. Curling, R. D. Ananjiwala, A. J. Norton and G. Newman, *J. Mater. Sci.*, 2011, **46**, 479.
26. S. Okubayashi, U. J. Griesser and T. Bechtold, *Carbohydr. Polym.*, 2004, **58**, 293.
27. S. Okubayashi, U. J. Griesser and T. Bechtold, *J. Appl. Polym. Sci.*, 2005, **97**, 1621.
28. S. Okubayashi, U. J. Griesser and T. Bechtold, *Cellulose*, 2005, **12**, 403.
29. K. Kachrimanis, M. F. Noisternig, U. J. Griesser and S. Malamataris, *Eur. J. Pharm. Biopharm.*, 2006, **64**, 307.
30. C. A. S. Hill, A. J. Norton and G. Newman, *Wood Sci. Technol.*, 2010, **44**, 497.
31. Y. Xie, C. A. S. Hill, Z. Xiao, Z. Jalaludin, H. Militz and C. Mai, *J. Appl. Polym. Sci.*, 2010, **117**, 1674.
32. X. Tang, M. R. De Roojj, J. Van Duynhoven and K. J. Van Breugel, *J. Microsc.*, 2008, **230**, 100.
33. P. S. Madamba, R. H. Driscol and K. A. J. Buckle, *Food Eng.*, 1996, **29**, 75.
34. M. S. Rahman, C. O. Perera and C. Thebaud, *Food Res. Int.*, 1998, **30**, 485.
35. Y. Xie, C. A. S. Hill, Z. Jalaludin and D. Sun, *Cellulose*, 2011, **18**, 517.
36. C. A. S. Hill and Y. Xie, *J. Mater. Sci.*, 2011, **46**, 3738.
37. J. M. Dinwoodie, *Timber: Its Nature and Behaviour*, Spon, London, 2nd edn., 2000.
38. H. W. Haslach, *Mech. Time-Depend. Mater.*, 1995, **4**, 169.
39. G. C. Davies and D. M. Bruce, *Textile Res. J.*, 1998, **68**, 623.
40. P. S. Mukherjee and K. G. Satyanarayana, *J. Mater. Sci.*, 1984, **19**, 3925.
41. L. Salmén, *C. R. Biol.*, 2004, **327**, 873.
42. J. Passard and P. Perre, *Ann. For. Sci.*, 2005, **62**, 823.
43. M. Hughes, J. Carpenter and C. A. S. Hill, *J. Mater. Sci.*, 2007, **42**, 2499.
44. C. A. S. Hill and M. Hughes, *J. Biobased Mater. Bioenergy*, 2011, **4**, 148.
45. Y. Xie, C. A. S. Hill, Z. Jalaludin and D. Sun, *Cellulose*, 2011, **18**, 517.
46. C. Hill and Y. Xie, *J. Mater. Sci.*, 2011, **46**, 3738.
47. H. A. Barnes, J. F. Hutton and K. Walters, *An Introduction to Rheology*, Elsevier, Amsterdam, 1989.
48. K. Krabbenhoft and L. Damkilde, *Mater. Constr. (Paris)*, 2004, **37**, 615.
49. R. Meredith, *J. Text. Inst. Trans.*, 1957, **48**, T163.
50. S. Matsuoka, *Relaxation Phenomena in Polymers*, Hanser, New York, 1992.
51. S. Matsuoka and A. Hale, *J. Appl. Polym. Sci.*, 1997, **64**, 77.
52. A. Bartolotta, G. Carini, G. Carini, G. Di Marco and G. Tripodo, *Macromolecules*, 2010, **43**, 4798.
53. C. A. Baillie, *Green Composites: Polymer Composites and the Environment*, Woodhead, Cambridge, 2004.
54. K. M. Mannan and Z. Robbany, *Polymer*, 1996, **37**, 4639.
55. R. Wimmer, B. N. Lucas, T. Y. Tsui and W. C. Oliver, *Wood Sci. Technol.*, 1997, **31**, 131.
56. W. Gindl and H. S. Gupta, *Composites, Part A*, 2002, **33**, 1141.
57. W. Gindl and T. Schöberl, *Composites, Part A*, 2004, **35**, 1245.

58. S. Ochi, *Mech. Mater.*, 2008, **40**, 446.
59. C. A. S. Hill, *Wood Modification: Chemical, Thermal and Other Properties*, Wiley, Chichester, 2006.
60. C. A. S. Hill, *Holzforschung*, 2007, **61**, 138.
61. C. A. S. Hill, H. P. S Abdul Khalil and M. D. Hale, *Ind. Crops Prod.*, 1998, **8**, 53.
62. C. A. S. Hill, S. F. Curling, J. H. Kwon and V. Marty, *Holzforschung*, 2009, **63**, 619.
63. C. A. S. Hill and H. P. S. Abdul Khalil, *J. Appl. Polym. Sci.*, 2000, **77**, 1322.
64. Y. Xie, C. A. S. Hill, Z. Xiao, H. Militz and C. Mai, *Composites, Part A*, 2010, **41**, 806.
65. Y. Xie, C. A. S. Hill, D. Sun, Z. Jalaludin, Q. Wang and C. Mai, *Bioresources*, 2011, **6**, 2323.

CHAPTER 8

Environmentally Friendly Coupling Agents for Natural Fibre Composites

R. CHOLLAKUP,*[a] W. SMITTHIPONG[a] AND
P. SUWANRUJI[b]

[a] Kasetsart University, Kasetsart Agricultural and Agro-Industrial Product Improvement Institute (KAPI), 50 Ngam Wong Wan Rd., Chatuchak, 10900, Bangkok, Thailand; [b] Department of Chemistry, Faculty of Science, Kasetsart University, 50 Ngam Wong Wan Rd., Chatuchak, 10900, Bangkok, Thailand
*Email: aaprmc@ku.ac.th

8.1 Introduction

In recent years, natural source materials have gained attention because they are renewable, widely distributed, locally accessible, environmentally recyclable, conveniently available in many forms and biodegradable. In particular, natural fibres, *i.e.* kenaf, flax, jute, hemp, sisal, pineapple, silk, *etc*, are abundant worldwide as biomaterial sources.[1–6] These fibres are based on either cellulose or protein. Cellulose is used to reinforce the structure of plant cells, whereas silk, a protein fibre produced from silkworms, is used to build the cocoon for protecting the silkworm from predators during its metamorphosis into a moth. Thus, these natural fibres have received much interest as possible reinforcing materials for plastic composites in many applications such as

RSC Green Chemistry No. 16
Natural Polymers, Volume 1: Composites
Edited by Maya J John and Thomas Sabu
© The Royal Society of Chemistry 2012
Published by the Royal Society of Chemistry, www.rsc.org

automative and package products.[7,8] The advantages of these biomaterial sources, compared to traditional reinforcing materials such as glass, aramid and carbon fibres, are their low cost, low density, toughness, acceptable specific strength properties, reduced tool wear, reduced dermal and respiratory irritation, good thermal properties, enhanced energy recovery and, especially, biodegradability.[9–11]

Fibre-reinforced composite materials are part of the general class of engineering materials called composite materials. A useful definition of composite materials is to state that composite materials are characterized by being multiphase in which the phase distribution and geometry have been deliberately tailored to optimize one or more properties.[12,13] This is clearly an appropriate definition for fibre-reinforced composites for which there is one phase, called the matrix, reinforced by material in the form of a fibre. Generally, the fibre is stronger than the matrix. When a composite is forced by external action, this force is transferred from the matrix to the fibre, which increases its strength compared to the matrix without the fibre. The size and shape of fibres play an important role in the efficiency of reinforcement. In addition to a wide choice of materials, there is the added factor of the manufacturing route to consider, because a valued feature of composite materials is the ability to manufacture the article at the same time as the material itself is being processed.

The use of natural fibres for the reinforcement of composites has received increasing attention both by the academic sector and industry. Currently, many types of natural fibres[14,15] have been investigated for use in plastics, including flax, hemp, jute straw, wood, rice husk, wheat, barley, oats, rye, cane (sugar and bamboo), grass, reed, kenaf, ramie, oil palm empty fruit bunch, sisal, coir, pennywort, kapok, paper mulberry, raphia, banana fibre, pineapple leaf fibre and papyrus. Thermoplastics reinforced with special wood fillers are enjoying rapid growth due to their many advantages: lightweight, reasonable strength and stiffness. Some plant proteins are interesting renewable materials because of their thermoplastic properties. Wheat gluten,[16] once plasticized, is unique among cereals and other plant proteins in its ability to form a cohesive blend with viscoelastic properties. For these reasons, wheat gluten has been utilized to process edible or biodegradable films and packaging materials. Hemp[17] is a bast lignocellulosic fibre, which comes from the plant *Cannabis sativa* and has been used as a reinforcement in biodegradable composites.

Generally, the properties of composites depend on the matrix and the fibre. The development of new composite materials based on natural fibres has been driven by interesting new applications. The role of the matrix in a fibre-reinforced composite is to transfer stress between the fibres, to provide a barrier against an adverse environment and to protect the surface of the fibres from mechanical abrasion. The matrix plays a major role in the tensile load-carrying capacity of a composite structure. The binding agent or matrix in the composite is of critical importance. Here, we summarize three types of matrix: (i) non-biodegradable synthetic polymers in Table 8.1; (ii) biodegradable synthetic polymers in Table 8.2; and (iii) biopolymers in Table 8.3.

Table 8.1 Examples of natural fibre composite materials based on non-bio-degradable synthetic polymers.

Matrix	Natural fibre	Improvement properties	Ref.
Epoxy	Sisal	Tensile strength	18
Phenol/formaldehyde (PF)	Palm	Tensile strength and modulus, dynamic mechanical properties	19,20
Polycarbonate	Pineapple	Mechanical properties	21
Polyethylene (PE)	Coir and palm	Young's modulus	22
	Pineapple	Dynamic storage and loss modulus, tensile strength	23–25
	Sisal	Tensile strength, electrical properties, degradation	26–28
Polypropylene	Palm	Tensile strength, flexural strength and modulus, water absorption	30,31
	Pineapple	Tensile strength	25,29,32
	Sisal	Tensile strength, dynamic mechanical properties, thermal degradation	33,34
Polystyrene (PS)	Palm	Modulus	35
	Sisal	Thermal stability	36
Poly(vinyl chloride) (PVC)	Palm	Flexural modulus	37
	Rice straw	Tensile strength	38

8.2 Interface between the Natural Fibre and the Matrix

Multicomponent materials and, in particular, reinforced materials have known a huge industrial development in the last few decades. There has also been an increasing interest in the scientific community to analyze and understand the physical and chemical nature of the interface between the reinforcing entity and the matrix. More precisely, good final mechanical performance or use properties of the resulting reinforced composites depend significantly on the quality of the interface that is formed between both solids. In other words, the adhesion established at this interface becomes one of the most important parameters in controlling the interfacial behaviour, and consequently, the mechanical properties of reinforced materials. Therefore, it is understandable that a better knowledge of the adhesion phenomena is required for practical applications of multicomponent materials. Consequently, the study of adhesion uses various concepts, depending on different special fields of expertise. This variety of approaches is emphasized by the fact that many theories of adhesion have been proposed which together are both complementary and contradictory, *i.e.* mechanical interlocking, electronic or electrostatic theory, theory of weak boundary layers, thermodynamic or adsorption model (also referring to wettability), diffusion or interdiffusion theory, and finally, chemical bonding theory.[69–72]

Table 8.2 Examples of natural fibre composite materials based on biodegradable synthetic polymers.

Matrix	Natural fibre	Improvement properties	Ref.
Poly(butylene succinate) (PBS)	Bamboo	Tensile properties, water resistance, interfacial adhesion and enzyme biodegradability	39
Poly(butylene adipate-*co*-terephthalate	Wheat straw	Mechanical properties	40
Poly-caprolactone (PCL)	Cellulose fibre	Electrical conductivity	41
Polyester amide (PEA)	Jute	Tensile strength	7
	Sisal	Mechanical properties	42
Polyester	Banana	Dynamic mechanical properties	43
Poly(hydroxy-butyrate) (PHB)	Flax	Bending modulus and dynamic mechanical properties	44
Poly(hydroxy-butyrate valerate) (PHBV)	Pineapple	Tensile and flexural strength and modulus	45,46
Poly(lactic acid) (PLA)	Bamboo	Impact strength, heat resistance, water resistance, interfacial adhesion and enzyme biodegradability	41,42,47
	Cotton with lignin	Tensile strength and modulus	48
	Flax	Impact, tensile strength, Young's modulus	49,50
	Hemp	Tensile strength, modulus and flexural strength and ease of thermal degradation	51,52
	Kenaf	Mechanical and thermal properties	53–55
	Reed	Mechanical properties	56
	Silk	Reinforcement for PLA for tissue engineering	57
	Wood	Fluxural modulus	58
Poly(vinyl alcohol) (PVA)	Baggase	Thermal stability	58
	Lignocellulosic fibre	Enhance PVA degradation	59
Unsaturated polyester (UPE)	Hemp	Impact and tensile strength	60

However, the thermodynamic model of adhesion, generally attributed to Sharpe and Schonhorn,[73] is certainly the most widely used approach in adhesion science at present. In this theory, it is considered that adhesion between two solids (more generally, between an adhesive and a substrate) is due to

Table 8.3 Examples of natural fibre composite materials based on biopolymers.

Matrix	Natural fibre	Improvement properties	Ref.
Thermoplastic starch	Flax	Tensile strength and modulus	61
	Sisal	Mechanical properties	62
	Cellulosic fibres from *Eucalyptus urograndis* pulp	Tensile strength and modulus	63
	Leafwood cellulose, paperpulp fibre	Mechanical properties (tensile tests), thermo-mechanical properties (DMTA) and thermal degradation	64
Soy protein isolate (SPI)	Ramie	Mechanical properties	65
Cross-linked soy flour (CSF)	Flax	Tensile and flexural properties	67
Corn gluten meal (CGM)	Wood	Mechanical properties	66
Modified acrylated expoxidized soy oil (AESO)	Flax and hemp	Tensile strength, flexural strength and modulus	68

interatomic and intermolecular forces established at the interface, provided that an intimate contact is achieved.[74] The most common interfacial forces result from van der Waals and Lewis acid–base interactions. The magnitude of these forces can generally be related to fundamental thermodynamic quantities, such as surface-free energies of both entities in contact. Generally, the formation of an assembly is obtained through a liquid–solid contact step. Therefore, criteria for good adhesion become, essentially, criteria for good wetting, although this is a necessary but not a sufficient condition. It is clear that when an intimate contact has been established at an interface, other adhesion phenomena can therefore take place, in particular the creation of interfacial chemical bonds, the molecular reorganization and reorientation near the interface, the interdiffusion of molecular or macromolecular chains across this interface in the case of contact involving polymers, *etc.* Each of these phenomena is able to enhance the level of adhesion.[75–82]

In fact, each of these adhesion theories is validated by some measurements, depending on the nature of the materials and the formation conditions of the adhesive junction. Although each theory is supposed to explain the adhesion in different ways, it in fact contributes to this complex phenomenon. Currently, six main mechanisms involving chemical or physical forces are considered. Their actions at the interface between two materials, the same or different, are shown in Figure 8.1. These mechanisms are only validated at different scales by the distance between the two materials: the atomic scale for the chemical bonds, a few angstroms for intermolecular bonds, a scale not known (but reasonably within the range of micrometers) in the case of mechanical bonding.

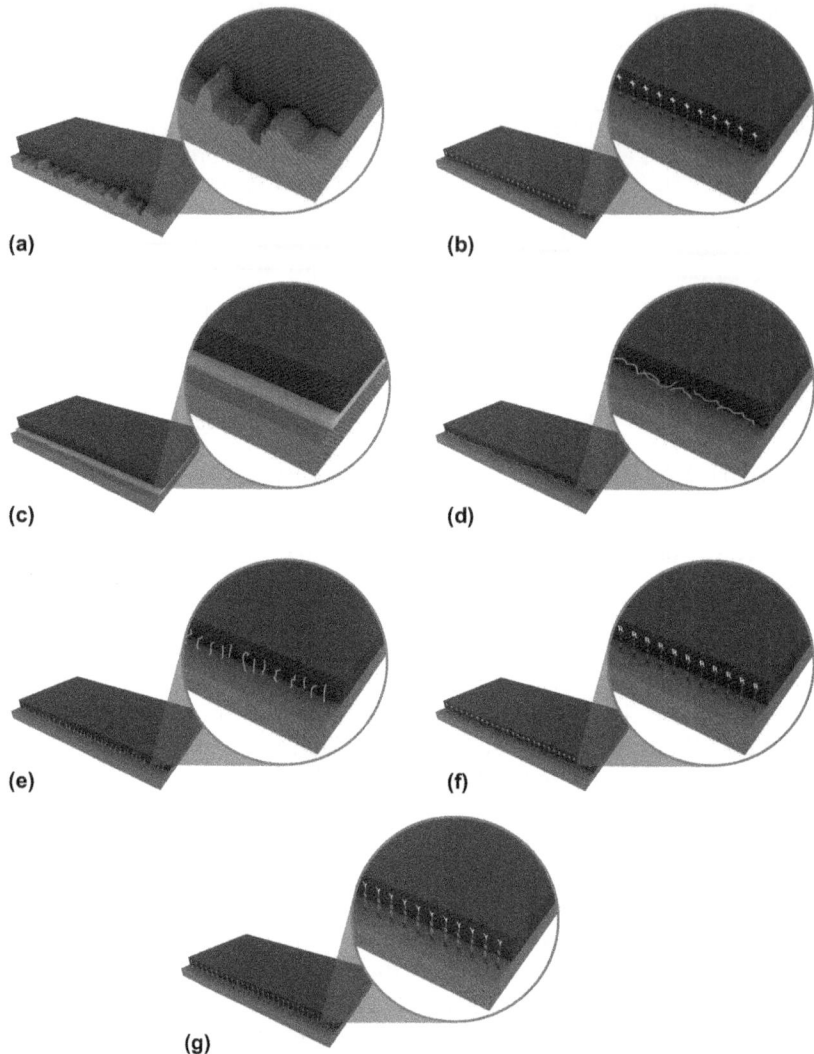

Figure 8.1 Model of adhesion theory: (a) mechanical bonding; (b) electrostatic
bonding; (c) interfacial layers of low cohesion; (d) adsorption; (e) diffusion
theory; (f) chemical bonding theory; and (g) chemical bonding of coupling
agents between the fibre surface and the polymer matrix.

8.3 Types of Coupling Agents

Coupling agents are compounds that function at the interface of natural fibres
and the polymer matrix to form a bridge of chemical and/or physical bonds
between the two phases. Generally, coupling agents possess two functions: the
first is to react with the fibre surface, *i.e.* at the polar hydroxyl (OH) group, and
the other is to react with functional groups of the polymer matrix. In this

heterogeneous system, coupling agents couple the two phases together *via* various interactions, such as covalent bonding, hydrogen bonding, acid–base interaction and chain entanglement, depending on the types of polymers and fibres and the functional groups of the coupling agents. The incorporation of coupling agents in fibre-reinforced polymer composites facilitates the formation of interfacial bonding between the fibres and the matrix, leading to the improvement of mechanical properties from effective stress transfer and bond distribution across the interface. In most cases, selection of a coupling agent for fibre-reinforced polymer composite is based on the combination of strength and toughness at the fibre/matrix interface. The most common coupling agents used for natural fibre-reinforced polymer composites are silane, isocyanate and maleated coupling agents.[10,83–85]

8.3.1 Conventional Coupling Agents

8.3.1.1 Silanes

Silane coupling agents are effectively used in natural fibre/polymer composites because of their bifunctional structures that can enhance the interfacial adhesion between the fibre reinforcement and the polymer matrix. The general chemical structure of a silane is $R–SiX_3$, where R represents an organofunctional group which can react with the polymer matrix and X are groups that can hydrolyze to form silanol groups in an aqueous solution.[10,83,84] Examples of silane coupling agents are shown in Table 8.4. Castellano *et al.* reported that silane itself does not react with the hydroxyl groups of cellulose fibres even at high temperatures, because of the low acidity of cellulose hydroxyl groups.[86]

Table 8.4 Silane coupling agents for natural fibre/polymer composites ($R–SiX_3$).

Name	R	X	Functionality
Hexadecyltrimethoxy-silane (HDS)	$-(CH_2)_{15}CH_3$	$-OCH_3$	Alkyl
Dichlorodiethylsilane (DCS)	$-(CH_2)_2CH_3$	$-Cl_2$	Alkyl
γ-(Aminopropyl) triethoxy-silane (APS)	$-(CH_2)_3NH_2$	$-OCH_2CH_3$	Amino
(Cyanoethyl) trimethoxy-silane (CES)	$-(CH_2)_2CN$	$-OCH_3$	Cyano
γ-(Mercaptopropyl)- trimethoxysilane (MPS)	$-(CH_2)_3SH$	$-OCH_3$	Mercapto (thiol)
γ-(Methacryloxypropyl)- trimethoxysilane (MCS)	$-(CH_2)_3-O-\overset{\overset{O}{\|\|}}{C}-\underset{\underset{CH_3}{\|}}{C}=CH_2$	$-OCH_3$	Methacryl
Vinyltri(2-methoxyethoxy)- silane (VTS)	$-CH=CH_2$	$-OCH_2CH_2OCH_3$	Vinyl

However, the hydrolyzable X groups can form the more reactive silanols in the presence of moisture (eqn 8.1), which then react with hydroxyl groups of natural fibre macromolecules:[87,88]

$$\underset{\text{silane}}{\text{R–SiX}_3} + 3\text{H}_2\text{O} \rightarrow \underset{\text{silanol}}{\text{R–Si(OH)}_3} + 3\text{XCl}$$

$$(8.1)$$

Owing to the high affinity for each other, silanol groups can also undergo self-condensation to form oligomers *via* –Si–O–Si bonds. Consequently, not only the silanol monomers but also the oligomers adsorb to the hydroxyl groups of the fibre by the formation of hydrogen bonds. Under thermal conditions, the hydrogen bonds formed between the adsorbed silanols and the hydroxyl groups of the fibre can be converted into covalent bonds of –Si–O–C linkages,[89–91] as shown in Figure 8.2.

The R groups at the other end of the silane can be a large number of functional groups, as shown in Table 8.4. These functional groups play an important role in the interaction between the silane and the polymer matrix used, depending on their functionality's reactivity towards the matrix. The non-reactive groups of silane, *i.e.* alkyl groups, would be expected to increase the compatibility with the hydrocarbon backbone of the polymer matrix, such as polyethylene (PE) and polypropylene (PP). On the other hand, reactive groups such as vinyl, methacryl and amino groups can form a chemical bond with a

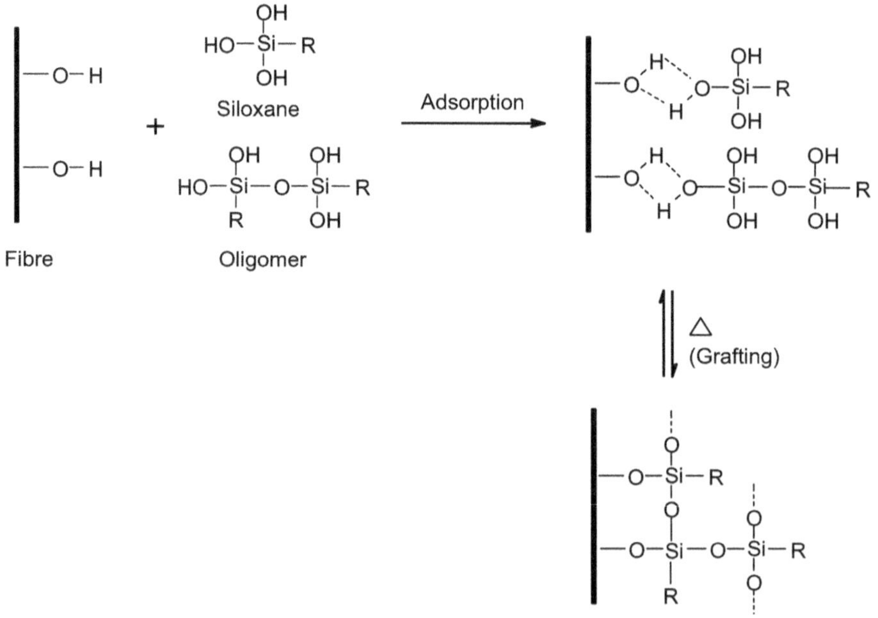

Figure 8.2   Mechanism of the interaction between silanes and cellulose fibres.

polymer matrix containing reactive functional groups.[88,90,92–98] Therefore, it is important to find a proper match between the silane functionality and the target matrix to improve the interfacial adhesion. The interaction between the silane group and the polymer matrix may be van der Waals forces[95] or by a grafting reaction.[93] Covalent bonds can also form between aminosilanes and epoxy resins.[97,98]

8.3.1.2 Isocyanates

Isocyanates function as useful coupling agents by the formation of covalent bonds between the hydroxyl groups of cellulose fibres and the isocyanate (–N=C=O) functional groups of the coupling agents.[85,88,99,100] The most common isocyanate coupling agents are shown in Table 8.5. The urethane or carbamate linkage between cellulose fibres and isocyanate coupling agents can be formed as shown in eqn (8.2). The formation of carbonyl groups on the fibre surface has been confirmed by Fourier transform infrared spectroscopy (FTIR).[88]

$$R-N{=}C{=}O \ + \ HO-Cell \ \longrightarrow \ R-\underset{\underset{H}{|}}{N}-\overset{\overset{O}{\|}}{C}-O-Cell \qquad (8.2)$$

Since natural fibres are highly hygroscopic with a great tendency to absorb water, isocyanate coupling agents are susceptible to hydrolysis. In the presence of even traces of moisture, isocyanates prefer the reaction with water to cellulosic

Table 8.5 Isocyanate coupling agents for natural fibre/polymer composites.

Name	Structure
Methylene diisocyanate (MDIC)	
Polymethylene polyphenyl isocyanate (PMPPIC)	
Toluene 2,4-diisocyanate (TDIC)	
Hexamethylene diisocyanate (HMDIC)	

hydrogen groups, producing polyurea, biuret and isocyanate dimers.[100] Therefore, the use of blocked isocyanates has been reported to prevent the hydrolysis of
isocyanates prior to reaction with the fibre and also its thermal dissociation. The
enhancement of mechanical properties and water resistance of wheat starch/PLA
blends was reported as a result of the addition of methylene diisocyanate
(MDIC).[101,102] George *et al.* claimed that the very high interfacial interaction
between pineapple-leaf fibre (PALF) and a PE matrix was because of the addition
of polymethylene polyphenyl isocyanate (PMPPIC). An effective water uptake
reduction of the composite was due to the improvement of interfacial bonding
from the coupling effect of PMPPIC.[88] The maximum retention in strength of
PMPPIC-treated PALF/linear density polyethylene (LDPE) composites at
higher temperatures was also reported.[32] The isocyanate groups in PMPPIC are
highly reactive with the hydroxyl groups of fibres and a urethane linkage can be
formed between them. The long-chain molecules in PMPPIC interact with the PE
matrix through a van der Waals interaction. The addition of toluene 2,4-diiso-
cyanate (TDIC) to short sisal fibre-reinforced polyester (PS) composites showed a
reduction of water uptake and an improvement of dimensional stability and
tensile properties. The π-electrons of benzene rings in TDIC, similar to PMPPIC,
lead to stronger interactions with the PS matrix.[103] By the reaction of cellulosic
hydroxyl groups and isocyanate groups of HMDIC, the polarity and hydrophilicity of cellulose fibres decreased and consequently improved the compatibility of the cellulose fibres and PP matrix.[100] Alkenyl isocyanates have also been
used in fibre/polymer composites, such as 3-isopropenyl-α,α'-dimethylbenzyl
isocyanate (TMIC) and 2-isocyanoethyl methacrylate (IEMC). The isocyanate
groups ensure the coupling reaction with the cellulosic hydroxyl groups while an
alkenyl moiety subsequently copolymerizes with the matrix monomers such as
styrene and methyl methacrylate.[104]

8.3.1.3 Maleated Coupling Agents

Maleic anhydride (MA) is an α,β-unsaturated carbonyl compound having a
C=C double bond on the heterocyclic ring conjugated with two carboxylate
groups. MA is extensively used to graft onto polymers, especially PE and PP, to
form maleated coupling agents such as maleated polyethylene (MAPE) and
maleated polypropylene (MAPP). The anhydride groups of the maleated
copolymer can react with the surface hydroxyl groups of the fibre reinforcement through esterification or hydrogen bonding, while the other end of the
copolymer can undergo entanglement interlocking with the polymer matrix as a
result of their similar polarity.[105,106] FTIR and electron spectroscopy for chemical analysis (ESCA) have presented evidence for chemical bridges between
the wood fibre and the polymeric matrix through esterification. The two primary covalent bonding products to link the wood fibre and polymer matrix at
the interface were succinic esters and half-esters.[106]

The inclusion of MAPP in wood-reinforced PP composites gives a significant
improvement in the strength properties of the composites.[107,108] The flexural

properties, impact strength and tensile modulus of PALF-reinforced PP composites were considerably improved in the presence of MAPP.[29] The mechanical properties of jute-reinforced PP composites were also reported to be improved by the application of MAPP, owing to the chemical bonds at the fibre/matrix interface established from MAPP coupling agents.[109,110] The improvement in the abrasive wear resistance of jute-reinforced PP composites with the addition of MAPP was also claimed by Chand and Dwivedi.[111]

8.3.1.4   Dichlorotriazines

Dichlorotriazines and their derivatives contain reactive chlorines on the heterocyclic ring that can react with the hydroxyl groups of natural fibres through an ether linkage. Moreover, the nitrogen in the heterocyclic ring may also link the hydroxyl groups *via* hydrogen bonding. Zedorecki and Flodin confirmed the reaction between the hydroxyl groups of cellulose fibres and triazine derivatives using FTIR and X-ray photoelectron spectroscopy (XPS). They reported that triazine compounds can improve the environmental ageing behaviour of cellulose fibre-reinforced polyester composites.[112–114] Joly *et al.* used dichlorotriazine substituted with an octadecyl chain as a coupling agent in cotton/PP composites. The chemical reaction of triazine with hydroxyl groups only occurred at the extreme surface of the fibre, while the octadecyl chain entangled with the PP matrix. There was no measurable effect on the water uptake of the composites.[115]

8.3.2   Bio-based Coupling Agents

With the world's increased environmental awareness, it would be good if coupling agents used in natural fibre/polymer composites were non-toxic and biodegradable to preserve the renewable and biodegradable character of the natural fibre reinforcement. In order to produce completely biodegradable composites without emitting toxic or noxious compounds, a bio-based coupling agent is needed. Therefore, many environmentally friendly coupling agents have been reported.

8.3.2.1   Cardanol Derivative of Toluene Diisocyanate (CTDIC)

Cardanol is the main component of cashew-nut-shell liquid obtained from the plant *Anacardium occidentale*, which is plentiful in tropical countries. CTDIC was prepared from the reaction between cardanol and toluene diisocyanate. The resulting product contained one free isocyanate group for further reaction with the hydroxyl groups of fibres. The formation of a covalent bond between isocyanate groups and fibres reduced the hydrophilic nature of the cellulosic fibre. The long-chain structure of cardanol in CTDIC helps the fibre have a higher hydrophobic character, better compatibility and dispersibility in a matrix. Owing to a strong interfacial bond between sisal fibres and LDPE, the

CTDIC-treated fibre composites exhibited superior tensile strength and modulus than the untreated composites. This strong fibre/matrix adhesion in the composites was supported by scanning electron microscopy (SEM) photomicrographs. The dielectric constant values of sisal/LDPE composites were found to decrease as a result of the addition of CTDIC.[27,116]

8.3.2.2   *Lysine Diisocyanate (LDIC)*

Conventional isocyanates have limited uses as coupling agents in view of environmental concerns because their ultimate hydrolysis products, such as 4,4′-methylenedianiline and 2,4-diaminotoluene, are suspected cancer agents and produce hepatitis in humans. LDIC prepared from lysine can be used as a coupling agent without producing toxic products during its degradation. Lysine is one of the essential amino acids, having two amino groups and one carboxyl group. LDIC can react with hydroxyl or carboxyl groups in poly(lactic acid) (PLA) or poly(butylene succinate) (PBS) and form hydrolyzable urethane bonds. Tensile properties and water resistance were improved by using LDIC in corn starch/PBS biocomposites. The enhancement of interfacial adhesion between corn starch and a PBS matrix was a result of the coupling effect of LDIC, even with a very small addition (<1 wt%).[117] Lee and Wang also reported the effect of LDIC as a coupling agent on the properties of biocomposites from PLA, PBS and bamboo fibre (BF). They concluded that the addition of LDIC improved the tensile properties, water resistance and interfacial adhesion of both PLA/BF and PBS/BF composites. The crystallization temperatures of both composites increased, whereas the enthalpy and heat of fusion decreased with increasing the amount of LDIC. However, the composites with LDIC exhibited higher thermal degradation temperatures than those without LDIC. Moreover, LDIC also delayed enzymatic degradation because stronger interfacial adhesion reduced the area exposed to enzyme hydrolysis.[39] In corn starch/PLA composites, isocyanate groups reacted with terminal hydroxyl or carbonyl groups of PLA and the hydroxyl groups of corn starch. The urethane linkage was confirmed by nuclear magnetic resonance spectroscopy (NMR).[118]

8.3.2.3   *Chitin and Chitosan*

Chitin and chitosan are long, linear, natural polysaccharide polymers composed of *N*-acetylglucosamine and glucosamine, respectively. Chitosan is thus the deacetylated form of chitin. The advantages of these two compounds are their wide availability, non-toxicity, biocompatibility and lower cost compared with many synthetic coupling agents. Chitin and chitosan have been used as coupling agents for wood flour/PVC composites. With the addition of chitin and chitosan coupling agents, the flexural strength, flexural and storage modulus of the composites increased compared with ones without coupling agents. The property improvement by the addition of chitin and chitosan was attributed to strong interfacial adhesion through acid–base interactions between the chlorine-containing PVC and the amino groups on the wood surface treated with chitin and chitosan. The optimum amount of chitin applied was 6.67 wt%. With up to 0.5 wt% of chitosan, a significant improvement in composite properties was attained. This difference in the amount applied was explained by the more reactive amino groups of chitosan over the less reactive acetylamino groups of chitin.[119]

8.3.2.4 Zein

Zein is an alcohol-soluble natural protein extracted from corn. Zein is microbial resistant and possesses the benefits of being renewable and biodegradable. The high molecular weight amphiphilic molecule contains an abundance of carboxyl, amino, phenyl and methyl groups. In 1938, zein became commercially available and quickly found applications in coatings, films, plastics, adhesives, inks and fibres. Zein in combination with vegetable oils and glycerine as plasticizers was used as a waxing or glaze, in order to enhance the shelf life of pharmaceutical tablets, nuts and candies by acting as a water and oxygen barrier. Zein has been used as a coupling agent for flax/PLA composites. The polar groups, *i.e.* carboxyl and amino groups, of the zein protein react with the hydroxyl groups of the fibres through hydrogen bonding, while aryl and alkyl groups react with alkyl groups of PLA through hydrophobic interaction.[120,121] The improvement of the storage modulus in flax/PP composites was due to enhanced interfacial addition from the addition of zein coupling agents.[122] The composites of kenaf fibre and PP with the addition of zein coupling agents also possess improved mechanical and viscoelastic properties.[120]

8.3.2.5 Fatty Acid Derivatives

Corrales *et al.* reported the use of oleoyl chloride, a fatty acid derivative, as a coupling agent for cellulose/polymer composites. Oleoyl chloride is able to react with cellulosic hydroxyl groups by an esterification reaction to form an ester functionality. Esterification is limited to the hydroxyl groups on the external surface of the fibre. The long hydrocarbon chain of oleoyl chloride induces hydrophobicity on the fibre surface and can improve the compatibility between the treated fibre and polymer matrix.[123]

8.3.2.6 Shellac

Shellac, which is a natural, biodegradable, renewable resource product made by bugs, is known to be compatible with cellulose materials like jute, wood, *etc.* Recently, jute/fibre-glass reinforced sheets were prepared with shellac. Their flexural strength values were found to be in the range of those reinforced sheets prepared using fibre glass alone. Flame retardance and thermal resistance of the jute/fibre-glass reinforced sheets were improved with shellac. Lamination of plywood and particleboard was also investigated with jute/fibre glass and shellac, and improved mechanical properties were obtained.[124] Shellac can be used as a matrix to prepare biocomposites made of natural fibre (jute) using biodegradable coupling agents (urea) to ensure superior mechanical properties.[125]

8.3.2.7 Fungal Modifications

A fungus can selectively degrade non-cellulosic compounds, particularly lignin, at a faster rate than they degrade wood cellulose. This reduction of non-cellulosic compounds is also believed to be responsible for the observed increase in both crystallinity and thermal stability of the fibre. In addition, fungal treatment is suggested to increase the surface roughness of the fibre. Microorganisms of *Ophiostoma ulmi* were used to modify the properties of hemp fibres for composite reinforcement. *O. ulmi* treatment slightly improved the moisture absorption characteristics of natural fibres, which should benefit the durability of the composites. Treated hemp fibres showed improved acid–base characteristics between fibre and resin, which are expected to improve the interfacial adhesion. Good interfacial adhesion will lead to both better performance and better durability of the composites by eliminating voids at the interface, which act as reservoirs for the accumulation of water.[126]

8.3.2.8 Furfuryl Alcohol Modifications

Recently, a new selective chemical modification of the surface of lignocellulosic natural fibres has been considered. This modification is based on the selective oxidation of guaiacyl and syringyl units of lignin, generating *ortho*- and *para*-quinones that are able to react by Diels–Alder reaction with furfuryl alcohol; the latter is commercially prepared by reduction of furfural, which in turn is obtained from agricultural residues.

A process was set for chemically modifying sugar cane bagasse and curaua fibres. A quite specific oxidation by chlorine dioxide of syringyl and guaiacyl phenols of the lignin polymer, creating quinones, was carried out. The latter were reacted with furfuryl alcohol, creating a coating around the fibre more compatible to phenolic resins. This modification favoured the fibre–matrix interaction at the interface for sugar cane and curaua fibre-reinforced composites, but caused some fibre degradation that affected the mechanical properties and decreased the mechanical resistance of the composites reinforced with them.[127] The fibre modified with furfuryl alcohol exhibited degradation of hemicelluloses, but cellulose maintained most of its crystallinity.[128] The chemical modification of coir fibres was also studied through oxidation with chlorine dioxide, followed by reaction with furfuryl alcohol. This surface modification reduced the hydrophilicity of coir fibres, leading to an improvement in their moisture resistance capability. The modified fibres were more compatible with the polymeric matrices.[129]

8.3.2.9 Guar Gum

Guar gum is a natural accumulated hydrocolloid stored in the endosperm of the seeds of the guar plant. The macromolecular structure of guar lies between a spherocolloid (like amylopectin) and a linear hydrocolloid (like cellulose). Like most polysaccharides, guar has two or three free hydroxyl groups on the mannose unit of the main chain or the galactose side chains. These are available for bonding and can be utilized in bonding with the numerous hydroxyl groups present in the jute fibre structure. The use of guar gum as a surface treating agent for jute yarns had an increasing effect on the performance of the treated jute/vinyl ester resin composites. However, the improvement was evident only at an optimum concentration of 0.2%. With a further increase in guar gum concentration (to 0.3%), the high viscosity of the solution might have caused a thicker coating on the jute fibre surface, which resulted in an insufficient penetration of the resin into the fibres, lowering the mechanical properties of the composites.[130]

8.3.2.10 Enzyme Treatments

Some published reports have described the effect of enzymatic degradation of non-cellulosic compounds of hemp fibres.[131] Pectinases degrade pectin and laccases modify lignin. However, waxes and other non-cellulosic compounds can be a barrier to these enzymes. In order to achieve good results, these enzyme treatments have been preceded by pre-treatments, such as chelator treatment with ethylenediaminetetraacetic acid (EDTA).[132] Chelating agents are organic compounds capable of forming covalent bonds with metals through two or more of their atoms. It has been reported that some chelators such as EDTA can remove calcium ions from pectin in plant cell walls such as in hemp fibres, resulting in the pectin becoming soluble in many liquids. This allows the hemp fibres to separate from their bundles. However, EDTA persists in the environment and, due to its strong metal chelating properties, enhances the mobility and bioavailability of contaminant heavy metals. An alternative is ethylenediaminetetramethylenephosphonic acid (EDTMPA), a phosphonated analogue of EDTA. EDTMPA has a very strong interaction with all mineral surfaces,[133] so it is easily removed from technical and natural systems. Owing to this strong adsorption, little or no remobilization of metals occurs. Therefore, compared to EDTA, EDTMPA has less impact on the environment.

Pectinases can be broadly classified into acidic and alkaline, based on their pH requirement for optimum enzymatic activity. Alkaline pectinases, which come mostly from bacterial sources, are capable of degrading pectin in the middle lamella of a fibre cell wall. They are widely used in the separation of bundles in crops such as flax, hemp and jute to obtain fibres.[134] Laccases are one of the most important lignin degrading enzymes. It has been reported that laccase alone cannot depolymerize lignin,[131] but when 1-hydroxybenzotriazole hydrate (HBT) is used as a mediator, a degree of delignification up to 40% has been obtained. The mechanism of laccase–mediator systems involves oxidation of the mediator by laccase, followed by oxidation of lignin by the low molecular weight oxidized mediator, which can diffuse into the structure to oxidize the lignin molecule.

8.4   Pros and Cons of Using Bio-based Coupling Agents

Using natural fibres as reinforcement in composites involves several challenges. The first and the most important problem is the low fibre/matrix adhesion. The role of the matrix in a fibre-reinforced composite is to transfer the load to the stiff fibres through shear stresses at the interface. This process requires good bonding between the polymeric matrix and the fibres. Poor adhesion at the interface means that the full capabilities of the composite cannot be exploited and leaves it vulnerable to environmental attack, thus reducing its life span. Natural fibres are hydrophilic and polar in nature, whereas common matrices are hydrophobic and non-polar. Natural fibres used in matrices are therefore dimensionally unstable and display insufficient adhesion between matrix and fibre, which results in poor composite mechanical properties. It is possible to improve bonding between the fibre and the matrix by modifying the surface of the fibre, or by modifying the matrix with the addition of a coupling agent or bio-based coupling agents.

Because bio-based coupling agents are inexpensive and highly eco-friendly materials, their use in either their raw form or some suitably modified form could be very effective for the composite industry. A fully biodegradable green composite from the combination of natural fibres and biodegradable resins could be achieved by rendering bio-based coupling agents. These green composites could be completely decomposed into innocuous products. Compared with conventional coupling agents, bio-based coupling agents themselves are considered as green chemicals because they are from renewable feedstocks, preserve efficacy of function while reducing toxicity and do not persist in the environment at the end of their use.[1] It is known that a silane coating of glass fibres is extensively used in the glass fibre composite industry for improved bonding between the glass fibres and the polymer resin. However, for natural fibres, such a universal coating material is not yet available that could be used effectively and extensively for specific natural fibres to make them more suitable as reinforcements in polymer composites. The use of inexpensive natural materials, as a coupling agent or surface treating agent of natural fibres, can be a techno-commercially viable one. In contrast, during the raw material storage required for the industrial production of composites, bio-based coupling agents would degrade due to standard bacterial and fungal processes. Assessing the degradation that occurs during storage, such that it could be taken account of in the composite processing, is recommended for future research. For example, in the case of fungal modifications, care must be exercised in choosing the treatment time since longer treatment times can give the enzymes secreted by a fungus an opportunity to penetrate deep into the crystalline region of the cellulose, which may affect the strength of the fibres. Moreover, the concentration of bio-based coupling agents may be optimized, depending on the natural fibre and the nature of the matrix.

8.5   Conclusions

Natural fibres, when used as reinforcement, compete with such synthetic fibres as glass fibre. The advantages of synthetic fibres are their good mechanical properties, while their disadvantage is the difficulty in recycling. By use of proper processing techniques, fibre treatments and coupling agents, several natural fibre composites reach the mechanical properties of glass fibre composites. Natural fibres are renewable, light in weight, low in cost and environmentally friendly. To enhance the interfacial adhesion between the fibre reinforcement and the polymer matrix, different types of natural and synthetic coupling agents have been used in natural fibre/polymer composites. Recently, there has been increasing interest in commercialization of natural fibre composites and their use, *e.g.* in the automobile and furniture industries. However, poor wettability and insufficient adhesion with resins limits the use of natural fibres in high-performance applications. In order to overcome this limitation, bio-based coupling agents like zein, fatty acid derivatives, enzyme treatments, *etc.*, have been successfully used. These bio-based coupling agents are effective and environmentally friendly methods to improve the interfacial bonding between the natural fibre and the matrix. The future trends of natural fibre-reinforced composites are to use bio-based coupling agents and new types of natural fibre which could be compatible with the environment and widen the applications of natural fibre composites, such as scaffolds for tissue engineering applications.

Acknowledgement

We gratefully acknowledge Kasetsart University Research and Development Institute (KURDI) and KAPI, Kasetsart University in Thailand, for their financial support.

References

1. M. J. John and S. Thomas, *Carbohydr. Polym.*, 2008, **71**, 343–364.
2. J. K. Pandey, S. H. Ahn, C. S. Lee, A. K. Mohanty and M. Misra, *Macromol. Mater. Eng.*, 2010, **295**, 975–989.
3. J. Ganster and H.-P. Fink, *Cellulose*, 2006, **13**, 271–280.
4. D. N. Saheb and J. P. Jog, *Adv. Polym. Tech.*, 1999, **18**, 351–363.
5. E. Franco-Marquès, J. A. Méndez, M. A. Pèlach, F. Vilaseca, J. Bayer and P. Mutjé, *Chem. Eng. J.*, 2011, **166**, 1170–1178.
6. R. Chollakup, W. Smithipong and K. Sriroth, *Green Biomaterials*, Manus Film, Bangkok, 2009.

7. A. K. Mohanty, M. A. Khan, S. Sahoo and G. Hinrichsen, *J. Mater. Sci.*, 2000, **35**, 2589–2595.
8. M. Hetzer and D. De Kee, *Chem. Eng. Res. Des.*, 2008, **86**, 1083–1093.
9. N. Reddy and Y. Yang, *Trends Biotechnol.*, 2005, **23**, 22–27.
10. J. George, M. S. Sreekala and S. Thomas, *Polym. Eng. Sci.*, 2001, **41**, 1471–1485.
11. F. Vilaplana, E. Strömberg and S. Karlsson, *Polym. Degrad. Stab.*, 2010, **95**, 2147–2161.
12. S. T. Peters, *Handbook of Composites*, Springer, Berlin, 2nd edn., 1998.
13. A. K. Mohanty, M. Misra and L. T. Drzal, *Natural Fibers, Biopolymers, and Biocomposites*, CRC, Boca Raton, FL, 2005.
14. A. K. Bledzki and J. Gassan, *Prog. Polym. Sci.*, 1999, **24**, 221–274.
15. S. Mishra, A. K. Mohanty, L. T. Drzal, M. Misra and G. Hinrichsen, *Macromol. Mater. Eng.*, 2004, **289**, 955–974.
16. M. Pommet, A. Redl, M.-H. Morel and S. Guilbert, *Polymer.*, 2003, **44**, 115–122.
17. L. Y. Mwaikambo and M. P. Ansell, *Compos. Sci. Technol.*, 2003, **63**, 1297–1305.
18. T. Paramasivan and A. P. J. A. Kalam, *Fibre Sci. Technol.*, 1974, **7**, 85.
19. A. K. Mohanty and M. Misra, *Polym.-Plast. Technol. Eng.*, 1995, **34**, 729–792.
20. M. S. Sreekala, S. Thomas and G. Groeninckx, *Polym. Compos.*, 2005, **26**, 388–400.
21. P. Threepopnatkul, N. Kaerkitcha and N. Athipongarporn, *Composites, Part B*, 2009, **40**, 628–632.
22. R. Chollakup, W. Smitthipong, W. Kongtud and R. Tantatherdtam, *J. Adhes. Sci. Technol.*, 2012, in press.
23. J. George, K. Joseph, S. S. Bhagawan and S. Thomas, *Mater. Lett.*, 1993, **18**, 163–170.
24. J. George, S. S. Bhagawan, N. Prabhakaran and S. Thomas, *J. Appl. Polym. Sci.*, 1995, **57**, 843–854.
25. R. Chollakup, R. Tantatherdtam, S. Ujjin and K. Sriroth, *J. Appl. Polym. Sci.*, 2011, **119**, 1952–1960.
26. K. Joseph, S. Thomas, C. Pavithran and M. Brahmakumar, *J. Appl. Polym. Sci.*, 1993, **47**, 1731–1739.
27. K. Joseph, S. Thomas and C. Pavithran, *Polymer*, 1996, **37**, 5139–5149.
28. A. Paul and S. Thomas, *J. Appl. Polym. Sci.*, 1997, **63**, 247–266.
29. U. Hujuri, S. K. Chattopadhay, R. Uppaluri and A. K. Ghoshal, *J. Appl. Polym. Sci.*, 2008, **107**, 1507–1516.
30. B. Wirjosentono, P. Guritno and H. Ismail, *Int. J. Polym. Mater.*, 2004, **53**, 295–306.
31. H. D. Rozman, C. Y. Lai, H. Ismail and Z. A. M. Ishak, *Polym. Int.*, 2000, **49**, 1273–1278.
32. J. George, S. Thomas and S. S. Bhagawan, *J. Thermoplast. Compos. Mater.*, 1999, **12**, 443–464.

33. P. V. Joseph, K. Joseph and S. Thomas, *Compos. Sci. Technol.*, 1999, **59**, 1625–1640.
34. P. V. Joseph, K. Joseph, S. Thomas, C. K. S. Pillai, V. S. Prasad, G. Groeninckx and M. Sarkissova, *Composites, Part A*, 2003, **34**, 253–266.
35. S. Zakaria and L. K. Poh, *Polym.-Plast. Technol.*, 2002, **41**, 951–962.
36. K. C. Manikandan Nair, S. Thomas and G. Groeninckx, *Compos. Sci. Technol.*, 2001, **61**, 2519–2529.
37. S. Joseph, M. S. Sreekala, Z. Oommen, P. Koshy and S. Thomas, *Compos. Sci. Technol.*, 2002, **62**, 1857–1868.
38. A. V. R. Prasad, K. M. M. Rao, K. M. Rao and A. V. S. S. K. S. Gupta, *Indian J. Fibre Text. Res.*, 2007, **32**, 399–403.
39. S.-H. Lee and S. Wang, *Composites, Part A*, 2006, **37**, 80–91.
40. F. Le Digabel, N. Boquillon, P. Dole, B. Monties and L. Averous, *J. Appl. Polym. Sci.*, 2004, **93**, 428–436.
41. M. Mičušík, M. Omastová, J. Prokeš and I. Krupa, *J. Appl. Polym. Sci.*, 2006, **101**, 133–142.
42. S. Mishra, S. S. Tripathy, M. Misra, A. K. Mohanty and S. K. Nayak, *J. Reinf. Plast. Compos.*, 2002, **21**, 55–70.
43. L. A. Pothan, Z. Oommen and S. Thomas, *Compos. Sci. Technol.*, 2003, **63**, 283–293.
44. R. A. Shanks, A. Hodzic and S. Wong, *J. Appl. Polym. Sci.*, 2004, **91**, 2114–2121.
45. S. Luo and A. N. Netravali, *Polym. Compos.*, 1999, **20**, 367–378.
46. S. Luo and A. N. Netravali, *J. Mater. Sci.*, 1999, **34**, 3709–3719.
47. R. Tokoro, D. Vu, K. Okubo, T. Tanaka, T. Fujii and T. Fujiura, *J. Mater. Sci.*, 2008, **43**, 775–787.
48. N. Graupner, *J. Mater. Sci.*, 2008, **43**, 5222–5229.
49. B. Bax and J. Müssig, *Compos. Sci. Technol.*, 2008, **68**, 1601–1607.
50. E. Bodros, I. Pillin, N. Montrelay and C. Baley, *Compos. Sci. Technol.*, 2007, **67**, 462–470.
51. R. Hu and J.-K. Lim, *J. Compos. Mater.*, 2007, **41**, 1655–1669.
52. R. Masirek, Z. Kulinski, D. Chionna, E. Piorkowska and M. Pracella, *J. Appl. Polym. Sci.*, 2007, **105**, 255–268.
53. M. S. Huda, L. T. Drzal, A. K. Mohanty and M. Misra, *Compos. Sci. Technol.*, 2008, **68**, 424–432.
54. M. Avella, G. Bogoeva-Gaceva, A. Bužarovska and M. E. Errico, G. Gentile and A. Grozdanov, *J. Appl. Polym. Sci.*, 2008, **108**, 3542–3551.
55. M. García, I. Garmendia and J. García, *J. Appl. Polym. Sci.*, 2008, **107**, 2994–3004.
56. M. S. Huda, L. T. Drzal, M. Misra and A. K. Mohanty, *J. Appl. Polym. Sci.*, 2006, **102**, 4856–4869.
57. H.-Y. Cheung, K.-T. Lau, X.-M. Tao and D. Hui, *Composites, Part B*, 2008, **39**, 1026–1033.
58. E. Grillo Fernandes, P. Cinelli and E. Chiellini, *Macromol. Symp.*, 2004, **218**, 231–240.

59. S. H. Imam, P. Cinelli, S. H. Gordon and E. Chiellini, *J. Polym. Environ.*, 2005, **13**, 47–55.
60. G. Mehta, A. K. Mohanty, M. Misra and L. T. Drzal, *Green Chem.*, 2004, **6**, 254–258.
61. G. Romhány, J. Karger-Kocsis and T. Czigány, *Macromol. Mater. Eng.*, 2003, **288**, 699–707.
62. V. A. Alvarez, A. N. Fraga and A. Vázquez, *J. Appl. Polym. Sci.*, 2004, **91**, 4007–4016.
63. A. A. S. Curvelo, A. J. F. de Carvalho and J. A. M. Agnelli, *Carbohydr. Polym.*, 2001, **45**, 183–188.
64. L. Averous and N. Boquillon, *Carbohydr. Polym.*, 2004, **56**, 111–122.
65. P. Lodha and A. N. Netravali, *Compos. Sci. Technol.*, 2005, **65**, 1211–1225.
66. M. D. H. Beg, K. L. Pickering and S. J. Weal, *Mater. Sci. Eng., A*, 2005, **412**, 7–11.
67. S. Chabba, G. F. Matthews and A. N. Netravali, *Green Chem.*, 2005, **7**, 576–581.
68. G. I. Williams and R. P. Wool, *Appl. Compos. Mater.*, 2000, **7**, 421–432.
69. A. J. Kinloch, *Adhesion and Adhesives: Science and Technology*, Chapman & Hall, London, 1987.
70. *Fundamentals of Adhesion*, ed. L. H. Lee, Plenum, New York, 1991.
71. *Adhesion Promotion Techniques: Technological Applications*, ed. K. L. Mittal and A. Pizzi, Dekker, New York, 1999.
72. W. Smitthipong, M. Nardin, J. Schultz and K. Suchiva, *Int. J. Adhes. Adhes.*, 2009, **29**, 253–258.
73. H. S. L.H. Sharpe, *Chem. Eng. News*, 1963, **15**, 67.
74. W. Smitthipong, M. Nardin, J. Schultz and K. Suchiva, *Int. J. Adhes. Adhes.*, 2007, **27**, 352–357.
75. W. Smitthipong, T. Neumann, S. Gajria, Y. Li, A. Chworos, L. Jaeger and M. Tirrell, *Biomacromolecules*, 2008, **10**, 221–228.
76. R. Chollakup, W. Smitthipong, C. D. Eisenbach and M. Tirrell, *Macromolecules*, 2010, **43**, 2518–2528.
77. M. Brogly, M. Nardin and J. Schultz, *Macromol. Symp.*, 1997, **119**, 89–100.
78. S. G. M.-F. Vallat and A. Coupard, *Rubber Chem. Technol.*, 1999, **72**, 701–712.
79. A. A. Roche, J. Bouchet and S. Bentadjine, *Int. J. Adhes. Adhes.*, 2002, **22**, 431–441.
80. S. S. Voyutskii, *Autohesion and Adhesion of High Polymers*, Interscience, New York, 1963.
81. W. Smitthipong, M. Nardin, J. Schultz, T. Nipithakul and K. Suchiva, *J. Adhes. Sci. Technol.*, 2004, **18**, 1449–1463.
82. P. G. de Gennes, *Scaling Concepts in Polymer Physics*, Cornell University Press, New York, 1979.
83. J. Lu, Q. Wu and H. McNabb, *Wood Fiber Sci.*, 2000, **32**, 88–104.

84. Y. Xie, C. A. S. Hill, Z. Xiao, H. Militz and C. Mai, *Composites, Part A*, 2010, **41**, 806–819.
85. X. Li, L. Tabil and S. Panigrahi, *J. Polym. Environ.*, 2007, **15**, 25–33.
86. M. Castellano, A. Gandini, P. Fabbri and M. N. Belgacem, *J. Colloid Interface Sci.*, 2004, **273**, 505–511.
87. S. Taj, M. A. Munawar and S. Khan, *Proc. Pak. Acad. Sci.*, 2007, **44**, 129–144.
88. J. George, S. S. Bhagawan and S. Thomas, *Compos. Sci. Technol.*, 1997, **58**, 1471–1485.
89. B.-D. Park, S. G. Wi, K. H. Lee, A. P. Singh, T.-H. Yoon and Y. S. Kim, *Biomass Bioenergy*, 2004, **27**, 353–363.
90. M. Abdelmouleh, S. Boufi, M. N. Belgacem and A. Dufresne, *Compos. Sci. Technol.*, 2007, **67**, 1627–1639.
91. M.-C. Brochier Salon, M. Abdelmouleh, S. Boufi, M. N. Belgacem and A. Gandini, *J. Colloid Interface Sci.*, 2005, **289**, 249–261.
92. L. A. Pothan and S. Thomas, *Compos. Sci. Technol.*, 2003, **63**, 1231–1240.
93. D. Maldas, B. V. Kokta and C. Daneault, *J. Appl. Polym. Sci.*, 1989, **37**, 751–775.
94. P. J. Herrera-Franco and A. Valadez-González, *Composites, Part B*, 2005, **36**, 597–608.
95. A. Valadez-Gonzalez, J. M. Cervantes-Uc, R. Olayo and P. J. Herrera-Franco, *Composites, Part B*, 1999, **30**, 321–331.
96. M. Bengtsson, N. M. Stark and K. Oksman, *Compos. Sci. Technol.*, 2007, **67**, 2728–2738.
97. M. Z. Rong, M. Q. Zhang, Y. Liu, G. C. Yang and H. M. Zeng, *Compos. Sci. Technol.*, 2001, **61**, 1437–1447.
98. L. M. Matuana, R. T. Woodhams, J. J. Balatinecz and C. B. Park, *Polym. Compos.*, 1998, **19**, 446–455.
99. J. Gironès, M. T. B. Pimenta, F. Vilaseca, A. J. F. de Carvalho, P. Mutjé and A. A. S. Curvelo, *Carbohydr. Polym.*, 2007, **68**, 537–543.
100. W. Qiu, F. Zhang, T. Endo and T. Hirotsu, *J. Mater. Sci.*, 2005, **40**, 3607–3614.
101. H. Wang, X. Sun and P. Seib, *J. Appl. Polym. Sci.*, 2001, **82**, 1761–1767.
102. H. Wang, X. Sun and P. Seib, *J. Appl. Polym. Sci.*, 2002, **84**, 1257–1262.
103. L. S. Nair and C. T. Laurencin, *Prog. Polym. Sci.*, 2007, **32**, 762–798.
104. V. R. Botaro and A. Gandini, *Cellulose*, 1998, **5**, 65–78.
105. J. M. Felix and P. Gatenholm, *J. Appl. Polym. Sci.*, 1991, **42**, 609–620.
106. J. Z. Lu, I. I. Negulescu and Q. Wu, *Compos. Interfaces*, 2005, **12**, 125–140.
107. M. Kazayawoko, J. J. Balatinecz, R. T. Woodhams and S. Law, *J. Reinf. Plast. Compos.*, 1997, **16**, 1383–1406.
108. A. R. Sanadi, D. F. Caulfield, R. E. Jacobson and R. M. Rowell, *Ind. Eng. Chem. Res.*, 1995, **34**, 1889–1896.
109. J. Gassan and A. K. Bledzki, *Composites, Part A*, 1997, **28**, 1001–1005.

110. S. Mohanty, S. K. Nayak, S. K. Verma and S. S. Tripathy, *J. Reinf. Plast. Compos.*, 2004, **23**, 625–637.
111. N. Chand and U. K. Dwivedi, *Wear*, 2006, **261**, 1057–1063.
112. P. Zadorecki and P. Flodin, *J. Appl. Polym. Sci.*, 1985, **30**, 2419–2429.
113. P. Zadorecki and P. Flodin, *J. Appl. Polym. Sci.*, 1986, **31**, 1699–1707.
114. P. Zadorecki and T. Rönnhult, *J. Polym. Sci., Part A: Polym. Chem.*, 1986, **24**, 737–745.
115. C. Joly, R. Gauthier and M. Escoubes, *J. Appl. Polym. Sci.*, 1996, **61**, 57–69.
116. A. Paul, K. Joseph and S. Thomas, *Compos. Sci. Technol.*, 1997, **57**, 67–79.
117. T. Ohkita and S.-H. Lee, *J. Appl. Polym. Sci.*, 2005, **97**, 1107–1114.
118. T. Ohkita and S.-H. Lee, *J. Adhes. Sci. Technol.*, 2004, **18**, 905–924.
119. B. L. Shah, L. M. Matuana and P. A. Heiden, *J. Vinyl. Addit. Technol.*, 2005, **11**, 160–165.
120. M. J. John, C. Bellmann and R. D. Anandjiwala, *Carbohydr. Polym.*, 2010, **82**, 549–554.
121. R. Kumar, M. K. Yakabu and R. D. Anandjiwala, *Composites, Part A*, 2010, **41**, 1620–1627.
122. M. J. John and R. D. Anandjiwala, *Composites, Part A*, 2009, **40**, 442–448.
123. F. Corrales, F. Vilaseca, M. Llop, J. Gironès, J. A. Méndez and P. Mutjè, *J. Hazard. Mater.*, 2007, **144**, 730–735.
124. D. N. Goswami, M. F. Ansari, A. Day, N. Prasad and B. Baboo, *Indian J. Chem. Technol.*, 2008, **15**, 325–331.
125. M. Khan, S. Ghoshal, R. Khan, S.-A. Pervin and A. Mustafa, *Chem. Chem. Technol.*, 2008, **2**, 231–234.
126. D. Gulati and M. Sain, *J. Polym. Environ.*, 2006, **14**, 347–352.
127. W. G. Trindade, W. Hoareau, J. D. Megiatto and I. A. T. Razera, A. Castellan and E. Frollini, *Biomacromolecules*, 2005, **6**, 2485–2496.
128. W. G. Trindade, W. Hoareau, I. A. T. Razera, R. Ruggiero and E. Frollini, and A. Castellan, *Macromol. Mater. Eng.*, 2004, **289**, 728–736.
129. S. K. Saw, G. Sarkhel and A. Choudhury, *Appl. Surf. Sci.*, **257**, 3763–3769.
130. D. Ray, A. K. Rana, N. R. Bose and S. P. Sengupta, *J. Appl. Polym. Sci.*, 2005, **98**, 557–563.
131. S. Camarero, O. García, T. Vidal, J. Colom, J. C. del Río, A. Gutiérrez, J. M. Gras, R. Monje, M. J. Martínez and Á. T. Martínez, *Enzyme Microb. Technol.*, 2004, **35**, 113–120.
132. T. Stuart, Q. Liu, M. Hughes, R. McCall, H. S. S. Sharma and A. Norton, *Composites, Part A*, 2006, **37**, 393–404.
133. B. Nowack, *Water Res.*, 2003, **37**, 2533–2546.
134. D. R. Kashyap, P. K. Vohra, S. Chopra and R. Tewari, *Bioresour. Technol.*, 2001, **77**, 215–227.

CHAPTER 9

Probing Interfacial Interactions in Natural Fibre Reinforced Biocomposites Using Colloidal Force Microscopy

G. RAJ,* E. BALNOIS,* C. BALEY AND Y. GROHENS

Université de Bretagne Sud (UBS-Ueb), LIMATB (Laboratoire d'Ingénierie des MATériaux de Bretagne), rue de St Maudé, 56321 Lorient, France
*Email: raj.gijo@gmail.com; eric.balnois@univ-brest.fr

9.1 Introduction

Growing environmental concerns have led recent research to focus on the exploitation of natural fibres as reinforcements in composite materials. Natural fibres such as flax, hemp, jute, sisal, and coir have already been used as reinforcements in thermoplastic composites for applications in various industries, mainly the automotive,[1] construction,[2] and packaging industries. Certain natural fibres, such as flax fibres, have demonstrated specific strength properties close to that of glass fibres and can potentially replace them in fibre-reinforced plastics (FRP).[3] Recently, new "eco-composites" or biocomposites with interesting mechanical properties have been developed from natural fibre reinforced biodegradable polymers such as poly(lactic acid) (PLA).[4] PLA has attracted much attention, not only because of its biodegradability and

RSC Green Chemistry No. 16
Natural Polymers, Volume 1: Composites
Edited by Maya J John and Thomas Sabu
© The Royal Society of Chemistry 2012
Published by the Royal Society of Chemistry, www.rsc.org

biocompatibility, but also because its price has become competitive due to the continuous increasing price of petroleum products.

Even though natural fibres present several advantages in terms of low cost, low density, acoustics, and biodegradability properties, their efficiency as reinforcement elements in composite materials depends on several key parameters. Firstly, the intrinsic strength of the fibres, which in turn is determined by the complex hierarchical organization of several polysaccharides that constitute the fibre's internal structure;[5,6] and secondly, the interfacial adhesion between the natural fibre and the polymer matrix. The last parameter plays an important role in efficient stress-transfer at the fibre/matrix interface region, and a clear understanding of this adhesion phenomenon is required to develop high-performance biocomposites suitable for structural applications.

Many attempts have been made to characterize interface adhesion in natural fibre reinforced composite materials. In most studies, an improvement in macroscopic material properties, e.g. tensile strength,[7–9] water absorption properties,[10,11] were attributed often indirectly to an improvement in fibre/matrix interface bonding. More recently, the adherence or "practical adhesion" between vegetal fibres and polymer matrix in biocomposite materials was analysed by estimating interfacial shear stress (IFSS) values obtained from different micromechanical tests, such as the pull-out test,[12] microbonding test,[13] and single fibre fragmentation tests.[14] For example, Arbelaiz et al.[7] used pull-out tests on the flax/polypropylene system to investigate the efficiency of their fibre surface treatments. The authors mentioned that the lack of homogeneity makes the interpretation of these tests difficult and reported relatively weak increases of IFSS values.

Nardin et al.[15] proposed a semi-empirical linear relationship between IFSS and the reversible work of adhesion established at the interface of model composite materials involving glass or carbon fibres in thermoplastic or thermosetting matrices. A remarkable feature of this relation is that, apart from the mechanical properties of the materials involved, the IFSS (a mechanical quantity) was completely dependent on the level of molecular interactions or the work of adhesion (a thermodynamic quantity) at the interface. In the case of biocomposites, a direct application of the Nardin–Schultz relation was difficult, as shown by earlier studies,[13] and was attributed to the complex composite structure of flax fibre. It should be noted that the work of adhesion estimated in these studies was obtained from the surface energy values based on the contact angle of different liquid probes on flax fibres.[13] Given the rough, heterogeneous, and porous build-up of flax fibres, the work of adhesion obtained from the contact angle measurements lacks homogeneity, and should be treated with caution. Complementary techniques are thus essential for a better understanding of the complex interactions between natural fibres and polymer matrix in a biocomposite.

Colloidal force microscopy (CFM) has emerged as a direct technique to quantitatively estimate intermolecular forces between two approaching surfaces by measuring force–distance curves.[16,17] The technique was, for example, extensively used to study the interaction between cellulose surfaces in the pulp and paper industries.[18–24] Most of the experiments were mainly focused on the preparation of ultra-flat cellulose surfaces and smooth, spherical, and non-porous cellulose microbeads in order to mimic a sphere–plane interaction. These experiments were performed in liquid or dilute electrolyte concentrations which essentially simulate the manufacturing conditions involved in the pulp and paper production. This approach forwarded a better understanding of the interactions between various materials and additives in paper making.

Recently, atomic force microscopy (AFM) was found to be a valuable complementary technique to characterize surface properties of natural fibres at the nanoscale.[25,26] It provides direct adhesion force measurements between a standard silicon tip and the fibre[25,27,28] or even directly between the polymer matrix and cellulose.[29] The quantitative interpretation of adhesion force measurements in AFM is nonetheless limited to a precise knowledge of the contact area between the two objects, and working with rough and hetero-geneous fibres remains a problem. In this work, we propose to decompose and study some of the main adhesion forces present in biocomposites. Molecularly smooth model surfaces, which mimic the three major polysaccharides in flax fibres, were prepared and colloidal force measurements between these surfaces and a PLA bead, assuming sphere–plane contact geometry, were performed and compared. Figure 9.1 represents a schematic description of the experi-mental set-up developed that allows us to gain information on some important adhesion mechanisms at the PLA/flax biocomposite interface.[30]

Force measurements were first performed under ambient conditions and recorded under various relative humidity conditions. From the AFM pull-off force values, the work of adhesion for the different interaction systems at the interface region were estimated and used to verify the Nardin–Schultz rela-tion[15] for the flax/PLA biocomposite.

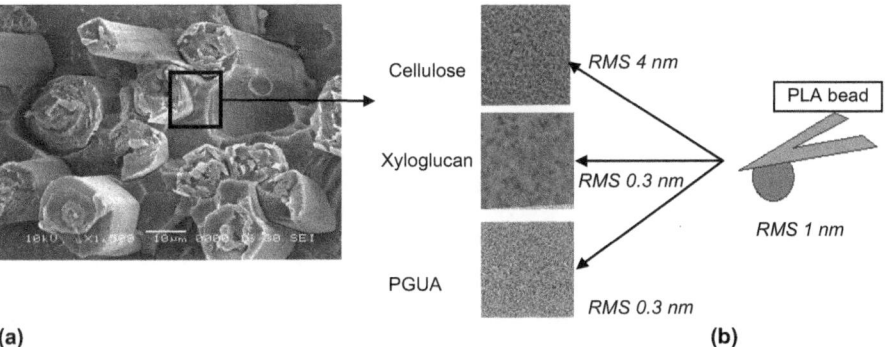

(a)   **(b)**

Figure 9.1   (a) SEM image of the flax/PLA biocomposite; (b) schematic representa-tion of the different systems "layer model" studied to probe interfacial interactions in the biocomposite using model surfaces and colloidal probes. (Reproduced with permission of Springer Science + Business Media from Raj *et al.*[30])

9.2 Materials and Techniques

9.2.1 Preparation of Cellulose Thin Films

One gram of microcrystalline cellulose powder ($M_W = 70000 \, \mathrm{g \, mol^{-1}}$) was suspended in 50 mL of Milli-Q water for 12 h. This step was followed by two exchanges for 45 min each with 25 mL of methanol, then four exchanges with 25 mL of dimethylacetamide (DMAc). The activation liquid was removed each time by using an aspirator pump for each solvent exchange. The filtered swollen cellulose was dried in a vacuum oven for 24 h at 60 °C. A cellulose solution (0.5%) was prepared by dissolving the activated microcrystalline cellulose in 9% LiCl/DMAc solvent system.[31] A drop of clear cellulose solution was spin-coated onto a freshly cleaved mica substrate at 3000 rpm for 1 min. The films were then dried in an oven at 160 °C for 15 min and washed in Milli-Q water to remove the excess LiCl on the surface and finally dried in a vacuum oven at 100 °C for 15 min. A thickness of $\sim 150 \, \mathrm{nm}$ was obtained for these films.

9.2.2 Preparation of Xyloglucan Thin Films

Xyloglucan was chosen as a model hemicellulose. Traditionally, hemicelluloses are defined as the alkali soluble materials after the removal of pectic substances from plant cell walls.[32] They are a heterogeneous class of polymers which may contain pentoses (D-xylopyranose, L-arabinofuranose), hexoses (D-glucopyranose, D-mannopyranose, D-galactopyranose), and uronic acids (D-glucuronic acid).[33,34] The main constituents of hemicelluloses[34] are shown in Figure 9.2.

Among the different types of hemicelluloses, xyloglucans (XGs) are the most predominant hemicellulosic polysaccharides in the primary cell wall, and can form up to 1.3% of the flax fibre mass.[35] Xyloglucans have a backbone chain of $(1 \rightarrow 4)$-linked β-D-glucopyranose units with a substitution of D-xylopyranose at position 6. Xyloglucans are classified as XXXG-type or XXGG-type,

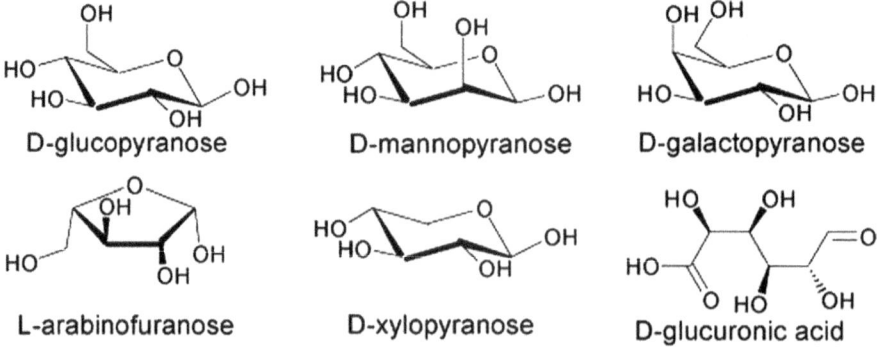

Figure 9.2 The main constituents of hemicelluloses. (Adapted with permission of the American Chemical Society from Hansen and Plackett.[34])

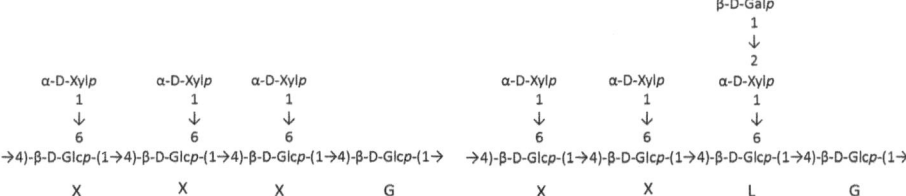

Figure 9.3 Chemical structure of the different repeating units of xyloglucan. Glc*p* = glucopyranose, Xyl*p* = xylopyranose, and Gal*p* = galactopyranose units.

depending on the distribution of xylopyranose side-chain units. The XXXGs have three consecutive backbone residues that are substituted with xylopyranose units and a fourth unbranched backbone residue, while the XXGG xyloglucans have two consecutive branched backbone residues and two unbranched backbone residues. Moreover, additional residues are attached to xylose, depending on the source of xyloglucan. This variation of the structure dominates the functionality and physicochemical properties. For instance, the galactose units substituted onto the xylose dominates the water solubility in the case of xyloglucan extracted from tamarind seed.[36] The structures of different types of xyloglucans[36] are shown in Figure 9.3.

A solution was prepared by dissolving xyloglucan ($M_W = 202\,000\,\mathrm{g\,mol^{-1}}$, $I = 1.8$) in Milli-Q water under continuous magnetic stirring during 12 h to make a $10\,\mathrm{g\,L^{-1}}$ solution. The solution was then filtered using a 0.1 µm membrane. A drop of solution was spin-coated onto freshly cleaved mica substrate at 2500 rpm for 1 min. The film was dried in a vacuum oven at 45 °C for more than 5 h.

9.2.3 Preparation of Pectin Thin Films

Poly(galacturonic acid) (PGUA) was chosen as a model of hydrophilic pectins. Pectin is a family of complex polysaccharides present in all plant primary cell walls. The chemical structure of pectin is shown in Figure 9.4.[37]

The most abundant constituent of pectin in flax fibre is homogalaturonan (HG), which is a (1 → 4) linked α-D-galacturonic acid and its methyl ester.[35] Some homogalacturonans contain α-D-xylose units. Another constituent present in pectin is rhamnogalacturonan I, which contains arabinan, galactan, and arabinogalactan side chains.[38] The PGUA used in this study is a polymer with a backbone chain containing repeating units of α-D-galacturonic acid (Figure 9.5).

PGUA in powder form (extracted from orange) was dissolved at a concentration of $10\,\mathrm{g\,L^{-1}}$ in Milli-Q water under continuous magnetic stirring for 12 h. The solution was filtered (0.1 µm) and spin-coated onto a freshly cleaved mica substrate at 2500 rpm for 1 min. The film was dried in a vacuum oven at 45 °C for more than 5 h.

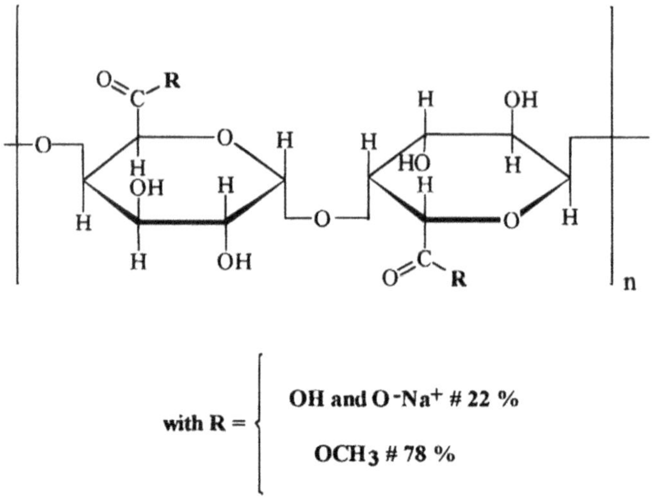

Figure 9.4 Chemical structure of high-methoxy pectin (DM = 78). (Adapted with permission of the American Chemical Society from Fischer *et al.*[37])

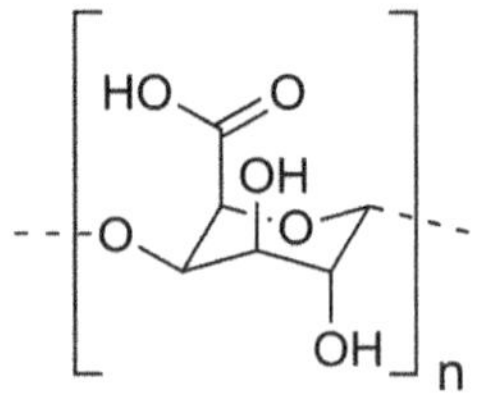

Figure 9.5 Chemical structure of the repeating unit in poly(galacturonic acid) (PGUA).

9.2.4 Preparation of PLA Colloidal Probes

Non-crystalline PLA microbeads (PDLLA) were manufactured by an emulsion–solvent evaporation process.[39] This enantiomer was used to ensure the lowest roughness at the bead surfaces thanks to its amorphous structure. One gram of PLA polymer was completely dissolved in 20 mL of CH_2Cl_2. The PLA/ CH_2Cl_2 solution (5%) was then poured into a special glass vessel containing 4 g of poly(vinyl alcohol) dispersant in 200 mL of Milli-Q water. The system was then continuously stirred for 24 h at room temperature. The PLA microbeads which had formed were then filtered, washed in deionized water, and vacuum dried at 45 °C for 3 days.

Colloidal probes were made by gluing a single microbead of PLA at the extreme end of an AFM cantilever tip. The glue was allowed to cure for 48 h under ambient conditions.

9.2.5 Atomic Force Microscopy

AFM experiments were performed using a Nanoscope IIIa multimode scanning probe microscope (Veeco, USA). Experiments under controlled humidity conditions were performed by placing the AFM inside a homemade glove box equipped with a hygrometer, with an accuracy of ~3%. The temperature inside the glove box was maintained at 25 °C. A relative humidity (RH) of less than 2% was achieved by keeping dried silica gel in the glove box and passing dry nitrogen gas through the glove box. The N_2 flux was stopped and the system was equilibrated prior to any AFM force measurements.

Surface imaging of the different polymer thin films was performed in tapping mode AFM (TM-AFM) under ambient conditions (23 °C and 56% RH). The silicon tapping mode cantilevers were of the LTESP model with a nominal tip radius of approximately 10 nm. Direct force measurements between model surfaces, glued on a magnetic stainless steel disk, and colloidal probes were performed by cycles of approach and retraction of the piezotube. The exact value of the cantilever spring constant k was determined to be 51 N m^{-1} by the thermal noise method that is based on the thermal fluctuation of the cantilever. The velocity of the probe was kept constant at 3.5 μm s^{-1} during all the experiments and the number of data points was set to 1500. The data analysis was done using the Nanoscope software V6.13. To improve the measurement accuracy, statistical analysis of at least 600 force–distance curves at 10 different arbitrary surface sites were performed. The adhesion force measured was then normalized according to Derjaguin's approxima-tion[40] by the radius of curvature of the colloidal probe, determined from SEM, and plotted as frequency distribution curves of normalized adhesion force (N m^{-1}).

9.2.6 Microbonding Test

The adherence strength between the flax fibres and the PLA matrix was estimated by calculating the apparent interfacial shear stress (IFSS) values obtained from microbond tests on a minimum of 10 pull-out experiments. A homogeneous PLA polymer microdroplet (length < 200 μm) was deposited on the surface of the flax fibre. In order to achieve this microdroplet, a microknot was made on the flax fibre with a microfilament of PLA. The set-up was placed in an oven preheated at 190 °C for 10 min. The sample was then immediately taken from the oven and quenched at room temperature. Prior to any tests, the diameter of the fibre near the PLA droplet, the embedded length of the fibre, and the drop height were measured using an optical microscope. Pull-out experiments were performed on a tensile testing machine. On the lower clamp, a homemade X–Y translator with two sharp knife edges was mounted. The microdroplet was brought just under these knife edges and the knife blades were brought close together so that the blades just touched the upper end of the droplet. Tensile loading was applied at the rate of 0.1 mm min^{-1}.

9.3 Results and Discussion

9.3.1 Model Surfaces

Surface force studies require polymer films that are smooth and homogeneous over a few micrometres square to ensure reproducible measurements and to allow for a quantitative determination of the adhesion force using a sphere–plane model. Figure 9.6 presents AFM images of the different polysaccharides thin films.

The root mean square (RMS) roughness of poly(galacturonic acid) and xyloglucan thin films were found to be around 0.5 nm (on a scan size of 25 μm^2), whereas cellulose films present a more important RMS, about 4 nm. The film thicknesses, as determined by AFM, were all around 100–200 nm. A SEM image of the PLA colloidal probe (Figure 9.7) shows that the PLA particle

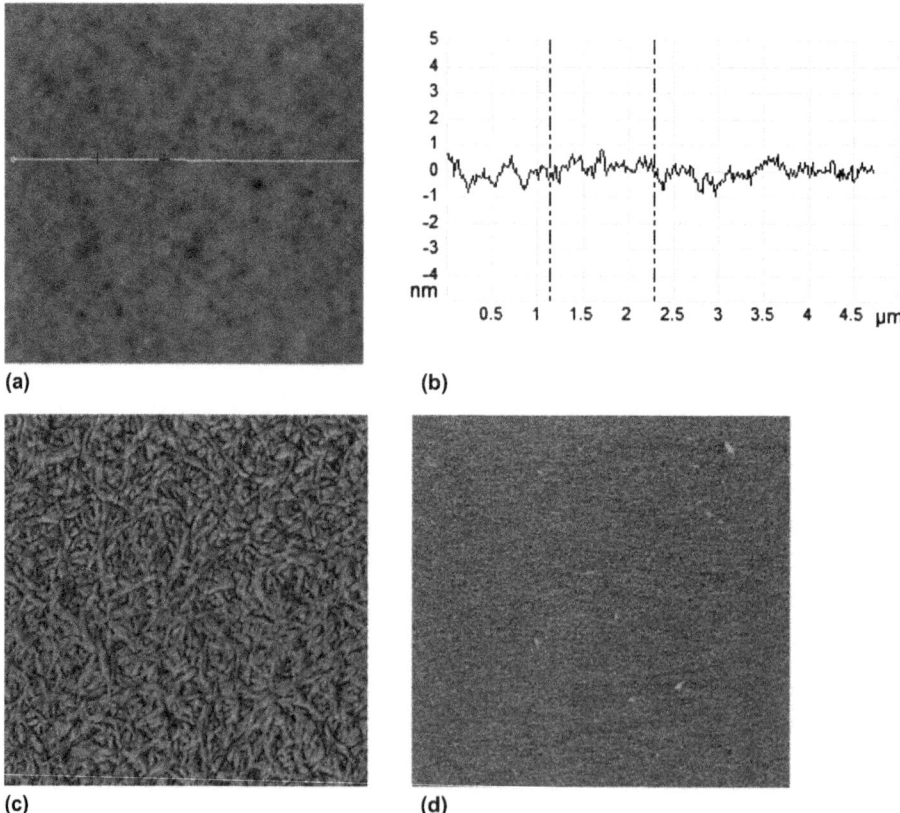

Figure 9.6 TM-AFM images (4 μm^2 scan size) of thin films of different poly-saccharides present in flax fibre: (a) xuloglucan, (b) cross-section analysis of (a), (c) cellulose, and (d) poly(galacturonic acid). (Reproduced with permission of Springer Science + Business Media from Raj *et al.*[30])

Figure 9.7 SEM image of a PLA colloidal probe. The scale bar in the picture is 15 μm. (Reproduced with permission of Springer Science + Business Media from Raj *et al.*[30])

is spherical, with an estimated diameter of $30 \pm 0,5\,\mu m$, non-porous, and smooth.[30]

9.3.2 Colloidal Force Microscopy

The principle of colloidal force microscopy has been described elsewhere.[17] An example of a force–distance curve measured with CFM is presented in Figure 9.8.[30]

In the region of large separation, where the probe and sample do not interact, the cantilever is not deflected by external forces and the resulting force is equal to zero (position A). This part is called the "non-contact region". On the approach stage, as the scanner Z position is increased, the cantilever and the sample start to interact and a cantilever deflection is induced. The attractive interaction between the two objects triggers an instability of the probe, which results in a jump of the probe onto the sample called "jump into contact" (position B). Below this point, for incompressible surfaces a repulsive force appears and linearly increases with the motion of the piezo-stage. This part (position C), which corresponds to a straight line in the deflection (or force) *versus* piezo displacement (or surface distance) curve, is called the "constant compliance region or hard-wall contact". After the Z scan has reached a specified value, the deflection of the cantilever is switched in the opposite direction (attractive). In the absence of plastic deformation, the deflection observed during retraction of the cantilever follows the constant compliance region of the approach (position D). When the position of the cantilever reaches point E, in which the cantilever spring constant overcomes the adhesion force, the

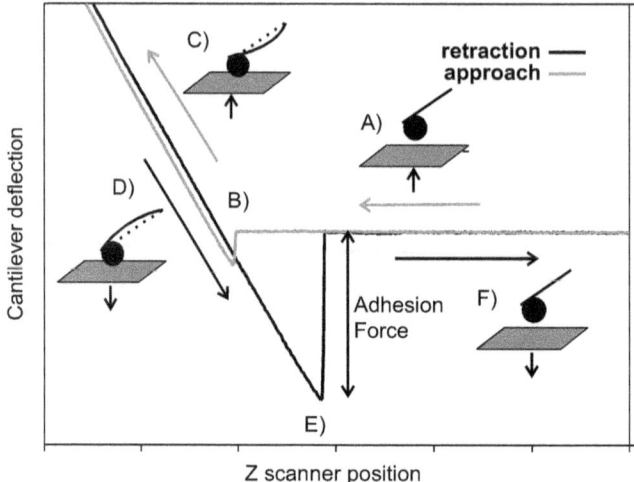

Figure 9.8 Example of force–distance curve showing characteristic features of interactions between two surfaces: (A) non-contact region, (B) jump into
contact, (C) constant compliance region, (D) adhesion event, (E) jump-
out, and (F) return to non-contact region. (Reproduced with permission of
Elsevier from Raj *et al.*[29])

particles and surface separate and the probe "jumps-out" from the surface to
its equilibrium position (position F). This transition is defined as the adhesion
force and can be calculated by simply converting the cantilever deflection to the
corresponding force value according to Hooke's law (eqn 9.1):

$$F = -k \times \Delta x \qquad (9.1)$$

where k is the accurate cantilever spring constant and Δx is the cantilever
deflection. For electrically neutral and chemically non-reactive surfaces,
the resulting adhesion force under ambient conditions is the contribution
from van der Waals and capillary forces. Capillary forces due to the formation of a thin layer of water molecules condensed on the surface can
be eliminated by performing the experiments under controlled humidity
conditions.

9.3.3 Direct Force Measurements

At first, the reproducibility of the method and set-ups developed here was
tested by measuring adhesion forces in different parts of different PGUA thin
films, as shown in Figure 9.9a.

It can be observed that good reproducibility is obtained for all the samples
studied, as proved by the small deviation of the centre of the different histo-
grams. For xyloglucan, similar results were obtained, whereas for cellulose a

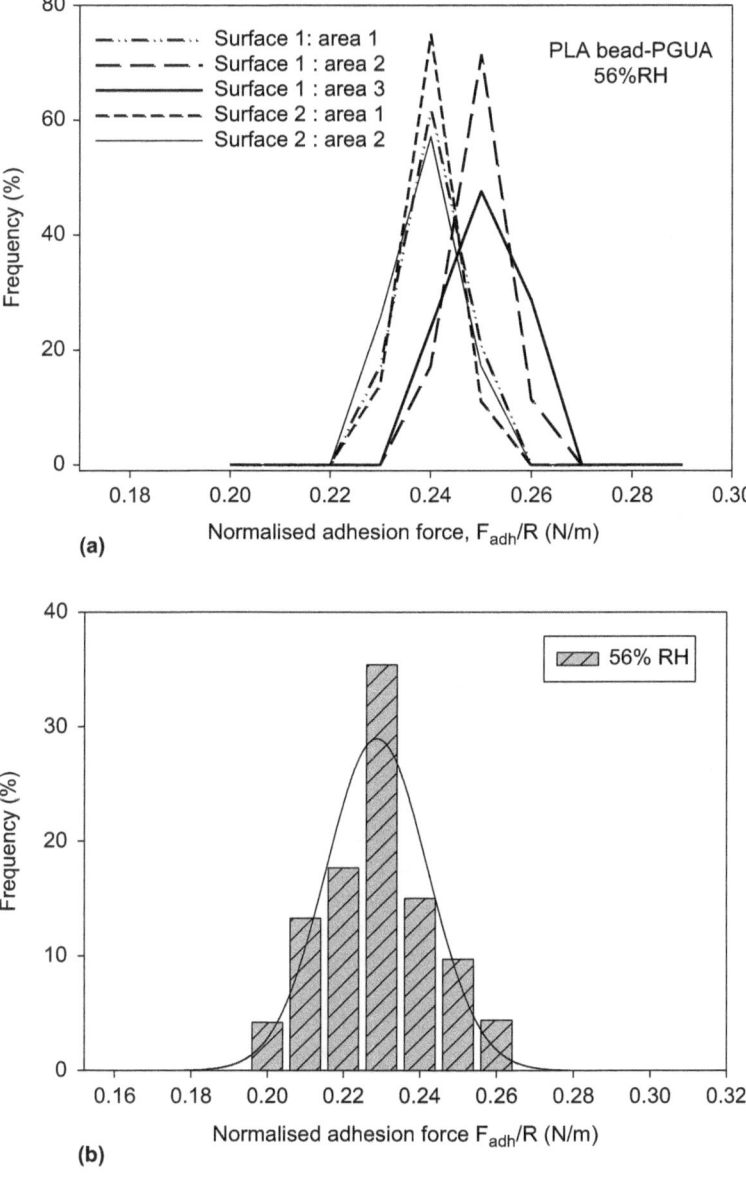

Figure 9.9 (a) Frequency distribution of the normalized adhesion force measurements between a PLA colloidal probe and a PGUA film under ambient
conditions measured on different areas of the PGUA film. (b) Frequency
distribution of normalized adhesion forces between a PLA colloid probe
and a PGUA film at 56% RH ($N = 600$). (Reproduced with permission of
Springer Science + Business Media from Raj *et al.*[30])

Table 9.1 RMS roughness calculated from AFM images of different biopolymers.

	PGUA, RMS (nm)	Xyloglucan, RMS (nm)	Cellulose, RMS (nm)	PLA, RMS (nm)
Thin films (scan area 25 μm²)	0.3	0.3	4	–
Microbeads (scan area 4 μm²)	–	–	–	1.5

broader adhesion force distribution was obtained. This phenomenon was attributed to the higher roughness of cellulose films. Once the reproducibility of the method was verified, a final histogram was plotted from at least 600 single force measurements on each sample (Figure 9.9b). A quantitative comparison of adhesion forces is only possible if the films' roughnesses are comparable, since adhesion force measurements using the colloidal probe technique are known to be largely influenced by the roughness of the surfaces or particles in contact.[17] The RMS roughness of all the surfaces used in this study is shown in Table 9.1.

The roughness parameter, even at the nanometre scale, can affect adhesion forces between particle and substrate by modifying the contact area and increasing the distance between the two objects.[41] In the present study, the same bead was used during all the experiments and regularly controlled to avoid any accidental contamination of the probe.

Figure 9.10a represents the distribution in the normalized adhesion force for the three different systems obtained under ambient conditions (56% RH).

Under such condition, the adhesion force results from the addition of van der Waals forces, H-bonds, and capillary forces.[17,29] We can observe that adhesion between PLA beads and the different surfaces is nearly the same, with peaks centred at around 0.22 N m⁻¹. With respect to the similar RMS characteristics of the films, the tendency is:

$$F_{adh}(PLA/xyloglucan) \approx F_{adh}(PLA/PGUA) \text{ at } 56\% \text{ RH}$$

As previously discussed, the adhesion force obtained for the PLA/cellulose system must be cautiously interpreted because of the difference in roughness with the other films. The results obtained here indicate that PLA interacts in a similar manner with pectins and hemicelluloses, and one might expect that capillary forces play a major role in adhesion between these materials under ambient conditions.

In the presence of humidity, it is well known that a condensed water meniscus between the probe and surface may increase the adhesion force. This additional force, called the capillary force, is related to the wettability of water to the substrate as well as to the probe surface. For hydrophobic surfaces, the thickness of a water film is limited and results in a relatively small capillary

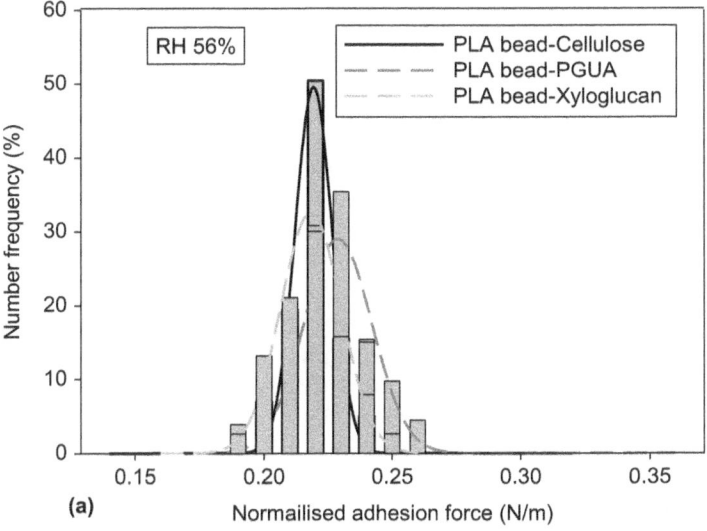

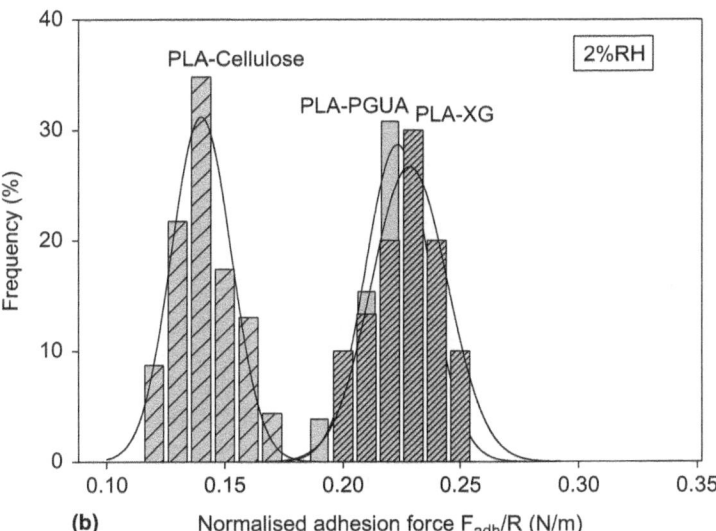

Figure 9.10 Comparison of normalized adhesion forces between a PLA probe and different polysaccharide surfaces such as PGUA, xyloglucan, and cellulose under (a) 56% RH and (b) 2% RH. (Reproduced with permission of Springer Science + Business Media from Raj *et al.*[30])

force. In opposition, if both surfaces in contact are hydrophilic, the water capillary bridge between the two surfaces is large and gives an important additional adhesion force. Experimentally, the contribution of the capillary force can be eliminated, from the overall adhesion force, by operating the AFM in vacuum or in a dry environment. Using such experimental conditions, Burnham *et al.*[42] found a good correlation between the calculated surface energy, derived from experimental adhesion forces measurements in dry air, and theoretical values. Similarly, Raj *et al.*[29] obtained a good agreement between the Hamaker constant for the cellulose/PLA system, as determined from AFM experiments at 2% RH, and the theoretical value as calculated by the Lifshitz theory.

Figure 9.10b shows the effect of RH on pull-off forces for the different systems. When the RH is decreased to 2%, the "PLA/cellulose" adhesion peak shifts to a lower value, whereas in the case of PLA/PGUA and PLA/xyloglucan, no variation is observed.[30] As already reported in the literature, when the RH tends to zero, the adhesion force remains simply ensured by van der Waals forces and H-bonds without capillary forces and therefore decreases.[43,44] Surprisingly, for PGUA and xyloglucan films, no such effect is observed. This failure to observe the humidity dependence has already been reported,[45] but it was specific for hydrophobic surfaces where the amount of water is limited. The roughness of the surface is also an important parameter to consider, since the contact angle of water on a rough surface is different from that on a smooth surface and consequently affects the magnitude of the capillary force.[46] In our case, the main contribution of the capillary force does not come from the difference in roughness between the films but rather in their characteristic surface energies, as revealed by contact angle measurements. Water contact angles on cellulose, xyloglucan, and PGUA films are respectively equal to $40 \pm 2°$, $25 \pm 1°$, and $13 \pm 1°$. Xyloglucan and PGUA are more hydrophilic surfaces and smoother than cellulose films and, as a consequence, are hydrated and retain water molecules even at low RH. Therefore, we believe capillary forces are important in these systems and several assumptions can be proposed to explain the non-decrease of the adhesion force with RH. The first one is based on the remaining water inside PGUA and xyloglucan films, which are known to influence the polymer film structure and properties. We are reminded here of the important role of water molecules in the fibre structure to ensure optimized cohesion between its various entities. Hydrogen bonds are known to be important in vegetal fibres.[47] Moreover, it is well known that water, acting as a plasticizer, affects the glass transition temperature (T_g) of amorphous hydrophilic polymers.[48,49] Water molecules yield a large increase in mobility, due to increased free volume and decreased local viscosity. Water molecules plasticize polysaccharide films and therefore decrease the polymer modulus through the reduction of its glass transition temperature, for example.[50] For instance, an amount of water of $0.1\,\mathrm{g\,L^{-1}}$ will decrease the T_g of pectin from 45 to 0 °C.

In the JKR theory, the contact area (a_c) is given by (eqn 9.2):

$$a_c = \left(\frac{3\pi R^2 W_{\mathrm{adh}}}{2K^*} \right)^{1/3} \qquad (9.2)$$

where R is the radius of the probe, W_{adh} is the work of adhesion, and K^* is the reduced elastic modulus. It can thus be observed that if K decreases, the contact area between the two objects will increase and so consequently will the measured adhesion force. Since water molecules are difficult to remove from polar PGUA and xyloglucan, the measured CFM adhesion force will not decrease from 52% to 2% RH. Some efforts have to be done to measure the water content inside polysaccharides films.

A second assumption is based on the fact that other adhesion mechanisms, independent of the RH, may also be present in the interaction mechanisms between PGUA, xyloglucan, and PLA. The "diffusion theory" of adhesion is accounted for by the time and temperature dependence of the adherence between miscible, soft polymers.[51,52] In practice, the inter-diffusion process requires sufficient chain mobility, which can, of course, be related to the glass transition temperature of the polymers.[53] There is, to the best of our knowledge, no evidence of such a mechanism in biocomposites and further experiments should be conducted to verify this hypothesis.

9.3.4   Interfacial Shear Stress and Work of Adhesion

Interfacial shear stress (IFSS) at the flax/PLA interface was estimated from the microbonding pull-out test using the Kelly–Tyson equation (eqn 9.3):

$$\tau_{i,\text{mean}} = \frac{F_{\max}}{\pi d l_e} \tag{9.3}$$

where $F_{\max}$ is the maximum pull-off force at debonding, d is the fibre diameter and l_e is the fibre embedded length inside the polymer droplet. The above equation assumes that the force $F_{\max}$ at the instant of debonding is predicted to be directly proportional to the joined surface area between the fibre and the matrix, and the droplet shears off from the fibre surface when the average shear stress at the interface, $\tau_{i,\text{mean}}$, becomes large enough to break the interface. The apparent shear stress was determined from the linear regression of the plot of debonding force *versus* bonding area.

The work of adhesion at the flax/PLA interface was estimated, on the one hand, from contact angle measurements according to the Young–Dupré equation (eqn 9.4):

$$W_{adh} = \gamma_l (1 + \cos \theta) \tag{9.4}$$

where γ_l is the interfacial energy of the PLA melt drop that makes a contact angle of θ with flax fibre's surface. Also, the work of adhesion was determined from colloidal force measurements between a PLA bead and different polysaccharide surfaces. The work of adhesion for two surfaces in sphere–plane geometry contact can be estimated according to the Johnson–Kendall–Roberts (JKR) theory (eqn 9.5):[54]

$$\frac{F_{adh}}{R} = \frac{3}{2}\pi W_{adh} \tag{9.5}$$

and the Derjaguin–Muller–Toporov (DMT) theory (eqn 9.6):[55]

$$\frac{F_{adh}}{R} = 2\pi W_{adh} \tag{9.6}$$

The JKR theory considers that adhesion takes place inside the area of contact and can be applied in cases of large tips and soft samples with a large adhesion. JKR theory is thus more suited to model colloid probe microscopy experiments where we use a large tip (colloidal probe) and a soft polymer film. On the other hand, the DMT theory takes into account the adhesion outside the contact area and is applied in cases of small tips and stiff samples with small adhesion.[17]

The Maugis parameter (λ) provides a useful measure of which model is appropriate and is given by eqn (9.7):[56]

$$\lambda = 2.06\sqrt[3]{\frac{R\gamma^2}{\pi E_{tot}^2 D_0^3}} \tag{9.7}$$

where E_{tot} is the reduced elastic modulus of the two surfaces and D_0 is the separation distance between the two materials. In cases where $\lambda > 1$, the JKR model is valid, whereas if $\lambda < 1$, the DMT model is used. In our case, we found the Maugis parameter, λ, close to 100, thus justifying the use of the JKR model.

The work of adhesion, as measured by AFM and contact angle measurements, are shown in Table 9.2.

Table 9.2   Comparison of work of adhesion (W_{adh}) estimated for different systems from contact angle, and CFM experiments.

System	CFM, JKR W_{adh} (56% RH) (mJ m^{-2})	CFM, JKR W_{adh} (2% RH) (mJ m^{-2})	W_{adh} contact angle (mJ m^{-2})
Flax fibre/PLA	–	–	60.78 ($\pm$2.8) ($\theta = 55.6 \pm 4°$)
PLA bead/cellulose film	40.3 ($\pm$3.0)	30.6 ($\pm$1.9)	–
PLA bead/xyloglucan film	42.5 ($\pm$1.7)	42.5 ($\pm$1.7)	–
PLA bead/PGUA film	46.7 ($\pm$1.9)	46.7 ($\pm$1.9)	–

9.3.5   Verifying the Nardin–Schultz Relation for Flax/PLA Biocomposites

The semi-empirical relationship between IFSS (τ) and the work of adhesion (W_{adh}) proposed by Nardin and Schultz[15] is shown in eqn (9.8):

$$\tau = \left(\frac{E_m}{E_f}\right)^{1/2}\frac{W_{adh}}{\delta}\tag{9.8}$$

where E_m and E_f are the elastic modulus of the polymer matrix and the fibres, respectively, and δ is the distance equal to 0.5 nm that corresponds to an equilibrium intermolecular centre-to-centre distance involved in molecular interactions.

We tried to verify the Nardin–Schultz model for flax fibre/PLA biocomposites, using our experimental data. To plot these experimental points, the elastic modulus of the fibre and PLA matrix were taken as $E_f = 65$ GPa[57] and $E_m = 3.6$ GPa,[58] respectively. These data, which represent macroscopic mean values for these materials, are well established in the literature. Figure 9.11 plots the Nardin–Schultz relationship between IFSS and W_{adh} for flax/PLA biocomposites, and compares that with the linear relation experimentally verified for model composite materials.

It can be observed that for flax/PLA composites the work of adhesion calculated from the contact angle technique did not fit the Nardin–Schultz linear

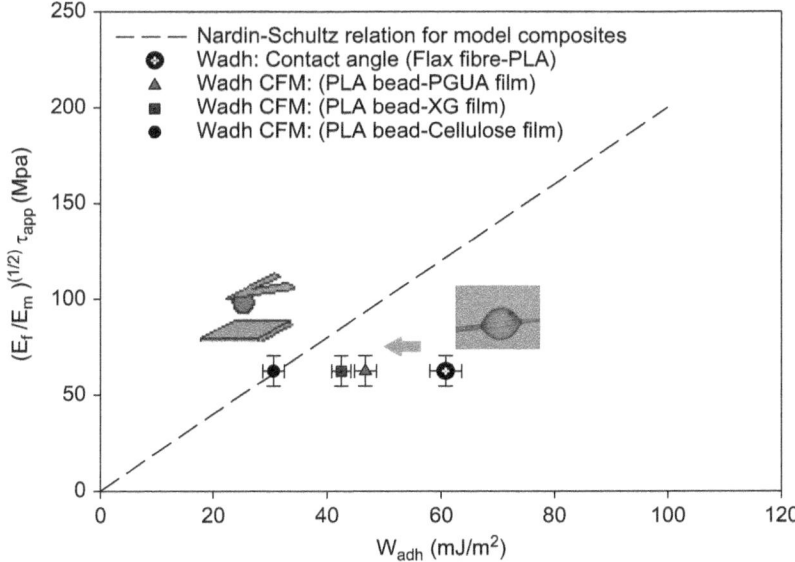

Figure 9.11   The Nardin–Schultz relationship between IFSS and W_{adh} obtained for flax/PLA biocomposites.

relation. However, when the work of adhesion was calculated from more local techniques such as CFM, we observed that the points were placed close to the Nardin–Shultz linear relation. It can be seen that the closest agreement was achieved for the work of adhesion calculated for the PLA/cellulose system. In an earlier work, we have shown the contribution of van der Waals forces for PLA/cellulose interactions by calculating the Hamaker constant from pull-off forces.[29] In the Nardin–Shultz model the work of adhesion between two solids is considered to be the contribution from dispersive interactions (London forces) and electron acceptor–donor interactions (Lewis acid–base interactions).[15] However, in cases where other adhesion mechanisms, *e.g.* covalent bonding, mechanical interlocking, or interdiffusion processes, may take part in interfacial adhesion, then the Nardin–Schultz relation cannot be directly applied. This was clearly evident in the case of the hydrated polysaccharides poly(galacturonic acid) and xyloglucan, where the work of adhesion with PLA was far from simple molecular interactions and involved complex mechanisms such as capillary forces and interdiffusion phenomena (see Section 9.3.3). The deviation from the Nardin–Schultz linear relation in the case of hydrated polysaccharides (Figure 9.11) could be attributed to the complimentary "adhesion mechanisms" that exist at the interface with these materials.

9.4   Conclusion

Colloid force measurements using AFM have been successfully tested to directly probe interfacial interactions in flax fibre/PLA biocomposite materials. These direct force measurements, using smooth polysaccharides films and PLA colloidal probes, allows for qualitative comparison between different systems and provides an insight into the nature of complex interactions involved at the biocomposite interface. Adhesion forces were shown to be dominated by capillary forces under ambient conditions. Force measurements under controlled humidity conditions show that PLA/cellulose is probably the weak interface in the biocomposite in terms of interactions. It also underlines the important role of water in the adhesion between the PLA and some hydrated polysaccharides such as pectin and hemicellulose. Even at low RH, the adhesion force was not reduced for these polymer films. We suggest that the large adhesion, and its insensitivity of as a function of RH, was due to the high amount of water within the film, which presents low elastic modulus, and hence may deform locally upon contact with the PLA bead to give an increased contact area. Interdiffusion mechanisms may also participate in the overall adhesion mechanism between these polymers. The importance of water was emphasized and is clearly interesting since it should be considered for real systems.

The Nardin–Schultz linear relation between work of adhesion and interfacial shear stress was tested for flax/PLA biocomposites. The work of adhesion estimated from colloidal force measurements using AFM was in better agreement with the Nardin–Schultz linear relation, when compared to the W_{adh}

calculated from global and averaging techniques such as contact angle measurements. A best linear fit was observed for the PLA/cellulose system, where the work of adhesion was dominated by molecular interactions such as van der Waals forces. However, for systems in which interfacial adhesion is dominated though complex interaction mechanisms, a slight deviation from the Nardin–Schultz linear relation was observed. This was typically the case observed for systems involving hydrated polysaccharides such as pectin and hemicellulose, where interfacial interactions were dominated through a combination of complex "adhesion mechanisms" involving capillary forces and interdiffusion phenomena.

Acknowledgement

Financial support for this work from Region Bretagne, Cap Lorient (France), is deeply acknowledged.

References

1. J. Holbery and D. Houston, *J. Mater.*, 2006, **58**, 80.
2. T. T. Nguyen, V. Picandet, P. Carré, T. Lecompte, S. Amziane and C. Baley, *Eur. J. Environ. Civ. Eng.*, 2010, **14**, 545.
3. E. Bodros, I. Pillin, N. Montrelay and C. Baley, *Compos. Sci. Technol.*, 2007, **67**, 462.
4. K. Oksman, M. Skrifvars and J. F. Selin, *Compos. Sci. Technol.*, 2003, **63**, 1317.
5. C. Morvan, C. Andème-Onzighi, R. Girault, D. S. Himmelsbach, A. Driouich and D. E. Akin, *Plant Physiol. Biochem.*, 2003, **41**, 935.
6. C. Baley, *Composites, Part A*, 2002, **33**, 939.
7. A. Arbelaiz, G. Cantero, B. Fernández, I. Mondragon, P. Gañán and J. M. Kenny, *Polym. Compos.*, 2005, **26**, 324.
8. J. T. Kim and A. N. Netravali, *Composites, Part A*, 2010, **41**, 1245.
9. M. Z. Rong, M. Q. Zhang, Y. Liu, G. C. Yang and H. M. Zeng, *Compos. Sci. Technol.*, 2001, **61**, 1437.
10. S.-H. Lee and S. Wang, *Composites, Part A*, 2006, **37**, 80.
11. S. Marais, F. Gouanvé, A. Bonnesoeur, J. Grenet, F. Poncin-Epaillard, C. Morvan and M. Métayer, *Composites, Part A*, 2005, **36**, 975.
12. M. Pommet, J. Juntaro, J. Y. Y. Heng, A. Mantalaris, A. F. Lee, K. Wilson, G. Kalinka, M. S. P. Shaffer and A. Bismarck, *Biomacromolecules*, 2008, **9**, 1643.
13. C. Baley, F. Busnel, Y. Grohens and O. Sire, *Composites, Part A*, 2006, **37**, 1626.
14. J.-M. Park, J.-W. Kim and D.-J. Yoon, *Compos. Sci. Technol.*, 2002, **62**, 743.
15. M. Nardin and J. Schultz, *Langmuir*, 1996, **12**, 4238.
16. W. A. Ducker, T. J. Senden and R. M. Pashley, *Nature*, 1991, **353**, 239.

17. H.-J. Butt, B. Cappella and M. Kappl, *Surf. Sci. Rep.*, 2005, **59**, 1.
18. R. D. Neuman, J. M. Berg and P. M. Claesson, *Nord. Pulp Paper Res. J.*, 1993, **8**, 96.
19. M. Holmberg, J. Berg, S. Stemme, L. Odberg, J. Rasmusson and P. Claessona, *J. Colloid Interface Sci.*, 1997, **186**, 369.
20. A. Carambassis and M. W. Rutland, *Langmuir*, 1999, **15**, 5584.
21. I. L. Radtchenko, G. Papastavrou and M. Borkovec, *Biomacromolecules*, 2005, **6**, 3057.
22. M. W. Rutland, A. Carambassis, G. A. Willing and R. D. Newman, *Colloids Surf., A*, 1997, **123**, 369.
23. J. Salmi, M. Osterberg and J. Laine, *Colloids Surf., A*, 2007, **297**, 122.
24. S. Zauscher and J. Klingenberg, *Colloids Surf., A*, 2001, **178**, 213.
25. E. Balnois, F. Bunel, C. Baley and Y. Grohens, *Compos. Interface*, 2007, **14**, 715.
26. A. Pietak, S. Korte, E. Tan, A. Downard and M. P. Staiger, *Appl. Surf. Sci.*, 2007, **253**, 3627.
27. M. Le Troëdec, A. Rachini, C. Peyratout, S. Rossignol, E. Max, O. Kaftan, A. Fery and A. Smith, *J. Colloid Interface Sci.*, 2011, **356**, 303.
28. G. Raj, E. Balnois, C. Baley and Y. Grohens, *J. Scanning Probe Microsc.*, 2009, **4**, 66.
29. G. Raj, E. Balnois, C. Baley and Y. Grohens, *Colloids Surf., A*, 2009, **352**, 47.
30. G. Raj, E. Balnois, M.-A. Helias, C. Baley and Y. Grohens, *J. Mater. Sci.*, 2011, **47**, 2175.
31. R. Sczech and H. Riegler, *J. Colloid Interface Sci.*, 2006, **301**, 376.
32. R. C. Sun, X. F. Sun and J. Tomkinson, in *Hemicellulose: Science and Technology*, ed. P. Gatenholm and M. Tenkanen, American Chemical Society, Washington, 2004, pp. 2–22.
33. F. M. Girio, C. Fonseca, F. Carvalheiro, L. C. Duarte, S. Marques and R. Bogel-Lukasik, *Bioresour. Technol.*, 2010, **101**, 4775.
34. N. M. L. Hansen and D. Plackett, *Biomacromolecules*, 2008, **9**, 1493.
35. C. Morvan, C. Andème-Onzighi, R. Girault, D. S. Himmelsbach, A. Driouich and D. E. Akin, *Plant Physiol. Biochem.*, 2003, **41**, 935.
36. K. Nishinari, M. Takemasa, S. H. Zhang and R. Takahashi, in *Comprehensive Glycoscience from Chemistry to Systems Biology*, ed. J. P. Kamerling, Elsevier, Amsterdam, 2007, p. 613.
37. A. Fischer, M. C. Houzelle, P. Hubert, M. A. V. Axelos, C. Geoffroy-Chapotot, M. C. Carré, M. L. Viriot and E. Dellacherie, *Langmuir*, 1998, **14**, 4482.
38. M. L. Fishman, P. H. Cooke, H. K. Chau, D. R. Coffin and A. T. Hotchkiss, *Biomacromolecules*, 2007, **8**, 573.
39. Y. Hong, C. Gao, Y. Shi and J. Shen, *Polym. Adv. Technol.*, 2005, **16**, 622.
40. J. N. Israelachvili, *Intermolecular and Surface Forces*, Academic Press, London, 1991.
41. Y. I. Rabinovich, J. J. Adler, A. Ata, R. K. Singh and B. M. Moudgil, *J. Colloid Interface Sci.*, 2000, **232**, 17.

42. N. A. Burnham, R. J. Colton and H. M. Pollock, *Nanotechnology*, 1993, **4**, 64.
43. D. L. Sedin and K. L. Rowlen, *Anal. Chem.*, 2000, **72**, 2183.
44. X. Xiao and L. Qian, *Langmuir*, 2000, **16**, 8153.
45. T. Eastman and D.-M. Zhu, *Langmuir*, 1996, **12**, 2859.
46. J.-A. Ko, H.-J. Choi, M.-Y. Ha, S.-D. Hong and H.-S. Yoon, *Langmuir*, 2010, **26**, 9728.
47. C. Baley, C. Morvan and Y. Grohens, *Macromol. Symp.*, 2005, **222**, 195.
48. C. Trotzig, S. Abrahmsén-Alami and F. H. J. Maurer, *Polymer*, 2007, **48**, 3294.
49. D. T. Turner and A. Schwartz, *Polymer*, 1985, **26**, 757.
50. M. Iijima, K. Nakamura, T. Hatakeyama and H. Hatakeyama, *Carbohydr. Polym.*, 2000, **41**, 101.
51. Y. M. Boiko and R. E. Prud'homme, *Macromolecules*, 1997, **30**, 3708.
52. Y.M. Boiko and R. E. Prud'homme, *J. Appl. Polym. Sci.*, 1999, **74**, 825.
53. Y. Grohens, M. Brogly, C. Labbe, M.-O. David and J. Schultz, *Langmuir*, 1998, **14**, 2929.
54. K. L. Johnson, K. Kendall and A. D. Roberts, *Proc. R. Soc. London, Ser. A*, 1971, **324**, 301.
55. B. V. Derjaguin, V. M. Muller and Y. P. Toporov, *J. Colloid Interface Sci.*, 1975, **53**, 314.
56. B. Cappela and G. Dietler, *Surf. Sci. Rep.*, 1999, **34**, 1.
57. G. Raj, E. Balnois, C. Baley and Y. Grohens, *Int. J. Polym. Sci.*, 2011, **2011**, 1.
58. A. Le Duigou, I. Pillin, A. Bourmaud, P. Davies and C. Baley, *Composites, Part A*, 2008, **39**, 1471.

CHAPTER 10

Zein: Structure, Production, Film Properties and Applications

NARPINDER SINGH,*[a] SANDEEP SINGH,[a]
AMRITPAL KAUR[a] AND MANDEEP SINGH BAKSHI[b]

[a] Department of Food Science of Technology, Guru Nanak Dev University, Amritsar, 143005, India; [b] Department of Chemistry, Wilfrid Laurier University, Science Building, 75 University Ave. W., Waterloo, Ontario N2L 3C5, Canada
*Email: narpinders@yahoo.com

10.1 Introduction

Alcohol-soluble storage proteins of corn are also known as prolamines and are generally referred to as zein. Zein is a naturally occurring protein polymer obtained from corn gluten meal (CGM), a co-product obtained during the wet milling of corn to produce cornstarch. Zein is also produced as a co-product of the ethanol industry. Whole corn grain contains around 39% zein and the endosperm portion has about 47%. CGM contains around 60–74% protein,[1] 20% starch and 4% oil. Zein constitutes one-half or more of CGM proteins; therefore CGM is the main raw material for both commercial and laboratory scale production of zein. Zein proteins account for about 60% of the total protein of corn grain. Zein contains 21.4% glutamine, 19.3% leucine, 9.0% proline, 8.3% alanine, 6.8% phenylalanine, 6.2% isoleucine, 5.7% serine and 5.1% tyrosine.[2] Zein is deficient in essential amino acids, such as lysine and tryptophan, which makes it poor in nutritional quality. Zein has a high proportion of non-polar amino acids (mainly leucine, alanine and proline),

RSC Green Chemistry No. 16
Natural Polymers, Volume 1: Composites
Edited by Maya J John and Thomas Sabu
© The Royal Society of Chemistry 2012
Published by the Royal Society of Chemistry, www.rsc.org

residues that are responsible for its poor solubility in water.[3] Both its insolubility in water and poor nutritional properties reduce its applications in human food formulations.

The present chapter deals with the isolation, structure, film forming properties and applications of zein.

10.2 Structural Characteristics

Zein is classified into four classes, *viz.* α, β, γ and δ, according to their solubility. Among the four fractions, α-zein with a molecular weight of 21–25 kD is the most abundant fraction. The β-, γ- and δ-zein fractions have molecular weights of 17–18 kD, 27 kD and 9–10 kD, respectively.[4,5] These fractions of zein differ in their amino acid sequences and surface charges. Paulis *et al.*[6] reported that 35% of the total zein was α-zein, which had two major bands with molecular weights of 24 kD and 22 kD and an amino acid and peptide composition similar to whole zein.[6–9] Zein varies in composition and concentration depending upon corn genotypes,[5,10] location of protein bodies in the grain[11,12] and maturity.[13,14] α-Zein has large amounts of hydrophobic residues such as leucine, proline, alanine and phenylalanine,[15] which contribute to its hydrophobic properties. The average hydrophobicity of zein was reported to be 50 times larger than albumin, fibrinogen, *etc.*, and hence a higher percentage of an alcohol/aqueous mixture is required for the dispersion of zein.[16] When dissolved in aqueous alcohol, the protein was reported to be rich in α-helical content (50–60%), with turn or random coil configurations comprising the remaining structure.[17] Several models have been proposed for the 3D structure of α-zein, including one composed of a series of anti-parallel helices clustered within a distorted cylinder.[17] Argos *et al.*[17] proposed a structural model for α-zein using optical rotatory dispersion and circular dichroism measurements (Figure 10.1). The model consists of repeat units forming an α-helix and having three polar segments on its surface (Figure 10.1a). The α-helices interact with the others by hydrogen bonds using two of the three polar segments (Figure 10.1b), while the third polar segment is utilized for intermolecular contacts (Figure 10.1c). α-Zein behaves in solution as asymmetric particles with an axial ratio from 7:1 to 28:1 as indicated by ultracentrifugation, birefringence, dielectric and viscosity experiments.[18] α-Zein has coiled-coil tendencies, resulting in α-helices with about four residues per turn in the central helical sections, with the non-polar residue side chains forming a hydrophobic face inside a triple superhelix.[18] The nine helical segments of the 19 kD protein were modeled into three sets of three interacting coiled-coil helices, with segments positioned end-to-end. The resulting structure lengthens with the addition of the N- and C-terminal sections, to give an axial ratio of ~6:1 or 7:1. Structural studies using different techniques such as circular dichroism, IR, NMR and optical rotatory dispersion suggested that Z19 (19 kDa) has ~35–60% helical character, made up of nine helical segments of about 20 amino acids with glutamine-rich "turns" or "loops".[17,18]

Garrett *et al.*[19] developed a model based on hydrophobic membrane propensities and helical "wheels." It was suggested that pairs of repeats formed

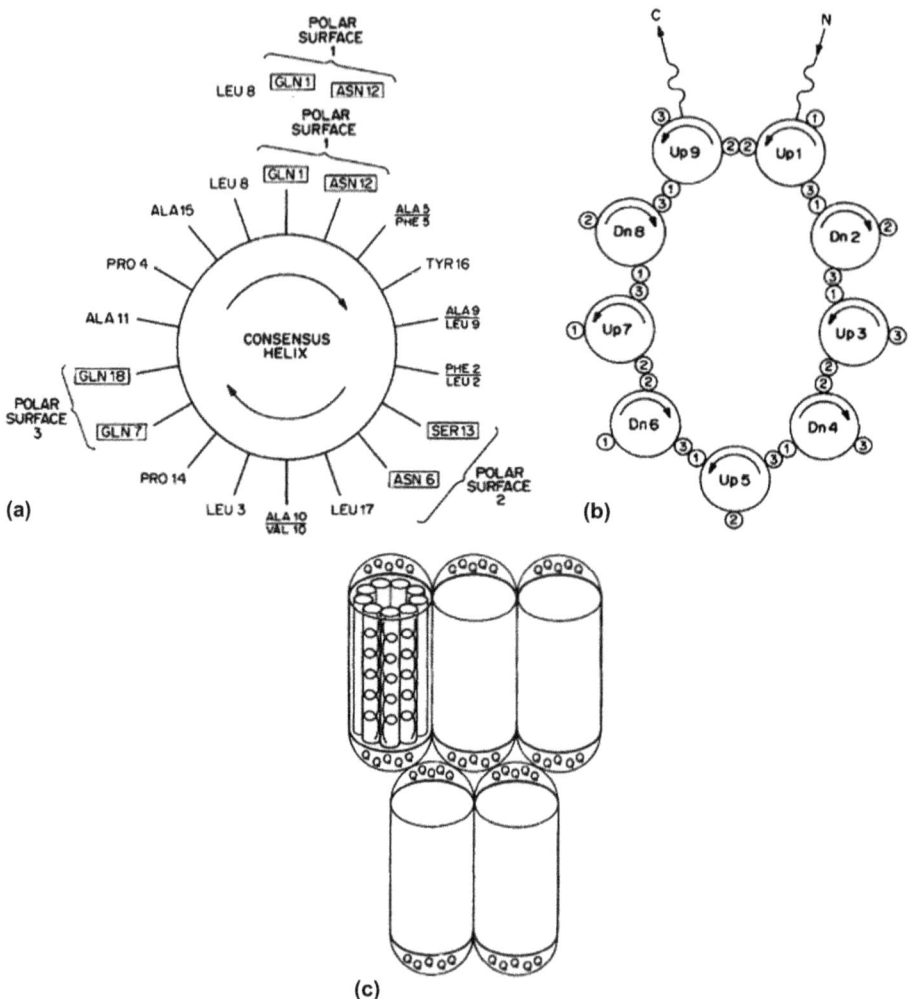

Figure 10.1   Proposed model of zein structure: (a) helical wheel for the 18-residue consensus repeat sequence showing three polar regions; (b) nine-helical zein protein structural model; (c) arrangement of zein proteins within a plane as well as for the stacking of molecular planes. Q = glutamine residues.[17]

anti-parallel α-helical hairpins and were arranged in a hexagonal net. X-ray scattering and viscosity studies have demonstrated that α-zein in solution is a relatively rigid asymmetric particle with a diameter between 0.7 and 1.5 nm and a length of ∼15 nm by small-angle X-ray scattering (SAXS) and 24 nm by viscosity.[20] Matsushima *et al.*[21] suggested another three-dimensional model consisting of nine (Z19) or ten (Z22) helical segments folded in anti-parallel fashion linked by glutamine-rich turns and held in place by hydrogen bonds. They proposed that the helical segments were aligned to form a compact elongated prism-like shape rather than an open ring. Such an arrangement allows α-zein molecules to contain different repeat units with greater flexibility and enhanced intramolecular hydrogen bonding. Momany *et al.*[18] favored the stick model (Figure 10.1) to represent the N-terminal segment of Z19. The resulting compact structure indicated that the cysteine residue (C27 or C6 in the renumbered sequence starting at residue 22) is exposed on the surface and therefore is available to form a disulfide bridge with the cysteine residue of the Z22 (22 kDa) protein, creating the naturally found dimer. The N- and C-termini are close to one another, a result of a large chain reversal in the middle of the segment and the disruption of secondary structure because of the nine proline residues in the segment.

10.3   Isolation and Production

Various methods are used for the manufacture of zein. These methods differ in terms of the raw material, the solvent used for extraction, purification and the method used for zein recovery. Almost the entire zein present in raw corn comes in the CGM, which is obtained as a co-product during the wet milling of corn. About one-half or more of the protein in CGM is zein. CGM is a low-cost raw material for the production of zein and is primarily used as an animal feed. The quality of CGM varies widely in protein content, depending upon the method of corn processing. The recovery of zein from CGM also depends upon its handling and processing conditions. For example, recovery of zein from CGM decreases with an increase in drying temperature.[22] Raw dry-milled corn gives a low zein yield when used as starting raw material[23] and leads to a higher cost of recovery. A low recovery of zein makes the process uneconomical, unless the methods used for solvent recovery and concentration of zein are efficient and cost effective.[24] Distiller's dried grains with "solubles" (DDGS) is another source of zein proteins and is obtained as a co-product during a dry-grind ethanol process. The dry-grind ethanol process involves corn grinding followed by saccharification and fermentation of glucose to produce ethanol, leaving concentrated fractions like cellulosic materials and proteins. DDGS has a protein content of around 27–30%. The quality and composition of the proteins present in DDGS varies widely, depending upon the drying and processing conditions. Drying at higher temperature can induce cross-linking of zein proteins and bring about changes in functional properties. DDGS is not a preferred raw material for producing high-quality zein since higher temperatures are used during drying.[25,26]

Zein contains predominantly nonpolar amino acids and therefore the solvents used for the extraction of zein need to possess wide characteristics, containing both ionic and non-ionic polar groups as well as nonpolar groups, either structure-wise (in the case of pure solvents) or composition-wise (for mixed solvents). Zein is considered "soluble" if more than 0.5% (w/v) of the protein dissolves in the solvent and produces a clear solution at room temperature. Most of the zein extraction processes involve the use of two solvents in succession: a polar solvent (aqueous solutions of ethanol or isopropanol) and a nonpolar solvent (hexane or benzene) for the removal of lipids and pigments. The solubility of zein in ethanol varies between 2 and 60% (w/w) at constant temperature, and varies with the ethanol concentration. At lower (<40%) and higher (>90%) concentrations of ethanol, two liquid phases appear; both contain zein, water and ethanol. This phenomenon is widely used to recover zein after extraction from CGM. Zein has two major fractions, α- and β-zein, as described by McKinney.[27] The α-zein is soluble in 95% ethanol while β-zein is soluble in 60% ethanol and insoluble in 95% ethanol. The α-zein has a lower content of histidine, arginine, proline and methionine than the β-zein. The β-zein is relatively unstable, and precipitates and coagulates readily, and hence is not a constituent of commercial zein.

A variety of laboratory-scale methods has been studied and described in the literature wherein zein has been fractionated by precipitation, cation exchange chromatography,[28,29] differential solubility,[5,30,31] cryo-precipitation,[32] charcoal and gel filtration.[33–35] Zein isolation by precipitation involves adding water to solutions of zein in ethanol or by using cellosolve.[31,36] Extraction of zein by the differential solubility method[37] from CGM gives yields up to ~90% with iso-propanol (80%), followed by a petroleum ether treatment. Cation exchange chromatography[28] involves use of defatted ground corn and extraction with 70% ethanol. The corn extract is refined using an Amberlite IRC-50 column with a NaCl gradient, which gives a yield of ~86%. Gel filtration chromatography[33] was used to separate zein from corn extract and processed by a combination of LH-20 purification and carbon adsorption, resulting in 90% purity.

Commercial methods for the production of zein from corn have been reviewed by Shukla and Cheryan[38] and are covered in several US patents. Most of the economically viable methods use aqueous solutions of ethanol or iso-propanol (IPA) for the first extraction of zein. IPA is the preferred solvent for zein extraction as it has more affinity than ethanol in terms of zein con-centration, *i.e.* a higher zein concentration at the same solvent-to-solids ratio. The subsequent separation (*e.g.* hexane extraction for pigment and oil removal) is better because the separation between zein and the hexane layer is easier. IPA dissolves in the hexane layer to a lesser extent, which results in lower distillation costs. These common solvents extract several other components (xanthophylls, polyamines, lipids, *etc.*) along with zein. Hence, various commercial processes or patents are distinguished based on methods of separation and purification to increase the purity of the zein.

Zein was manufactured on a large scale by Corn Products Corporation (CPC) from 1939 to 1967 and the method of purification by precipitation of zein has been described in patents.[39,40] In the CPC process (Figure 10.2.),[41] dried or wet CGM was contacted with hot 86–88% IPA or 93–95% ethanol at high pH and temperature (50–60 °C) in either a batch or continuous extractor for about 1.5–2 h. The use of reducing agents has been mentioned in the literature,[42] but are not commonly used in any industrial process. The extract was filtered and/or centrifuged and the filtrate, containing zein (~6% w/v) and impurities, was then clarified by standing or vacuum filtration[43] and followed by cooling (15 °C). Swallen[44] patented a process of extraction of zein from CGM using 85% IPA in a

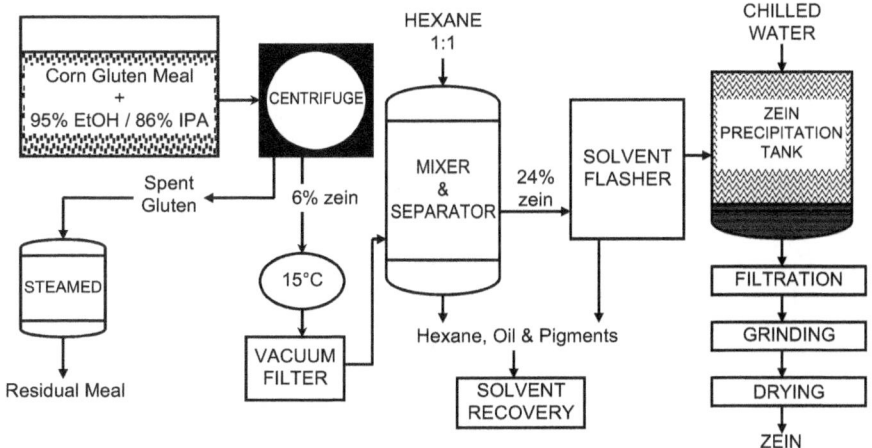

Figure 10.2 CPC process for production of zein from corn gluten meal.

solute-to-solvent ratio of 1:3.5 at 60 °C. Fats and colour pigments were extracted from zein by using solvents like hexane or benzene. The zein was finally precipitated either using excess amounts of cold water or at low temperatures (–15 to –25 °C). It was then vacuum dried and ground to produce zein of a light yellow colour. The CPC process was modified to include alkali treatment, where the pH was raised to 12 with NaOH and held for 30 min for deamidation of amino acid residues.[2] The pH was subsequently lowered with HCl and zein precipitated in cold water. This procedure improved the stability and gelation properties of zein.[45] The CPC process had some major disadvantages, such as high operating costs due to its complex solvent recovery systems (primarily distillation), low yield and high solvent losses during extraction.[46,47] Gelation of zein due to the variation in the pH of the solution and the quality of final product were also some of the other shortcomings of this process.

A process by Carter and Reck[48] known as the "Nutrilite process" involved extraction of zein from CGM using 88% aqueous IPA in a solute-to-solvent ratio of 1:4, with 0.25% NaOH for 1 h at 55–65 °C (Figure 10.3). Precipitation of the zein solution was carried out by chilling to –15 °C. Zein produced by this method contains 2% oil and xanthophylls, which can be partially removed by re-extraction with 88% IPA and subsequent chilling. Zein prepared in this way contains less oil (0.6%) and a lighter colour. This process appears to control gelation problems because low temperatures impart stability against denaturation of zein. However, the process was expensive since large volumes of solvent and low temperature precipitation were required.

Cook *et al.*[49] patented a process for the purification of zein and removal of pigments for pharmaceutical use. The CGM was made free of starch and washed several times with absolute ethanol to remove pigments and oil. The CGM was then washed with water and extracted with 80% ethanol. The extract was treated with activated carbon to remove further flavors and

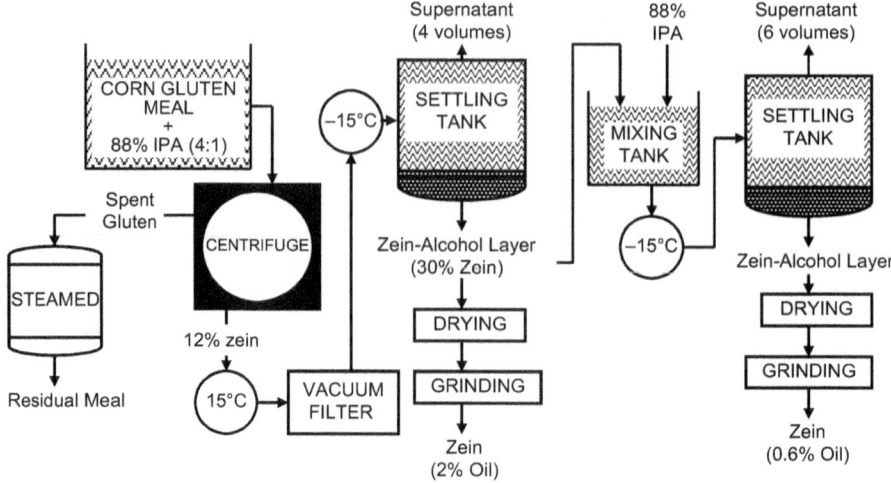

Figure 10.3   Nutrilite process for the production of zein from corn gluten meal.

pigments, precipitated with water and finally dried. The method extracts zein devoid of flavor and pigments; however, a large amount of ethanol is used with about ~2% of zein lost during the extraction of pigments. The removal of odor improves zein's marketability for applications such as gum.[50]

A method for refining zein after extraction was patented by Takahashi and Yanai.[51] Zein was extracted from CGM using aqueous acetone (70% v/v) in a respective ratio of 1:5 at 40 °C for 4 h. The solution was extracted and concentrated by evaporation to a syrup-like consistency, followed by precipitation using absolute acetone. The method gave higher zein yield with minimal colour than the zein produced by the method of Carter and Reck.[48]

In most of these methods, organic solvents, high pH, enzymes, *etc.*, were used for zein extraction and had some effect on the native conformation and composition of the proteins, which limit its food and non-food applications.

10.4   Film Forming

Zein is chemically inert and its globular structure poses difficulties in molding into different articles. Zein film formation occurs through development of hydrophobic, hydrogen and limited disulfide bonds between zein chains.[52] Films made from zein alone are brittle and plasticizers are incorporated to make them soft and permanently flexible.[45,53–57] Parris and Coffin[58] prepared unplasticized casting zein films in aqueous ethanol or acetone. The tensile properties indicated that films prepared in acetone were stronger but less flexible than those prepared in ethanol. Both types of films, however, were too brittle for most applications. Plasticizers are low molecular mass organic compounds added to soften rigid polymers.[59] These compounds act by reducing the glass transition temperature of polymers, reducing their crystallinity or melting temperature.[60] Plasticizers dissolve in the polymer that separates the chains from each other and thus facilitate molecular movement. These are applied to increase workability, flexibility and extensibility of polymers.[61] Plasticizers also act as internal lubricants by reducing frictional forces between polymer chains,[62] and their action may bring changes in the mechanical properties of polymers.[63,64] Plasticizers cover a wide range of chemical compounds, including esters, hydrocarbons, water, alcohols, glycols, phenols, ketones and ethers.[62,63] A plasticizer is considered ideal if miscible and compatible in wider proportions with plastic components. Generally, compatibility and permanence are said to follow opposite trends; therefore, the more compatible and efficient is a plasticizer, the more rapidly it can diffuse out of the polymer.[63] Lai and Padua[65] indicated that the addition of palmitic acid at a ratio of $0.5\,g\,g^{-1}$ of zein or stearic acid to $0.25\,g\,g^{-1}$ of zein substantially increased the tensile strength of the zein sheets. Glycerol has also been tried as a plasticizer in zein films.[66,67] However, Park *et al.*[68] reported that the interaction between protein molecules and glycerol is weak, and excess glycerol easily migrates through the film matrix. The processing techniques such as kneading, blowing and/or extrusion, used for film formation, could strengthen the plasticization effect of free fatty acids in zein films.[65,69] Therefore, some input energy to promote the interactions between fatty acid plasticizers and zein molecules is required during processing.[70]

Zein films synthesized through an acylation reaction with lauryl chloride were reported to be more flexible and tougher than pure zein. The hydrophobicity on the surface was further increased because of the attached alkyl chains, which could potentially improve the water barrier properties of these zein-based biomaterials. This study suggested a great potential of acylated zein as food packaging and coating materials, with improved properties. Ghanbarzadeh *et al.*[71] reported that the films plasticized by galactose had relatively higher tensile strength than films containing fructose and glucose. This could be attributed to a more hygroscopic nature of zein films containing glucose and fructose than those containing galactose. Zein films are generally formed by a wet-and-dry process. Film formation involves dissolving zein, plasticizers and other agents in an appropriate solvent and the solution is cast on a flat nonstick surface. The films are peeled off after the solvent has evaporated. Another method involves the preparation of a moldable resin of zein and plasticizers, which are later formed into a film by a heat process.[65,72] Resin zein films are obtained by stretching the resin over circular rims or pressing in a hot press.

10.5 Film Properties

Formation of zein films for food packaging has been extensively reported. The tensile strength of zein films was reported to be comparable with wheat gluten films.[73] Singh *et al.*[66] reported tensile strength and strain at failure of 27 MPa and 0.029, respectively, for zein–glycerol films. Alcohol-soluble proteins, such as zein, wheat gluten and fish myofibrillar proteins, form films of relatively low water vapor permeability (WVP) when compared to other proteins.[74] Such proteins are required to be modified by different methods to improve their functional properties. Films made from zein acetate have higher water resistance, strength and flexibility.[75] The major problem encountered during modification of zein is its lower solubility. Zein readily dissolves in an ethanol/water (90:10 w/w) mixture. However, this solution poses difficulty during chemical modification, since alcohol/water readily reacts with electrophilic reagents before zein does. To overcome this, the use of N,N-dimethylformamide as an inert solvent was suggested for zein modification.[76] Plasticization affects the WVP: zein films without plasticizers had lower WVP values than those plasticized with glycerol or mixtures of glycerol with poly(ethylene glycol) or poly(propylene glycol).[53,68,77] Protein films have low oxygen permeability at low to intermediate relative humidity. Oxygen permeability values were comparable to those of modest oxygen barriers, such as polyesters, and approaching those of the best oxygen barriers, *i.e.* ethylene–vinyl alcohol copolymer and poly(vinylidene chloride).[74] Low molecular weight molecules such as some polyols and fatty acids could be used as plasticizers to improve the mechanical properties of zein films.[78] However, phase separation as a result of immiscibility between zein and such plasticizers take place; this usually results in instability in the mechanical properties and often heterogeneous surfaces. Another disadvantage of using polyol plasticizers is that they exacerbate the sensitivity of zein-based materials to moisture.[79–81] Thus, new techniques to improve the water resistance and mechanical properties of zein-based materials need to be developed. Oleic acid has also been used to plasticize zein-based films.[45,65,82] Zein films plasticized with oleic acid were reported to exhibit tensile and moisture barrier properties that make them potentially useful as biodegradable packaging materials.[65,83] Zein films prepared with drying flax or tung oils showed higher resistance to water vapor.[84] Santosa and Padua[69] also evaluated the plasticizing effect of oleic acid on zein film characteristics. The WVP of zein films was observed to be lower than that of ethyl- and methyl-cellulose films. Zein films have lower storage stability and become discoloured and oxidized. Zein sheets prepared using oleic and linoleic acids as plasticizers resulted in an increase in the elongation percent, while Young's modulus and water absorption decreased.[69] Linoleic acid was more effective than oleic acid in reducing the water absorption of sheets, possibly due to the filling up of the pores and gaps in the structure and preventing it from swelling.[69]

10.6   Modifications

The zein structure is modified by physical and chemical methods to enhance its functional properties such as water solubility, foaming and emulsification.[85,86] Moderate modification of zein structures by enzymatic and chemical treatments is desired for partially unfolding zein molecules to improve the functional properties. The modifications of zein are done to the desired level because excessive and uncontrolled modifications lead to reduce functionality[87,88] due to fragmentation and truncation of the zein backbone.[89] Protein structures are influenced by many factors such as temperature, concentration, salt type, pH and solvent type. High temperature (90 °C) results in denaturation of secondary and tertiary structures, along with oligomerization of the zein.[85] Cross-linking induced by chemical reagents such as formaldehyde, glutaraldehyde, epichlorohydrin, citric acid, butane-1,2,3,4-tetracarboxylic acid, polymeric dialdehyde starch, 1,2-epoxy-3-chloropropane and dialcohols between zein molecules has also been reported. The cross-linking of zein molecules caused a significant increase in tensile strength of zein films.[58,89–91] It was also reported that cross-linking of zein molecules with 1-(3-dimethylaminopropyl)-3-ethylcarbodiimide hydrochloride and *N*-hydroxysuccinimide improved film forming properties while the aggregation phenomenon in solution was suppressed.[92] Wu *et al.*[93] used polycaprolactone (PCL) to modify zein, where the hydrophobic chain of PCL besides cross-linking the zein molecule also serves as a plasticizer and significantly increases the elasticity of zein sheets.

Different cations (*i.e.* Na^+, K^+, NH_4^+ and Mg^{2+}) have also been reported to strengthen zein structure selectivity and interaction with non-ionic surfactants.[94] Methanol, ethanol and isopropanol have also been tried to modify the morphology, diameter and distribution of fibres from the electrospinning of zein.[95] Cabra *et al.*[85] reported that different pH values (0.5–2 N HCl or NaOH), with or without applying heat, would induce zein deamidation, resulting in an improved emulsifying property. Under similar pH conditions (0.05 N HCl with heat), Chiue *et al.*[87] found a decrease in zein's fat-binding capacity, leading to the loss of radical scavenging ability. Bakshi *et al.*[96] blended zein films with selfassembled (SA) bovine serum albumen (BSA) conjugated nanoparticles (NPs) to improve the mechanical properties of zein films. BSA conjugated NPs were produced through biomineralization of a gold salt by BSA. SA NPs promote the mechanical properties of zein films, as hydrophobic proteins of zein have preferential interactions with predominantly hydrophobic SA NPs. Colloidal SA NPs in the form of soft films thus provide better orientations for unfolded zein to produce strong and flexible protein films.

10.7 Applications

Zein is generally recognized as safe (GRAS) and is biodegradable[97,98] and biocompatible.[99,100] Some applications include free-standing packaging material,[101] emulsifier,[85,86] chewing gum base,[102] antioxidant,[87,103] *etc.* Many of these applications (chewing gum base and antioxidant) are due to higher aliphatic indexes, surface hydrophobicity[17,85] and the high fatty acid-binding capacity[88] of zein. In the food and confectionary industries, zein is being used for coating enriched rice, candies such as chocolates and jellybeans, dried fruits, nuts, nutmeats, and for the encapsulation of flavors and sweeteners. It is used as a coating on bakery products as a vegetable protein. Other packaging applications include the use for frozen foods and ready-to-eat chicken.[104]

Some other applications include tablet coating, drugs encapsulation[105,106] and as a tissue scaffold.[107,108] Singh *et al.*[66] reported that zein iodine solution could be used as a film forming medication (ointment) when high anti-microbial activity against pathogenic microorganisms or infections is required. Such ointments have the potential to replace substantially oily ointments that are inefficient and might spoil or stain clothes. Zein films containing salicylic acid and acetylsalicylic acid with or without glycerol for structural, mechanical and dissolution properties were also evaluated.[67] This study reported that random coils, α-helix and β-sheets mainly governed the secondary structure of zein, depending on glycerol and the level of model molecules. It was suggested that, from a bio-medical perspective, zein films could act as drug release devices provided a thorough understanding of the interactions between zein proteins and the active ingredients be established.

Zein has a wide variety of other non-food applications, such as the manufacture of plastics, fibres, adhesives, coatings, binders, *etc.* Films produced using zein have excellent water barrier properties because of its hydrophobic characteristics and stability in acid and alkaline solutions.[52] Therefore zein has been employed as a raw material for waterproofing paper, coating wood, adhering plywood, damp-proofing and tablets.[45] The fibres made from zein are used for garments, hats and other commercial applications.[109]. Zein fibres produced by electrospinning have been suggested to have better applications in the fibre market.[95,110]

Chen *et al.*[111] produced zein nanofibres by the electrospinning of aqueous alcohol solutions. The electrospinning process produced nanoparticles, nanofibre mats or ribbon-like nanofibre mats and they concluded that fibre morphology was primarily affected by solution concentration. Zein electrospun nanofibre mats were flexible and lustrous but showed poor mechanical properties. Cross-linking of zein by hexamethylene diisocyanate significantly increased the tensile strength of the zein electrospun nanofibre mats. The mechanical properties of zein nanofibres can be improved by combining electrospinning and cross-linking. Zein can be further processed into resins and other bioplastic polymers, which can be extruded or rolled into a variety of plastic products.[112] With increasing environmental concerns about synthetic coatings and the ever increasing price of hydrocarbon-based petrochemicals, there is a great potential for utilizing zein as a raw material for a variety of non-toxic and renewable polymer applications, particularly in the paper industry.[113,114] Other reasons for the renewed interest in zein include concern about the landfill costs of plastics, and the increased demand by consumers for natural substances.

Acknowledgements

The financial assistance from CSIR to NS is acknowledged.

References

1. S. W. Wu, D. J. Myers and L. A. Johnson, *Cereal Chem.*, 1997, **74**, 258.
2. A. F. Pomes, in *Encyclopedia of Polymer Science and Technology*, ed. H. Mark, Wiley, New York, 1971, vol. 15, p. 125.

3. J. L. Kokini, A. M. Cocero, H. Madeka and E. de Graaf, *Trends Food Sci. Technol.*, 1994, **5**, 281.
4. A. Esen, *Plant Physiol.*, 1986, **80**, 623.
5. A. Esen, *J. Cereal Sci.*, 1987, **5**, 117.
6. J. W. Paulis, C. James and J. S. Wall, *J. Agric. Food Chem.*, 1969, **17**, 1301.
7. J. W. Paulis and J. S. Wall, *Cereal Chem.*, 1977, **54**, 1223.
8. J. S. Wall and J. W. Paulis, in *Advances in Cereal Science and Technology*, ed. Y. Pomeranz, American Association of Cereal Chemists, St. Paul, MN, 1978, vol. 2, p. 135.
9. J. W. Paulis, *Cereal Chem.*, 1981, **58**, 542.
10. J. Landry, C. Damerval, R. A. Azevedo and S. Delhaye, *Plant Physiol. Biochem.*, 2005, **43**, 549.
11. A. Larkins and W. J. Hurkman, *Plant Physiol.*, 1978, **62**, 256.
12. R. Lending and B. A. Larkins, *Plant Cell*, 1989, **1**, 1011.
13. Y. M. Woo, D. W. N. Hu, B. A. Larkins and R. Jung, *Plant Cell*, 2001, **13**, 2297.
14. R. Holding and B. A. Larkins, *Maydica*, 2006, **51**, 243.
15. E. Gianazza, V. Viglienghi, P. G. Righetti, F. Salamini and C. Soave, *Phytochemistyry*, 1977, **16**, 315.
16. H. D. Belitz, R. Kieffer, W. Seilmeier and H. Wieser, *Cereal Chem.*, 1986, **63**, 336.
17. P. Argos, K. Pedersen, M. D. Marks and B. A. Larkins, *J. Biol. Chem.*, 1982, **257**, 9984.
18. F. A. Momany, D. J. Sessa, J. W. Lawton, G. W. Selling, S. A. H. Hamaker and J. L. Willett, *Food Chem.*, 2006, **54**, 543.
19. R. Garratt, G. Oliva, I. Caracelli, A. Leite and P. Arruda, *Proteins: Struct. Funct. Genet.*, 1993, **15**, 88.
20. A. S. Tatham, J. M. Field, V. J. Morris, K. J. I. Anson, L. Cardle, M. J. Dufton and P. R. Shewry, *J. Biol. Chem.*, 1993, **268**, 26253.
21. N. Matsushima, G. Danno, H. Takezawa and Y. Izumi, *Biochim. Biophys. Acta*, 1997, **1339**, 14.
22. S. Wu, D. J. Myers, L. A. Johnson, S. R. Fox and S. K. Singh, *Cereal Chem.*, 1997, **74**, 268.
23. R. Shukla, M. Cheryan and R. E. DeVor, *Cereal Chem.*, 2000, **77**, 724.
24. M. Cheryan, *US Pat.* 6 433 146, 2002.
25. Y. V. Wu, K. R. Sexson and J. S. Wall, *Cereal Chem.*, 1981, **58**, 343.
26. W. J. Wolf and J. W. Lawton, *Cereal Chem.*, 1997, **74**, 530.
27. L. L. McKinney, in *The Encyclopedia of Chemistry*, ed. G. L. Clark, Reinhold, New York, 1958, suppl., p. 319.
28. M. Craine, D. V. Freimuth, J. A. Boundy and R. J. Dimler, *Cereal Chem.*, 1961, **38**, 399.
29. J. Landry and P. Guyon, *Biochimie*, 1984, **66**, 451.
30. T. B. Osborne, *Vegetable Proteins*, Longmans Green, New York, 2nd edn., 1924.
31. R. A. Gortner and R. T. MacDonald, *Cereal Chem.*, 1944, **21**, 324.

32. B. Jain, *Purification of Zein from Ethanol Extracts of Corn*, MS thesis, University of Illinois, 2002.
33. L. A. Danzer and E. D. Rees, *Cereal Chem.*, 1971, **48**, 118.
34. J. Mossé and J. Landry, in *Cereals for Food and Beverages: Recent Progress in Cereal Chemistry and Technology*, ed. G. E. Inglett and L. Munck, Academic Press, New York, 1980, p. 255.
35. J. Landry and P. Guyon, *Biochimie*, 1984, **66**, 461.
36. C. Watson, S. Arrhenius and J. W. William, *Nature*, 1936, **137**, 322.
37. D. Evans, R. J. Foster and C. B. Croston, *Ind. Eng. Chem.*, 1945, **37**, 175.
38. R. Shukla and M. Cheryan, *Ind. Crops Prod.*, 2001, **13**, 171.
39. L. C. Swallen, *US Pat.* 2 120 946, 1938.
40. R. A. Reiners, J. C. Pressick and L. Morris, *US Pat.* 3 840 515, 1974.
41. J. F. Walsh, S. M. Kinzinger and W. L. Morgan, *US Pat.* 2 360 381, 1944.
42. C. Y. Tsai, *Cereal Chem.*, 1980, **57**, 288.
43. L. C. Swallen, *US Pat.* 2 221 560, 1940.
44. L. C. Swallen, *US Pat.* 2 287 649, 1942.
45. R. A. Reiners, J. S. Wall and G. E. Inglett, in *Industrial Uses of Cereals*, ed. Y. Pomeranz, American Association of Cereal Chemists, St. Paul, MN, 1973, p. 285.
46. R. H. Manley and C. D. Evans, *US Pat.* 2 354 393, 1944.
47. L. Morris, L. G. Unger and A. L. Wilson, *US Pat.* 2 733 234, 1956.
48. R. Carter and D. R. Reck, *US Pat.* 3 535 305, 1970.
49. R. B. Cook, F. M. Mallee and M. L. Shulman, *US Pat.* 5 580 959, 1996.
50. D. J. Sessa and D. E. Palmquist, *Bioresour. Technol.*, 2008, **99**, 6360.
51. H. Takahashi and N. Yanai, *US Pat.* 5 342 923, 1994.
52. A. Gennadios and C. L. Weller, *Trans. ASAE*, 1994, **37**, 535.
53. M. Mendoza, *MS Thesis, University of Massachusetts*, Amherst, 1975.
54. C. Andres, *Food Process.*, 1984, **45**, 48.
55. K. N. Wright, in *Corn Chemistry and Technology*, ed. S. A. Watson and P. E. Ramstad, American Association of Cereal Chemists, St. Paul, MN, 1987, p. 447.
56. T. P. Aydt and C. L. Weller, presented at the ASAE Meeting, St. Joseph, MI, 1988, paper 88-6522.
57. T. A. Trezza and P. J. Vergano, *J. Food Sci.*, 1994, **59**, 912.
58. N. Parris and D. R. Coffin, *J. Agric. Food Chem.*, 1997, **45**, 1596.
59. I. M. Ward and D. W. Hadley, *An Introduction to the Mechanical Properties of Solid Polymers*, Wiley, New York, 1993.
60. L. H. Sperling, *Introduction to Physical Polymer Science*, Wiley, New York, 2nd edn., 1992.
61. J. D. Ferry, *Viscoelastic Properties of Polymers*, Wiley, New York, 3rd edn., 1980, p. 486.
62. J. H. Briston and L. L. Katan, *Plastics in Contact with Food*, Food Trade Press, London, 1974, ch. 4 and 5.
63. H. W. Chatfield, in *Varnish Constituents*, Leonard Hill, London, 1953.
64. Y. H. Roos, *Phase Transitions in Foods*, Academic Press, San Diego, 1995, p. 109.

65. M. Lai and G. W. Padua, *Cereal Chem.*, 1997, **74**, 771.
66. N. Singh, D. M. R. Georget, P. S. Belton and S. A. Barker, *J. Agric. Food Chem.*, 2009, **57**, 4334.
67. N. Singh, D. M. R. Georget, P. S. Belton and S. A. Barker, *J. Cereal Sci.*, 2010, **52**, 282.
68. H. J. Park, J. M. Bunn, C. L. Weller, P. J. Vergano and R. F. Testin, *Trans. ASAE*, 1994, **37**, 1281.
69. F. X. Santosa and G. W. Padua, *J. Agric. Food Chem.*, 1999, **47**, 2070.
70. K. Shi, Y. Huang, H. Yu, T.-C. Lee and Q. Huang, *J. Agric. Food Chem.*, 2011, **59**, 56.
71. B. Ghanbarzadeh, A. R. Oromiehie, M. Musavi, Z. E. D-Jomeh, E. R. Rad and J. Milani, *Food Res. Int.*, 2006, **39**, 882.
72. M. W. Andrianaiov and G. W. Padua, *Starch*, 2003, 25.
73. A. Gennadios, H. J. Park and C. L. Weller, *Trans. ASAE*, 1993, **36**, 1867.
74. J. M. Krochta, *Protein-Based Films and Coatings*, CRC Press, Boca Raton, FL, 2002.
75. C. Veatch, *US Pat.* 2 236 768, 1941.
76. A. Biswas, D. J. Sessa, J. W. Lawton, S. H. Gordon and J. L. Willett, *Cereal Chem.*, 2005, **82**, 1.
77. H. J. Park and M. S. Chinnan, *J. Food Eng.*, 1995, **25**, 497.
78. J. Lawton, *Cereal Chem.*, 2002, **79**, 1.
79. B. Ghanbarzadeh, M. Musavi, A. R. Oromiehie, K. Rezayi, E. R. Rad and J. Milani, *LWT–Food Sci. Technol.*, 2007, **40**, 1191.
80. B. Ghanbarzadeh, A. R. Oromiehie, M. Musavi, P. M. Falcone, Z. E. D-Jomeh and E. R. Rad, *Packag. Technol. Sci.*, 2007, **20**, 155.
81. J. W. Park, R. F. Testin, P. J. Vergano, H. J. Park and C. L. Weller, *J. Food Sci.*, 1996, **61**, 401.
82. J. L. Kanig and H. Goodman, *J. Pharm. Sci.*, 1962, **51**, 77.
83. H. M. Lai and G. W. Padua, *Cereal Chem.*, 1998, **75**, 194.
84. Q. Wang and G. W. Padua, *J. Agric. Food Chem.*, 2005, **53**, 3444.
85. V. Cabra, R. Arreguin, R. Vazquez-Duhalt and A. Farres, *J. Agric. Food Chem.*, 2007, **55**, 439.
86. Y. H. Yong, S. Yamaguchi, Y. S. Gu, T. Mori and Y. Matsumura, *J. Agric. Food Chem.*, 2004, **52**, 7094.
87. H. Chiue, K. Iwami, T. Kusano and F. Ibuki, *Biosci. Biotechnol. Biochem.*, 1994, **58**, 198.
88. H. Chiue, T. Kusano and K. Iwami, *Food Chem.*, 1997, **58**, 111.
89. M. Zhang, C. A. Reitmeier, E. G. Hammond and D. J. Myers, *Cereal Chem.*, 1997, **74**, 549.
90. K. Yamada, H. Takahashi and A. Noguchi, *Int. J. Food Sci. Technol.*, 1995, **30**, 599.
91. Y. Yang, L. Wang and S. Li, *J. Appl. Polym. Sci.*, 1996, **59**, 433.
92. S. Kim, D. J. Sessa and J. W. Lawton, *Ind. Crops Prod.*, 2004, **20**, 291.
93. Q. Wu, T. Yoshino, H. Sakabe, H. Zhang and S. Isobe, *Polymer*, 2003, **44**, 3909.

94. T. Cserhati, E. Forgacs and Z. Illes, *J. Liq. Chromtogr. Relat. Technol.*, 2003, **26**, 2751.
95. G. Selling, A. Biswas, A. Patel, D. Walls, C. Dunlap and Y. Wei, *Macromol. Chem. Phys.*, 2007, **208**, 1002.
96. M. S. Bakshi, H. Kaur, T. S. Banipal, N. Singh and G. Kaur, *Langmuir*, 2010, **26**, 13535.
97. B. A. McGowan, G. W. Padua and S. Y. Lee, *J. Food Sci.*, 2005, **70**, S475.
98. J. Sessa, G. W. Selling, J. L. Willett and D. E. Palmquist, *Ind. Crops Prod.*, 2006, **23**, 15.
99. P. Hurtado-Lopez and S. Murdan, *J. Microencapsulation*, 2006, **23**, 303.
100. Q. Wang, W. J. Xian, S. F. Li, C. Liu and G. W. Padua, *Acta Biomater.*, 2008, **4**, 844.
101. V. M. Hernandez-Izquierdo and J. M. Krochta, *J. Food Sci.*, 2008, **73**, R30.
102. J. Liu and W. W. Lee, *US Pat.* 6 733 578, 2004.
103. L. Zhu, J. Chen, X. Tang and Y. Xiong, *J. Agric. Food Chem.*, 2008, **56**, 2714.
104. M. E. Janes, S. Kooshesh and M. G. Johnson, *J. Food Sci.*, 2002, **67**, 2754.
105. P. Hurtado-Lopez and S. Murdan, *J. Drug Delivery Sci. Technol.*, 2005, **15**, 267.
106. X. Liu, Q. Sun, H. Wang, L. Zhang and J. Wang, *Biomaterials*, 2005, **26**, 109.
107. S. Gong, H. Wang, Q. Sun, S. Xue and J. Wang, *Biomaterials*, 2006, **27**, 3793.
108. Q. Wang, L. Yina and G. W. Padua, *Food Biophys.*, 2008, **3**, 174.
109. O. Sturken, *US Pat.* 2 361 713, 1944.
110. T. Miyoshi, H. Toyoharaa and H. Minematsu, *Polym. Int.*, 2005, **54**, 1187.
111. Y. Chen, L. Xinsong and T. Song, *J. Appl. Polym. Sci.*, 2007, **103**, 380.
112. J. W. Lawton, Jr., *Cereal Chem.*, 2004, **81**, 1.
113. A. J. Jabar, M. A. Bilodeau, A. Michael, D. J. Neivandt and J. Spender, *US Pat. (pending)*, 2005.
114. N. Parris, M. Sykes, L. C. Dickey, J. L. Wiles, T. J. Urbanik and P. H. Cooke, *Prog. Pap. Recycl.*, 2002, **11**, 24.

CHAPTER 11

Silk Fibre Composites

PANYA WONGPANIT,[a]
ORATHAI PORNSUNTHORNTAWEE[b]
AND RATANA RUJIRAVANIT*[b,c]

[a] Faculty of Agricultural Product Innovation and Technology,
Srinakharinwirot University, Bangkok 10110, Thailand; [b] The Petroleum and
Petrochemical College, Chulalongkorn University, Bangkok 10330, Thailand;
[c] Center for Petroleum, Petrochemicals, and Advanced Materials,
Chulalongkorn University, Bangkok 10330, Thailand
*Email: ratana.r@chula.ac.th

11.1 Introduction to Silk Fibres and their Protein Components

Natural fibres can be divided into two main categories based on their origins:
(i) plant-based lignocellulose fibres, which predominantly contain cellulose,
hemicellulose, and lignin; and (ii) animal-based protein fibres, including silk
and wool.[1] Humans have utilized natural fibres as textiles for several thousand
years; however, a recent increase in environmental concerns has led to the use
of natural fibres, which are renewable resources, for the development of
environmentally friendly materials, especially in the form of composites.
Among both types of natural fibres, the fabrication of silk-based composites is
of great interest. In this chapter, the structures and important characteristics of
silk fibres and their protein components are briefly discussed, followed by an
overview of the fabrication of silk-based materials, both in the form of raw silk
and regenerated silk. Finally, the preparation of silk-based composites and
their potential applications are intensively reviewed.

RSC Green Chemistry No. 16
Natural Polymers, Volume 1: Composites
Edited by Maya J John and Thomas Sabu
© The Royal Society of Chemistry 2012
Published by the Royal Society of Chemistry, www.rsc.org

11.1.1 Silk Fibres and their Sources

Silk fibres are produced by arthropods, especially by members of the class *Arachnida*, including mites and more than 30 000 species of spiders, and those of the class *Insecta*, which are the larvae of some *Lepidoptera*, such as butterflies and moths.[2–4] These animals produce the silk fibres for various purposes. For example, silkworm moths produce the silk fibres to provide shelters in the form of cocoons during their metamorphosis, while spiders use the silk fibres in the form of aerial nets to capture prey, to provide structural support in the form of egg sag suspension threads, to aid dispersal of neonate spiderlings in the wind, and to escape from predators.[2–5]

In silk production, proteinaceous materials are first produced by either special glands, which are generally modified salivary glands in the larvae of some *Lepidoptera*, or by other structures in some immatures and adults of mites and spiders.[3] The silk proteins are initially stored in those structures as a liquid before being spun into fibres during secretion. Although there are many types of silk in nature, each species of silk producer, except spiders, produce only one silk type.[3,6] In the case of spiders, they can produce at least seven different silk types, including major ampullate or dragline silk, minor ampullate silk, flagelliform silk, pyriform silk, aggregate silk, cylindriform or tubuliform silk, and aciniform silk.[5–7]

Among various kinds of silk producers, silkworms and spiders are intensively studied for their silk production. Utilization of silk fibres by mankind has been done at least since 2600 BC by harvesting *Bombyx mori* silkworms, a member of the family *Bombycidae* in the order *Lepidoptera*, on mulberry trees.[3] Up to now, the mulberry *B. mori* silk fibres are of the most commercial importance, especially in the textile industry, due to their large-scale production and unique luster appearances.[5,8,9] Smaller quantities of silk fibres can also be produced by other species fed on a variety of their natural hosts, such as the productions of tasar silk, muga silk, and eri silk by silkworms in the family *Saturniidae*.[3] Silk production by silkworms is also known as sericulture.[3,8] For spider silks, there has been an attempt to use them as either fishing nets or wound closures for thousands of years.[3,6] However, in contrast to silkworms, it is impossible to harvest spiders for large-scale silk production because of the territorial and cannibalistic nature of most spiders.[3,6] As a result, the utilization of spider silks is now restricted, although much effort is used to enable their industrialization, especially by biotechnological production.[6] Based on their widespread use, this chapter will mainly focus on the silk fibres produced by *B. mori* silkworms. For spider silks and other silk types, a number of comprehensive reviews have already been published.[2,3,5–7,9]

11.1.2 Physical Structure and Chemical Composition of Silk Fibres

Silk fibres produced by *B. mori* and other silkworms possess a "sheath-around-two-cores" structure, containing two protein monofilaments with triangular sections, named brins, embedded in a sericin layer, which is a water-soluble protein responsible for the gum-like sticky coating.[7,8,10,11] The brins contain several macrofibrils, which are bundles of nanofibrils made of fibroin filaments.[7] The silk fibres are composed of fibroin proteins in the range of 70–80 wt% and sericin proteins in the range of 20–30 wt%, depending on their origin and culture conditions.[10,12,13] Other minor components, which can be exclusively found in the sericin layers, include waxes and fats (1.5 wt%), carotenoids and minerals (1 wt%), and small amounts of polysaccharides, known as glycosylated sericin fractions.[10]

For the chemical composition, the major amino acids that can be found in the fibroin proteins of the silk fibres produced by *B. mori* silkworms are glycine (43%), alanine (30%), serine (12%), tyrosine (5%), and valine (3%).[14,15] Therefore, the silk fibroin predominantly contains hydrophobic uncharged amino acid residues. The fibroin consists of three components: two protein chains, named as a heavy chain (H-chain, with molecular weight of about 350 kDa) and a light chain (L-chain, with molecular weight of about 25 kDa), which are linked together *via* disulfide bonds, and a glycoprotein, named P25, with molecular weight of about 25 kDa, that is associated *via* non-covalently hydrophobic interactions.[16–18] The molar ratio of H-chain:L-chain:P25 is 6:6:1.[16–18] The H-chain is more hydrophobic and is responsible for the formation of crystalline structures, while the L-chain is more hydrophilic and relatively elastic.[16–18] The fibroin possesses three possible conformations, including random-coil, α-helix (silk I), and β-sheet (silk II).[19–21] The first conformation is an amorphous state, while the α-helix and β-sheet are crystalline structures of the fibroin. The random-coil conformation and α-helical structure can easily convert to the β-sheet structure by mechanical agitation or by exposure to heat or to hydrophilic organic solvents, like methanol.[2,14,19–21] The β-sheet structure is thermodynamically stable and is water-insoluble due to the presence of extensive inter- and intra-chain hydrogen bonding as well as hydrophobic interactions.[14] In general, the fibroin proteins of the silk fibres produced by *B. mori* silkworms contain α-helical and β-sheet structures of about 13% and 56%, respectively.[22] Thus, the fraction of crystalline structures of the silk fibroin is more than 60%.[15] In the case of the sericin, it contains at least six proteins with molecular weights in the range of 20–400 kDa.[10,23] The silk sericin predominantly contains nearly 40% serine, with smaller amounts of other amino acid components, such as glycine, aspartic acid, glutamic acid, threonine, and tyrosine.[12,24,25] The hydrogen bonding between hydroxyl groups of these amino acids are responsible for the glue-like properties of the sericin layer.[10] However, unlike the silk fibroin, sericin possesses only two conformations: random coil (about 63%) and β-sheet (about 35%).[24] A low degree of crystallinity and a large content of strongly polar side group-containing amino acid components make the sericin coating of the silk fibres become water-soluble.[10,12,25]

11.1.3 Characteristics of Silk Fibres and their Protein Components

The silk fibres exhibit outstanding mechanical properties in terms of toughness, tensile strength, and extensibility (or strain at break) because of the extensive hydrogen bonding and hydrophobic interactions of their protein components, as well as a high degree of crystallinity of the silk fibroin.[7,8,10] For the *B. mori* silk fibres, their average toughness, tensile stress, tensile modulus, and strain at break are 70 mJ m^{-3}, 700 MPa, 10 GPa, and 20%, respectively.[2,7,8,10,13,26] These mechanical parameters are comparable to those of both other natural fibres and synthetic fibres, including hemp, jute, ramie, coir, sisal, flax, cotton, glass fibre, and Kevlar.[7,8,13,26] The silk fibres can withstand heat up to nearly 200 °C, with a glass transition temperature (T_g) of 175 °C, before the side-chain groups of the amino acid residues and the main chain of the silk molecules begin to degrade at above 200 °C and 300 °C, respectively.[10,19] Moreover, compared to other protein fibres (wool), silk fibres have lower flammability.[10] Silk fibres are biodegradable at a slow predictable rate and show excellent biocompatibility after the removal of sericin.[7-9,27] The unique mechanical properties, slow bio-degradabilities, and biocompatabilities of silk fibres are valuable for tissue engineering applications, which require support and transfer of load from the scaffolding material to the developing tissue.[7] As a proteinaceous material, silk fibres are amphoteric with an isoelectric point (p*I*) in the range of 4–5.[10,15] Compared to wool, silk fibres are more resistant to alkali but more sensitive to acid.[10] The presence of the side-chain groups of amino acid residues also provides reactive sites for chemical modification of the silk fibres, such as by graft copolymerization which can further improve the processabilities of the silk-based composites (see Section 13.3).

11.2 Fabrication and Potential Applications of Silk-based Materials

Silk fibres have been used as biomedical sutures for decades and as textiles for centuries.[8] However, the original form of silk is unsuitable for use in some applications. To expand their potential exploitation, the silk proteins have to be denatured by dissolving the silk fibres in a solvent system prior to processing into other alternative forms, such as microtubes, films, spheres, hydrogels, or three-dimensional porous materials. The fabrication of silk proteins into new morphologies with desirable shapes and structures is called regeneration. Normally, the regenerated silk-based materials require post-treatments in order to induce the recrystallization of the silk proteins.

11.2.1 Raw Silk Threads and Spun Silks

After the mature silkworms completely spin their cocoons, which are made of a single continuous filament of silk, the cocoons are collected and the larvae are killed by overheating, either by boiling in hot water or exposure to steam.[5,9,10] Since the sericin is water soluble, it can be partially removed at this step. When the sticky sericin coating dissolves in water, the cohesion between the silk fibres begins to reduce. Hence, the silk fibres in the outer region of the cocoons are loosened and can be caught on a turning spool.[3] Since a single silk filament is very fine, it has to be wound together, or reeled, to form a continuous silk thread.[3] After being reeled, the raw silk threads, or silk yarns, are obtained. Normally, a single silk thread is still too fine and too fragile for end-use, so about nine threads (each thread containing two silk fibres) are twisted together in order to obtain a silk thread with enough strength to handle.[9] In general, a good quality cocoon can produce silk fibre of length in the range of 300–1000 m, and about 1500 cocoons are required to manufacture a kilogram of silk.[3] It should be noted that the raw silk threads are often called raw silk fibres in most research. Furthermore, not all silk fibres are successfully wound. The remaining unreeled silk fibres as well as shorter or broken filaments can be used to produce spun silks.

In the textile industry and in some specific applications (especially in bio-related fields), the raw silk has to be further subjected to a degumming step. In biomedical applications, the remaining sericin in silk sutures can cause wound inflammation and allergic reactions, consequently limiting their clinical use.[8,9] The degumming step takes advantage of the solubility of the silk sericin in boiling aqueous solutions containing different degumming agents, such as surfactants, alkalis, and organic acids.[10] However, aqueous alkaline solutions are the most widely used to degum the raw silk fibres.[7,9,10] The degummed silk fibres are sometimes called the fibroin fibres because it is considered that the sericin coating is almost completely removed.

Currently, unlike the fibroin which is the core of the silk fibres, the sericin is a major waste material in raw silk processing. It is estimated that average cocoon production is one million tonnes worldwide and raw silk processing from this amount of cocoons can yield sericin waste as high as 50 000 tonnes.[25] Although the sericin protein is unutilized in the textile industry, extensive research has been carried out to find possible exploitation in other fields, especially in pharmacological, cosmetic, biomedical, and biotechnological applications. An important reason why sericin has gained attention in recent years is that the discarded sericin in wastewater can cause extreme environmental problems, so the recovery and recycling of sericin are expected to yield economic and social benefits.[12,25] The potential uses of silk sericin have been intensively discussed in a number of review articles.[12,25] Although the sericin proteins can be prepared in different forms, such as films, hydrogels, and porous materials, the majority of the sericin-based material is classified as polymer blends;[12,25] only a few are composite materials. Therefore, this chapter will focus on the fabrication of the regenerated silk fibroin.

11.2.2 Regenerated Silk-based Materials

Figure 11.1 shows an overview of the fabrication of regenerated silk-based materials into various forms. The degummed silk fibres are first dissolved in a solvent system to prepare silk fibroin solution for reprocessing into desired shapes and structures. However, the silk fibres are insoluble in most aqueous solutions, including dilute acids and bases, due to the highly ordered orientations of silk fibroin chains in combination with intermolecular hydrogen

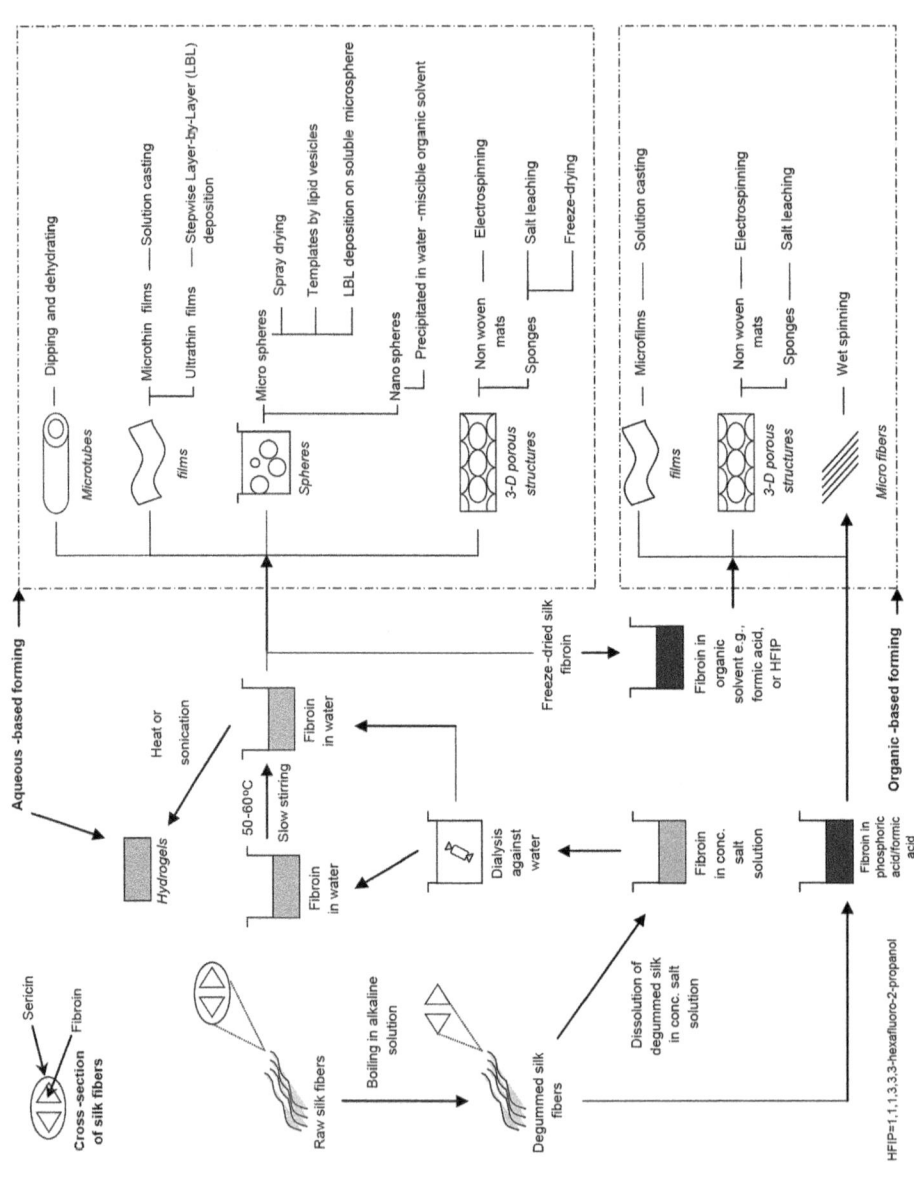

Figure 11.1   An overview of the fabrication of regenerated silk-based materials into various forms.

bonds and hydrophobic interactions.[10,11] Therefore, to effectively dissolve the silk proteins, the compact fibrous structure should be swollen and the hydrogen bond network as well as the hydrophobic interaction should be interrupted. Moreover, an ideal solvent should penetrate into the silk fibres without inducing any undesirable reactions, such as depolymerization and derivatization.[10]

Normally, the most common solvents used to dissolve the silk fibres can be categorized into two main groups: aqueous-based and organic-based solvent systems. The aqueous-based solvent systems are concentrated aqueous salt solutions, such as lithium bromide[28–33] and a ternary solvent of calcium chloride/water/ethanol.[34–38] After complete dissolution, the metal ions have to be removed from the as-prepared silk fibroin solution, perhaps by dialysis against distilled water, because the remaining metal ions can further hinder the regeneration process of the silk fibroin.[9] In general, the fibroin proteins in aqueous condition should be kept at a low temperature (4 °C) in order to maintain the solution lifetime before gelation occurs (which can be accelerated by heat and high fibroin concentration under static conditions). Practically, to increase the concentration of aqueous silk fibroin solution by water evaporation, the solution is slowly stirred in the temperature range of 50–60 °C for the disruption of inter- and intramolecular forces of the fibroin proteins without providing strong shear force.[39] The presence of physical shear can consequently induce β-sheet formation, which is the water-insoluble form of silk fibroin.[19,21] For organic-based solvent systems, the silk-based materials (either the degummed silk fibres or the freeze-dried silk fibroin) are able to be dissolved in organic acids, such as formic acid,[40–42] or fluorinated solvents, like 1,1,1,3,3,3-hexafluoropropan-2-ol (HFIP),[43–46] prior to fabrication into desired forms. It should be noted that the freeze-dried fibroin materials dissolve in these organic solvents easier than the degummed silk fibres. Another effective solvent system for the preparation of the fibroin solution is ionic liquid-based, including 1-butyl-3-methylimidazolium chloride.[47]

After the degummed silk fibres are completely dissolved in the selected solvent, the insoluble materials are removed from the as-prepared silk fibroin solution by centrifugation and the clear supernatant is subsequently used to regenerate the silk fibroin into diverse forms by various processing techniques, as shown in Figure 11.1. For example, the silk fibroin microtubes can be prepared by the dipping and dehydrating method,[48] while the micro- and ultrathin silk fibroin films are prepared using the solution casting procedure[38,49] and spin-assisted layer-by-layer (LBL) assembly,[47] respectively. The silk fibroin micro- and nanospheres are fabricated by spray drying,[50] vesicle templating,[51] LBL deposition on soluble templates,[52] and precipitation in a water-miscible organic solvent.[52] Regenerated silk fibroin hydrogels can be formed with the aid of heat and sonication.[53] For three-dimensional porous structures of the regenerated silk fibroin, nonwoven mats are prepared by electrospinning,[39,54,55] while sponges, or scaffolding materials, are fabricated with the use of either salt leaching[43–46] or freeze drying.[37] Other alternative forms of the regenerated silk fibroin are microfibres, which can be prepared by wet spinning.[56,57]

Since the dissolution of the silk proteins is achieved at the expense of their crystalline structure, the as-prepared regenerated silk-based materials are water soluble. With the use of an aqueous-based solvent system, the silk fibroin possesses either the random coil conformation (at low fibroin concentrations) or the α-helical structure (at high fibroin concentrations and in the absence of physical shear).[19,21] Therefore, after being regenerated, post-treatments are often required to induce the recrystallization (the formation of the β-sheet structure) of the silk fibroin. An example of the widely used post-treatments is methanol treatment. After the post-treatment, the regenerated-silk based materials become water insoluble because of a change in crystallinity and can be effectively used as biomaterials in a wide range of applications, such as controlled-release drug delivery systems and scaffolds in tissue engineering applications. Interestingly, regenerated silk-based materials are now being used in clinical trials as an anterior cruciate ligament (ACL) replacement. It is also expected that such material should achieve medical approval for clinical use in the near future and will be commercially available in both Europe and the USA within 2014.[9]

11.3   Silk-based Composites: Fabrication and their Potential Applications

Conventionally, composites are defined as materials consisting of two or more distinct and identifiable components, which subsequently provide desirable properties that cannot be found in the original materials.[58,59] However, this definition is quite ambiguous because, in the case of a polymer/polymer system, it also covers copolymers and blends. Hence, composites have recently been defined as materials that can be prepared by aligning strong and stiff constituents, such as fibres and particulates, in a continuous phase that acts as a binder.[59] In general, a filler phase is responsible for the mechanical strength of the composites, so it is called a reinforcing phase, while a binder phase is called a matrix. Materials that can be used as reinforcements include fibres, particles, microspheres, whiskers, and flakes, while those serving as a composite matrix are metals, ceramics, and polymers. It should be noted that the fillers can be incorporated into the matrix not only for providing mechanical strength but also for specific purposes, such as for providing either thermal stability or electrical conductivity to the resulting composite.

Since the properties of composite materials are determined by the properties of their components, by the morphology or shape of the filler phase, and by the nature of the interface between two or more distinct phases, the composite performance can be greatly varied by simply altering their components. However, in some cases, the composite fabrication has to be compromised and the processabilities of the composites should be taken into account because the combination of some components can possibly lead to undesirable properties. Another important aspect is that the interfacial adhesion between two or more discrete phases can play an important role in the mechanical performance of the fabricated composites.

Composites can be divided into many categories, based on the morphology or shape of the incorporated fillers, but two intensively studied composites are particulate-filled composites, which contain discrete particles as a discontinuous filler phase in a continuous matrix phase, and fibre-filled composites. The fabrication of the particulate-filled composites involves suspension of rigid particles in liquids or molten polymers, while the fibre-filled composites can be prepared by several techniques, including impregnation of fibre mats, lamination of impregnated sheets of oriented fibres or woven fabrics, impregnation of three-dimensional woven fabric, and injection or compression molding of polymers containing reinforcing fibres.[58]

Silk materials can be used to fabricate both types of composites, particulate-filled and fibre-filled. Moreover, in silk-based composites, both silk fibres and regenerated silk fibroin can serve as either reinforcement or matrix. Different kinds of fillers can be incorporated into silk materials in order to provide not only mechanical strength but also new favorable properties, such as antimicrobial activities, biological activities, optical properties, magnetism, and electrical conductivity. Industrial scale production of silk and its unique characteristics, especially its degradability and high biocompatibility, are the key factors that promote intensive study on the development of silk-based composites for possible exploitation in various fields, ranging from traditional clothing to highly innovative devices and biomedical end-uses.

Figure 11.2 shows an overview of the fabrication of silk-based composites. Among a variety of processing techniques, the most widely used methods for the fabrication of silk-based composites include injection molding, compression molding, freeze drying, and salt leaching.

The first two methods (injection and compression molding) frequently involve molten polymers or heated viscous materials. In a conventional injection molding, either a pre-blended or a physical mixture of polymer pellets/granules and reinforcing fibres is fed into the hopper before being conveyed by the flights of a rotating screw to the heated barrel.[60,61] As the screw rotates, the fed materials are accumulated at the front of the screw. At a specific time interval, the screw stops rotating and moves forward in order to inject the heated viscous material under high pressure into the mold cavity. After solidification, the mold is open and the fabricated products are ejected.[60,61] In compression molding, a mixture of either polymer pellets/granules or a resin dough and reinforcing fibres are generally placed in the fixed bottom mold half before being compressed at an elevated temperature as the top mold half moves down, usually with the aid of a hydraulic press. After solidification, the top mold moves up and the composite product is removed from the bottom mold with the aid of ejector pins.[62] Compared to injection molding, compression molding shows a number of advantages.[62] For example, it produces a lower amount of waste because of the use of a simply designed mold. Moreover, since compression molding involves a much low shear rate during the composite fabrication, no severe degradation is observed in the resulting composites. In addition, high reinforcing fibre volume fractions can be used to fabricate the composite materials with compression molding, so the obtained compression

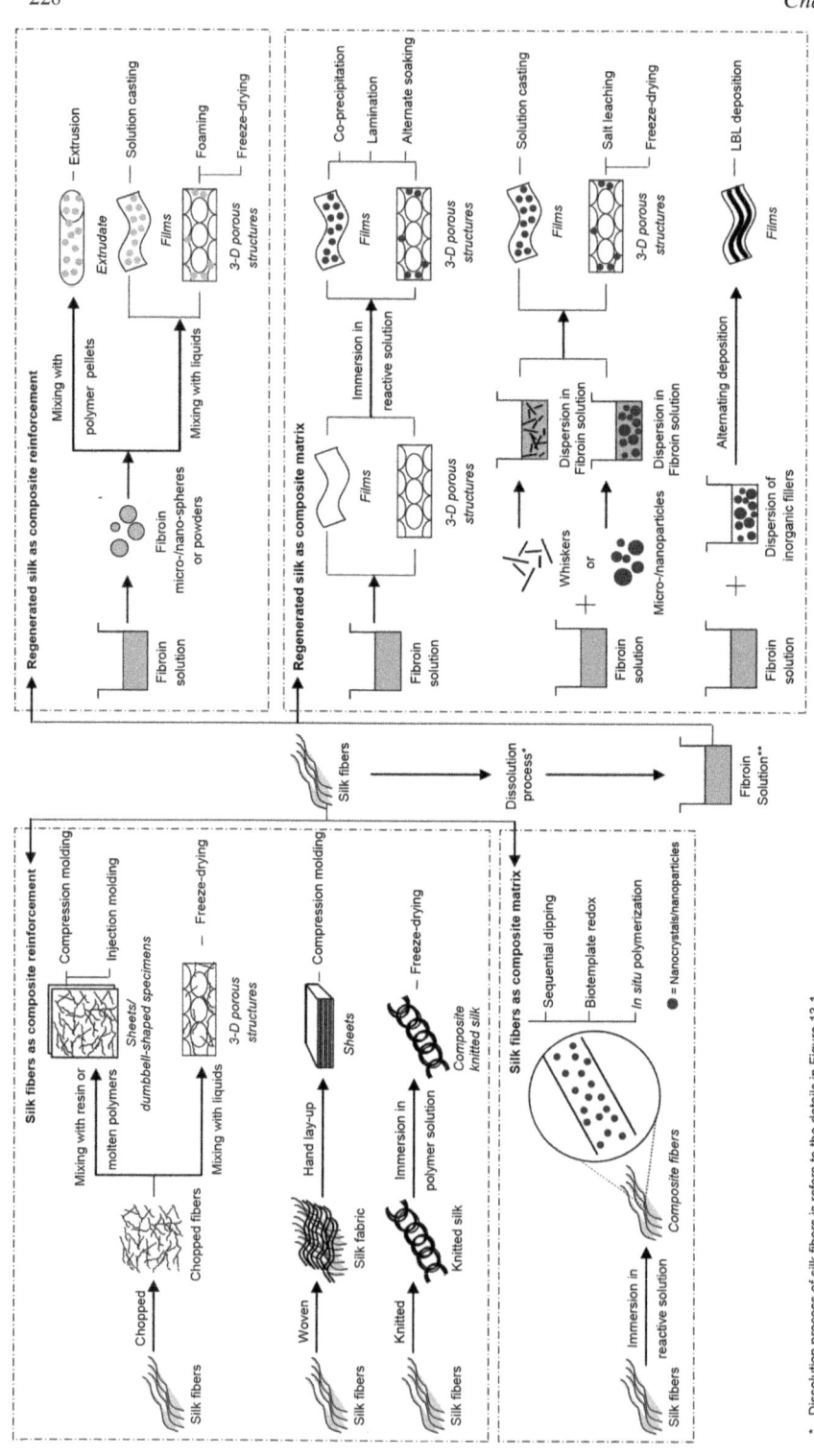

Figure 11.2   An overview of the fabrication of silk-based composites.

* Dissolution process of silk fibers is refers to the details in Figure 13.1.
** Fibroin solution means fibroin proteins dissolved in either water or organic solvents depending on a pair of matrix and reinforcement and/or method of composite fabrication.

molded products possess better physical and mechanical properties.[62] The compression molding is also more suitable for the production of composite materials with a large surface area.[62]

Freeze drying and salt leaching techniques are generally used for the fabrication of composite materials with three-dimensional porous structures, and both of them involve either the dispersion of solid particles or the impregnation of structured materials in a polymer solution. In the freeze-drying technique, as an aqueous polymer solution is frozen at a temperature below its freezing point, ice crystals begin to form; this phenomenon is called nucleation.[63] These ice crystals in the frozen polymer solution are subsequently removed by sublimation, yielding the three-dimensional porous structures. Typically, porosity, pore size, and pore structure of the freeze-dried materials depend on several factors, such as freezing temperature, freezing rate, and polymer concentration. Normally, the lower the freezing temperature, the smaller the obtained pore size. Another important aspect is that the pore shape can be affected by heat transfer during freezing, because uniform heat dissipation in all directions causes uniform ice crystal formation in the frozen polymer solution.[63] In the case of salt leaching, salt granules are used as a porogen instead of ice crystals in the freeze drying. A desired amount of salt is added to the polymer solution before the solvent is evaporated, resulting in the embedment of salt granules in a solid polymeric material. After that, the salt granules, a water-soluble porogen, are leached out with an excessive wash with water. Thus, the pore size and shape of the final product are dependent on the size and shape of salt granules. Compared to the freeze-drying technique, the salt leaching has some limitation due to the leaching step, especially when small molecules, like bioactive compounds or drug molecules, have to be retained in the final products. However, no expensive instrument is required in the salt leaching technique.

11.3.1 Silk Fibres as Composite Reinforcement

Recent emphasis on environmental awareness has led to the consideration of natural fibres as potential alternatives to synthetic fibres for being used as composite reinforcement. Among two types of natural fibres, namely plant-based lignocellulose fibres and animal-based protein fibres, the silk fibres possess several interesting characteristics, such as mechanical properties, thermal stability, biodegradability, and high biocompatibility. As a result, the silk fibres are used to reinforce various kinds of matrices, including synthetic polymers, biopolymers, polymer blends, and inorganic matrices. For the silk fibre-reinforced polymer matrices, the fabricated composites can be called biocomposites, which are defined as composite materials consisting of natural fibres as reinforcing elements and biodegradable or non-biodegradable polymers as matrices.[13,64,65] The reinforcing silk fibres can be either the raw silk fibres, which are partially degummed during fibre processing, or the degummed silk fibres, which are sometimes called the fibroin fibres. The incorporation of silk fibres in the composite matrices can be in the form of a complex structure, like knitted silks or woven fabrics. Moreover, in some researches, waste silk fibres obtained from the textile industry are also used as composite reinforce-ment. The silk fibres can even be combined with other fibres, *e.g.* plant-based fibres or synthetic fibres, before being used as reinforcement because it is believed that a combination of two or more different types of fibres in a matrix offers a wide range of properties that cannot be obtained from a single kind of fibre reinforcement. Since short fibres can be aligned in the composite matrix with a higher degree of orientation, leading to greater composite performance, short silk fibre-reinforced composites are extensively studied. Regardless of the fibre length, the silk fibre content is another key factor affecting the composite performance. The silk fibre-reinforced composites are good candidates to be potentially used in diverse fields, from general-purpose plastics to scaffolding materials in tissue engineering applications. Table 11.1 shows examples of silk-based composites with silk fibres as composite reinforcement in various matrices, their improved properties, and their proposed applications.

11.3.1.1 *Silk Fibre-reinforced Synthetic Polymers*

Waste silk fibres obtained from cocoon reeling, which are undesirable in the textile industry, can be chopped prior to incorporation into nitrile rubber.[66] To improve the adhesion between the chopped silk fibres and the rubber matrix, a "resorcinol/hexamethylenetetramine/silica" bonding system was used. All components were mixed by a conventional laboratory open mill before being compression molded. It was reported that the composite performance was strongly dependent on the fibre content. An increase in the silk fibre content increased hardness, tensile strength, tear strength, modulus, heat build-up, abrasion loss, restriction to solvent swelling, and processability of the fabricated composite, but simultaneously decreased resilience and elongation at break. After ageing at 100 °C for 48 h, the silk fibre-reinforced nitrile rubber composite was able to retain its mechanical strength in terms of tensile strength, tear strength, and elongation at break. In contrast, the nitrile rubber without silk fibre reinforcement deteriorated and its strength was expected to reduce after ageing.

Chopped natural silk fibres were used to reinforce poly(butylene succinate) (PBS), a biodegradable, thermoplastic, and aliphatic polyester.[65] The bio-composite was prepared using the compression molding method and its mechanical and thermal properties were investigated. The chopped raw silk fibres were found to markedly improve tensile and flexural properties and thermomechanical stability of the PBS, while the thermal stability of the silk fibre/PBS biocomposite was intermediate between the reinforcement and the matrix. Therefore, the natural silk fibre was an effective reinforcement and might be a potential candidate for enhancing the performance of biodegradable polymer matrix resins. These findings are also consistent with the work of Han *et al.*, who reported that the use of chopped waste silk fibres, which are industrially available, as reinforcement improved both static and dynamic mechanical properties of the PBS matrix.[67] Moreover, above the glass transi-tion temperature, the storage modulus of the composite was significantly higher than that of the PBS matrix.

Table 11.1 Examples of silk-based composite with silk fibres as composite reinforcement.

Silk-based composite[a]	Fabrication method	Improved properties	Proposed applications	Ref.
Silk fibre-reinforced synthetic polymers				
Silk/PP	Compression molding	Mechanical properties, degradabilities	Semi-structural and structural applications	1
Silk/nitrile rubber	Compression molding	Mechanical properties	–	66
Silk/PBS	Compression molding	Mechanical properties, thermal properties, water absorption abilities, degradabilities	Automobile applications	67,68
Silk/PLA	Extrusion and injection molding	Mechanical properties, thermal properties, dimensional stabilities, water absorption abilities, degradabilities	Biomedical and bioengineering applications, environmental engineering applications	69–71
Silk/PCL	Melt mixing	Mechanical properties	Biomedical materials	72–76
CNTs-coated silk/PBSA	Compression molding	Mechanical properties	Commodity materials	77
Silk/epoxy resin	Microbond method	Interfacial strength	Tough composites	78
	Hand lay-up	Interfacial strength, mechanical properties, chemical resistance	Applications in chemical surroundings	79,80
Silk/PMMA, silk/PC, silk/ABS	Hand lay-up	Mechanical properties, chemical resistance	Applications in chemical surrounding	81–83
Silk/sisal fibres/unsaturated polyester resin, silk/coir fibres/unsaturated polyester resin, silk/glass fibres/epoxy resin	Hand lay-up	Mechanical properties, chemical resistance	Materials for making chemical storage tanks	84–87
Silk fibre-reinforced biopolymers				
Silk/collagen	Freeze drying	Mechanical properties, biocompatibilities	Ligament and tendon tissue engineering	88,89
Silk/collagen/hyaluronic acid	Freeze drying	Biocompatibilities	Ligament and tendon tissue engineering	90
Silk/gelatin	Compression molding	Mechanical properties, degradabilities	Single-used operating theater trays	91

Table 11.1 (*Continued*)

Silk-based composite*a*	Fabrication method	Improved properties	Proposed applications	Ref.
Silk fibre-reinforced polymer blends				
Silk/PP/NR	Extrusion and compression molding	Mechanical properties, degradabilities	Semi-structural and structural applications	92
Silk fibre-reinforced inorganic matrices				
Silk/MPAHC	Sol/gel method	Mechanical properties	Biocements for ostepedic, ortho-paedic, or dental surgery	93
Silk/PVOH/MMT	Freeze drying	Mechanical properties	Environmentally benign materials	94

*a*PP = polypropylene; PBS = poly(butylene succinate); PLA = poly(lactic acid); PCL = poly(ε-caprolactone); CNTs = carbon nanotubes; PBSA = poly[(butylene succinate-*co*-(butylene adipate)]; PMMA = poly(methyl methacrylate); PC = polycarbonate; ABS = acrylonitrile/butadiene/styrene; NR = natural rubber; MPAHC = magnesium-bearing phosphoaluminate/hydroxyapatite; PVOH = poly(vinyl alcohol); MMT = montmorillonite.

A subsequent study further investigated the long-term stability of the silk fibre/PBS biocomposite in a wet environment in a comparison to that of the PBS reinforced with chopped henequen fibres, lignocellulose-based fibres.[68] Both PBS biocomposites were fabricated using the compression molding technique and the hygrothermal effects on silk fibre- or henequen-reinforced PBS biocomposites were observed at 60 °C and 85% relative humidity (RH). At the exposure duration of 1000 h, it was found that different degradabilities of these two reinforcing fibres led to different mechanical performances of the biocomposites. Although the incorporation of the silk fibres in the PBS enhanced the water absorption of the biocomposite and consequently caused a higher hydrolysis degradation of the PBS matrix, the silk fibre/PBS biocomposite still showed superior mechanical properties because the reinforcing silk fibres did not degrade. On the contrary, although the henequen/PBS biocomposite exhibited much lower water absorption, the reinforcing henequen fibres degraded rapidly and the mechanical properties of the reinforced biocomposite could not be maintained. As a result, the origin of the natural fibres significantly affected the biocomposite degradability in the long-term durability, so the selection of natural fibres as reinforcements must be carefully carried out and depends on the intended biocomposite applications. These findings are in accordance with another study which revealed that the origin of the natural fibres not only affected the degradability of the fabricated composite but also influenced its mechanical properties in terms of bending strength, bending modulus, tensile strength, tensile modulus, and impact strength.[1] In that work, either silk fibres or jute fibres, lignocellulose-based fibres, were used to reinforce polypropylene (PP). Compared to the neat PP, the incorporation of either silk fibres or jute fibres into the PP matrix remarkably improved its mechanical properties. However, when the two composites were compared, after 12 weeks of the soil degradation test the silk fibre/PP composite still possessed its excellent mechanical properties while the jute fibre/PP composite degraded much faster. Again, the results showed that the silk fibre-reinforced composite retained its mechanical performance longer than the lignocelluloses-based fibre-reinforced composite because of its slower degradation.

Natural silkworm silk fibres were gently cut into short fibre fractions before being used as a reinforcement of poly(lactic acid) (PLA), a synthetic biodegradable polymer which is widely used as an implant in medical applications, to prepare a biodegradable biocomposite.[69] The silk fibre/PLA composite was fabricated using extrusion and injection molding techniques and its mechanical and thermal properties were studied. It was observed that the silk fibres were well bonded with the polymer matrix. Compared to the pure PLA, the mechanical properties, in terms of elastic modulus and ductility, and thermal stability of the biocomposite were substantially improved. To evaluate the potential use of the silk fibre/PLA biocomposite in tissue engineering applications, its biodegradability was subsequently studied *in vitro* at specific time periods.[70] In that study, the reinforcement of silk fibres increased both stiffness and ductility of the PLA, and a biodegradation rate of the silk fibre/PLA composite was faster than that of the pure PLA, indicating that the incorporation of silk fibres not only improved the mechanical properties of the PLA but also altered its biodegradability.

Regardless of the tissue engineering applications, the silk fibre/PLA biocomposite is potentially useful as a high-performance material for environmental engineering applications.[71] Compared to pure PLA, although the incorporation of the silk fibres in the PLA matrix reduced the thermal stability of the fabricated composite, the stiffness at high temperature (in the range of 70–160 °C) and dimension stability were markedly enhanced. The biodegradability of the biocomposite was carried out with the use of a well-known protein-digesting enzyme: proteinase K. The silk fibres increased the water absorption of the biocomposite, so the silk fibre reinforcement enhanced the hydrolysis of the PLA matrix. Moreover, the enzymatic degradation of the reinforcing fibres also promoted the biocomposite degradability. Since the degradation of the silk fibre/PLA biocomposite increased with increasing the silk fibre content, it was suggested that the degradability of the PLA matrix should be controlled by altering the amount of the silk fibres in the fabricated composite.

The degummed silk fibres were used to reinforce a film of poly(ε-caprolactone) (PCL), a biodegradable polyester.[72–76] The silk fibre-reinforced PCL composite was prepared by melt mixing. The influence of the reinforcing silk fibres on the mechanical properties, viscoelastic properties, and crystallization behavior of the resulting biocomposites was investigated. The results showed that the incorporation of the silk fibres into the PCL matrix improved the mechanical properties of the biocomposites in terms of tensile and flexural strengths. Moreover, both the storage modulus and loss modulus also increased with increasing fibre content. The silk fibres improved the crystallization, both isothermal and non-isothermal processes, of the PCL matrix. Since the silk fibre-reinforced PCL biocomposite has a potential use as a biomedical material, the effect of electron beam irradiation (widely used sterilization in the medical industry) on the properties of the biocomposite was further studied. It was observed that the irradiation induced both the cross-linking and the degradation of the polymer components and the mechanical properties of the bio-composite were directly dependent on the irradiation dose. The tensile and flexural strengths moderately increased with increasing irradiation dose before dramatically decreasing as the irradiation dose increased above 150 kGy. The irradiation was also found to reduce the overall crystallization rate of the PCL matrix due to the cross-linking. This negative effect became more dominant in the non-isothermal crystallization, compared to isothermal crystallization.

Biodegradable composites containing poly[(butylene succinate)-*co*-(butylene adipate)] (PBSA), an aliphatic polyester, as a polymer matrix and silk fibres coated with carbon nanotubes (CNTs) as a reinforcement were prepared by the compression molding method.[77] Compared to the uncoated silk fibres, the CNT coating provided better interfacial interaction between the matrix and the reinforcement, leading to greater mechanical performance of the resulting CNT-coated silk fibre/PBSA composite, which should be beneficial for use as a commodity material, such as a general-purpose plastic.

A degummed silk fibre/epoxy composite was prepared and the interaction between the silk fibre and the resin matrix was studied.[78] With the use of the microbond test, the mean interfacial shear strength was found to be about 15 MPa, indicating that the silk fibre efficiently reinforced the epoxy resin, one of the most important thermosetting polymers. Although epoxy resin is widely used as either a structural material or a structural adhesive, it is brittle and has poor resistance to crack propagation. Since the silk fibres possesses high tensile strength and high extensibility, comparable to those of glass fibres or synthetic polymers, the determined interfacial shear strength suggests another exploitation of the silk fibre in tough composites.

Not only silk fibres but also waste silk fabric obtained from the textile industry has been used as composite reinforcement.[79–83] The waste silk fabric-reinforced epoxy resin was prepared by a hand lay-up method. The incorporation of the waste silk fabric into the epoxy resin increased the mechanical performance of the fabricated composite in terms of tensile, flexural, impact, and compression strength. The waste silk fabric-reinforced epoxy resin showed good chemical resistance to acids, alkalis, water, and organic solvents. More-over, good interfacial adhesion between the reinforcing silk fabric and the epoxy resin also resulted in a lower void content of the fabricated composite. However, it was suggested that interfacial bonding should be improved by surface treatment of the silk fabric, probably with the use of an appropriate coupling agent.

In later work,[81–83] the influence of tougheners and waste silk fabric content on mechanical properties and chemical resistance of the fabricated composites were studied. In the presence of toughener, the epoxy resin was prior mixed with three thermoplastics, including poly(methyl methacrylate) (PMMA),[81] polycarbonate (PC),[81] and an acrylonitrile/butadiene/styrene (ABS) copolymer,[82,83] before being reinforced with the waste silk fabric. Both in the absence and presence of toughener, the mechanical performance of the composites in terms of tensile, flexural, and impact properties were found to increase with increasing in the fabric content. For their resistance to moisture, the water absorption capacity of the silk fabric-reinforced epoxy resin decreased in the presence of toughener but increased with an increase in the fabric content. In addition, the toughened epoxy resin composites also displayed good chemical resistance to acids, alkalis, and organic solvents, indicating good efficacy for applications in chemical surroundings.

Recently, it has been reported that either the silk fibres or the silk fabric can be combined with other types of reinforcing fibres, including sisal fibres,[84,85] coir fibres,[86] and glass fibres,[87] prior to use as reinforcement of resin matrices, either unsaturated polyester resin[84–86] or epoxy resin.[87] The composites were prepared with the use of the hand lay-up technique. Again, the fabricated composites exhibited an increase in mechanical performance in terms of tensile, flexural, and compressive properties, and with good chemical resistance, indicating their potential use for making chemical storage tanks. It was also reported that the interfacial adhesion between the natural fibres and the resin matrix was able to be improved by treating the fibre surface with an alkaline solution.[84–86]

In conclusion, although the silk fibres are effectively used as reinforcement for various kinds of synthetic polymer matrices, the silk fibre content in the polymer matrix should be optimized in order to obtain the desired composite performance. Another important aspect is that the hydrophilic nature of the silk fibres can enhance the water absorption of the fabricated composite and subsequently promote hydrolysis of the biodegradable synthetic polymer matrix. Compared to plant-based fibres, the silk fibres (animal-based fibres) are better composite reinforcements from the viewpoint of long-term stability because of their slower biodegradability.

11.3.1.2 Silk Fibre-reinforced Biopolymers

Since both silk fibres and some biopolymers possess excellent biocompatibilities, most silk fibre-reinforced biopolymers are prepared for possible use in the biomedical field, particularly for tissue engineering applications. However, a slow *in vivo* degradation of the silk fibres makes the fabricated composites valuable for longer regenerating engineered tissues, such as bones, cartilages, ligaments, and tendons.

The incorporation of knitted silk fibres in a collagen matrix has been reported.[88,89] The raw silk fibres were used to prepare a warp knitted scaffold with the use of a knitting machine. After being degummed, the knitted silk was immersed in an acidic collagen (type I) solution and then freeze dried. The fabricated knitted silk/collagen composite scaffold exhibited optimal internal space, good biocompatibility, and good mechanical strength. The composite scaffold not only increased structural and functional ligament repair, but also enhanced the differentiation of human embryonic stem cell-derived mesenchymal stem cells (hESC-MSCs), the most widely used stem cells in tendon tissue engineering. Thus, the knitted silk/collagen composite scaffold should be a promising material for practical utilizations in both ligament and tendon tissue engineering.

In another study, the knitted silk-based composite scaffold was prepared by immersing the knitted silk in an acidic mixture solution of collagen and hyaluronic acid, or hyaluronan, one of the glycosaminoglycans.[90] The composite scaffold was then fabricated by the freeze-drying technique. The biocompatibility of the composite scaffold was evaluated *in vitro* by assessing anterior cruciate ligament (ACL) and T-lymphocyte cultures. The utilization of the composite scaffold as artificial ligaments in ACL reconstruction and its influence on angiogenesis was investigated *in vivo*. The initial attachment and proliferation of human ACL cells on the composite scaffold were higher than those on the knitted silk; however, the *in vitro* immune responses between the two scaffolds were not different. After being grafted in the knee joint of a dog for 6 weeks, both granulation tissue and new blood vessels were observed in the dog grafted with the composite scaffold, while implantation of the knitted silk scaffold did not provide the formation of reparative tissues. These findings indicated that the knitted silk scaffold with a lyophilized collagen/hyaloronic acid mixture was biocompatible *in vitro* and promoted the formation of new blood vessels as well as the cell migration *in vivo*, suggesting its potential for ligament and tendon tissue engineering.

A natural silk fibre-reinforced gelatin composite was fabricated in order to prepare a biodegradable composite to serve as a single-used operation theater (OT) tray because traditional OT trays, which are made of steel, require labor- and time-consuming processes for cleaning and sterilization after operations.[91] The composite was fabricated by undirectionally arranging three layers of silk fibres among four layers of gelatin film before being compression molded. Compared to the pure gelatin matrix, the silk fibre-reinforced gelatin composite showed higher tensile strength, tensile modulus, bending strength, bending modulus, and impact strength. The fabricated composite rapidly degraded after being buried in compost soil by losing more than 50% of its original weight within 24 h.

11.3.1.3 Silk Fibre-reinforced Polymer Blends

Short silk fibres were used to reinforce a thermoplastic elastomer blend of polyethylene (PE) and natural rubber (NR).[95] To obtain a uniform dispersion of silk fibres in the PE/NR blend, the silk fibres were first added to the molten thermoplastic before blending with the NR. With a proper mixing sequence, it was observed that each silk fibre was coated with a thermoplastic layer. The "resorcinol/hexamethylenetetramine/silica" bonding system was also used to promote the adhesion between the reinforcing fibres and the matrix. Although the bonding agent was able to improve composite performance, a longer curing time was required for full adhesion development. An increase in the silk fibre content substantially increased both the tensile and the tear properties of the fabricated composite; however, its elongation sharply reduced. Further examination revealed that the silk fibre acted as a mechanical anchor between the thermoplastic phase and the NR. Nevertheless, the effect of the silk fibre content on the mechanical properties of the composite became less dominant at high thermoplastic blend ratios.

A recent study reported on the preparation of a silk fibre-reinforced PP/NR blend.[92] The NR lumps were cut into small pieces and blended with PP in an extruder before being cold pressed. The obtained PP/NR blend film was used to fabricate a silk fibre-reinforced composite by arranging four layers of the blend films among three layers of the silk fibres before being compression molded. The silk fibre-reinforced PP/NR blend exhibited good degradability, but the NR in the blend dramatically decreased the mechanical properties of the fabricated composite in terms of tensile strength, tensile modulus, bending strength, bending modulus, and hardness (except impact strength). Hence, γ-radiation was employed to improve the mechanical properties of the composite. It was found that the mechanical properties of the composite increased with increasing radiation dose to a certain limit before starting to decrease. At low radiation doses, the radiation not only removed moisture from the composite but also induced cross-linking between the silk fibre and the blend, leading to better fibre/matrix adhesion and an increase in mechanical properties of the composite. In contrast, high radiation doses possibly induced polymer chain scissions of silk fibre, PP, and NR, hence consequently reducing composite performance. The silk fibre-reinforced PP/NR composite was degradable so it was an environmentally friendly material which should be suitable for semi-structural as well as structural applications.

11.3.1.4 Silk Fibre-reinforced Inorganic Matrices

Raw silk fibres were used as reinforcement for a composite cement of magnesium-bearing phosphoaluminate/hydroxyapatite (MPAHC).[93] The silk fibres were treated with nitric acid and subsequently soaked in an aqueous ethanol solution. After drying, the treated silk fibres were mixed with the cement before molding. Among the test specimens, the composite cement containing 0.5 wt% of the treated silk fibres possessed the highest splitting strength and Vickers micro-hardness. The incorporation of treated silk fibres in the cement enhanced the mechanical strength of the studied system while maintaining its biocompatibility. As a result, MPAHC reinforced with the treated silk fibres should be a preferential candidate for biocement purposes, which might find applications in the fields of osteopedic, orthopedic, or dental surgery.

The silk fibres and a number of biologically based fibres, such as hemp, ramie, and bamboo top, were used as reinforcements of clay aerogel to fabricate environmentally benign materials for a wide range of end-uses.[94] Poly(vinyl alcohol) (PVOH) was also incorporated in the clay aerogel to improve its mechanical performance. An aqueous PVOH solution was mixed with a suspension of sodium montmorillonite (Na^+-MMT) before adding chopped fibres to the PVOH/clay mixture at various fibre contents. The fibre-reinforced PVOH/clay composites were prepared *via* the freeze-drying procedure. The chopped fibres formed a woven-like structure in the clay aerogel. The mechanical properties in terms of compressive strength and moduli of the fabricated composites depended on the type of reinforcing fibres, but increased remarkably with increasing the fibre content.

11.3.2   Silk Fibres as a Composite Matrix

The unique fibril structure, the good mechanical properties, and the presence of amino acid components containing strongly charged and reactive side groups make silk fibres an attractive substrate for the embedment of different kinds of inorganic nanoparticles. The embedded nanoparticles provide new biological activities and physical properties to the silk fibres, such as antimicrobial activities, optical properties, magnetism, and electrical conductivity, hence further expanding their possible use as smart textiles in various fields. Table 11.2 lists examples of silk-based composites with silk fibres as a composite matrix containing various types of fillers, their improved properties, and their proposed applications.

Cubic silver chloride nanocrystals were synthesized on silk fibres *via* sequential dipping of the silk fibres in alternating solutions of silver nitrate and sodium chloride.[96] It was proposed that the negative charge at the surface of the silk fibres provided the immobilization of Ag^+ prior to reaction with Cl^-, leading to the formation of an initiator seed of AgCl, and the sequential dipping further resulted in the growth of nanocrystals. The AgCl-incorporated silk fibres should be used as either a photocatalyst in water-splitting applications or as an anti-bacterial agent. Another attempt to develop silk nanocomposites with anti-bacterial properties as potential textiles for hygienic clothing, wound healing, and

Table 11.2   Examples of silk-based composite with silk fibres as composite matrix.

Silk-based composite	Fabrication method	Improved properties	Proposed applications	Ref.
AgCl/silk	Sequential dipping	Optical properties, antimicrobial activities	Photocatalyst in water-splitting applications, antibacterial materials	96
AgBr/silk	Sequential dipping	Antimicrobial activities	Hygienic clothing, wound healing, and medical applications	97
Ag/silk	Biotemplate redox	Optical properties	Chemical sensing, separation, catalysis, data storage, and photonic materials	98,99
CdS/silk	Bio-inspired process	Optical properties	Compartment in nanodevices for bio-related fields	100

medical applications was the incorporation of silver bromide nanoparticles in silk yarns *via* sequential dipping in alternating solutions of silver nitrate and potassium bromide under ultrasound irradiation.[97] Compared to the traditional synthesis procedure, the sonochemical method yielded smaller AgBr nanoparticles with narrower particle size distribution on the silk yarns.

Silver nanoparticles were successfully incorporated into degummed silk fibres using a biotemplate redox procedure at room temperature, a green and mild technique to fabricate functional hybrid nanocomposites for possible utilization in the fields of chemical sensing, separation, catalysis, data storage, and photonic materials.[98,99] The tyrosine in the silk fibroin was found to mediate the reduction of Ag(I) ions to Ag(0) under alkaline conditions, resulting in the *in situ* formation of Ag nanoparticles in the silk fibroin fibres, an inorganic/organic hybrid nanocomposite.

Regardless of the nanoparticles of Ag and its complexes, there is a report on the synthesis of cadmium sulfide (CdS) strings and hexagons on the degummed silk fibres through a room temperature bio-inspired process.[100] Unlike the Ag which provided an antibacterial effect, the CdS, a typical semiconductor, gave optical properties to the fabricated composites. In that study,[100] the silk fibres which acted as reactive substrates provided *in situ* formation sites of CdS nanocrystals and simultaneously controlled the arrangement of the resulting nanocrystals into specific patterns. The CdS/silk nanocomposite displayed photoluminescence, suggesting its utilization as a compartment in nanodevices for bio-related fields.

Although most studies focused on the use of the silkworm silk fibres as a composite matrix, the preparation of spider silk fibre composite was also reported.[101,102] Dragline spider silks from *Nephila edulis* were impregnated with different kinds of inorganic nanoparticles, such as superparamagnetic magnetite (Fe_3O_4), gold, and CdS, by immersing the spider silk fibres in the as-prepared nanocolloidal sol.[101,102] Compared to bare dragline spider silk fibres, the impregnation of inorganic nanoparticles showed a negligible effect on the mechanical properties of the resulting composite fibres in terms of tensile strength, elongation at break, and tensile modulus. However, the combination of the natural strength and elasticity of the spider silk with physical properties of the added inorganic nanoparticles (including magnetism, electrical conductivity, and semiconductivity) should be useful as smart structural fabrics in a wide range of applications.

11.3.3 Regenerated Silk as a Composite Reinforcement

Several research studies aimed to use the regenerated silk fibroin either in the form of microspheres or microparticles as reinforcement for various kinds of biodegradable synthetic polymers in order to develop new composites with desirable properties, such as biocompatibility, biodegradability, and environmentally friendly characteristics, for specific purposes. Since silk fibroin powder is already commercially available, the widespread utilization of the regenerated silk as a composite reinforcement is now being promoted. Table 11.3

Table 11.3   Examples of silk-based composites with regenerated silk as composite reinforcement.

Silk-based composite[a]	Fabrication method	Improved properties	Proposed applications	Ref.
Silk fibroin/PLA	Supercritical CO$_2$	Foam cell size, foam cell density	Bioplastic foams	103
	Freeze drying	Biocompatibilities	Hepatic tissue engineering applications	104,105
Silk fibroin/poly-(ε-caprolactone-co-D,L-lactide)	Solution casting	Mechanical properties	–	106
Silk fibroin/WPU	Solution casting	Mechanical properties	–	107
Silk fibroin/chicken egg shells/SEBS	Melt processing	Thermal properties	Biodegradable applications	108

[a]PLA = poly(lactic acid); WPU = waterborne polyurethane; SEBS = poly(styrene-b-ethylene/butylene-b-styrene).

lists some examples of silk-based composites with regenerated silk as composite reinforcement in various matrices, their improved properties, and their proposed applications.

Microcellular biodegradable silk fibroin/PLA composite foam was fabricated using supercritical carbon dioxide to develop a new bioplastic.[103] The silk fibroin powders were mixed with a PLA solution of various powder content prior to foaming. Compared to the neat PLA foam, the composite foam possessed a larger cell size and a higher cell density. In addition, both the cell size and the cell density increased with increasing powder content. Regardless of the silk fibroin powder content, the cell size and the cell density of the composite foam were also affected by the saturation temperature and pressure during the fabrication. Both the cell size and the cell density were simultaneously reduced with an increase in the saturation temperature and pressure. Therefore, the microcellular structure of the composite was able to be controlled by altering either the silk fibroin powder content or the processing conditions.

A composite scaffold composed of silk fibroin, either in the form of microspheres or microparticles, and a PLA matrix was prepared using the freeze-drying technique.[104,105] The hepatocellular compatibility and the inflammatory response of the silk fibroin/PLA composite scaffold were subsequently evaluated with the use of human hepatocellular liver carcinoma cells (HepG2) and mouse RAW264.7 macrophage cell lines, respectively. The scaffold possessed interconnected pore structures and the silk fibroin uniformly dispersed in the PLA. The HepG2 cells attached, spread, and proliferated on the composite scaffold much better than those on the PLA scaffold, indicating that the embedded silk fibroin improved the cell–sponge interactions. The inflammatory response of the macrophages to the composite scaffold was also found to be lower than that to the PLA scaffold. Thus, the silk fibroin/PLA scaffold was a good candidate for hepatic tissue engineering due to its higher hepatocellular compatibility with lower inflammatory response.

The silk fibroin particles were also incorporated into poly(ε-caprolactone-*co*-D,L-lactide), which is a biodegradable polymer.[106] The composite film was fabricated using the solution casting procedure and its thermal and mechanical properties were investigated. The silk fibroin particles were homogeneously dispersed in the polymer matrix. Although the decomposition temperature and the degradation rate of the fabricated composite decreased with an increase in the silk fibroin content, its mechanical properties (such as storage modulus and hardness) were improved.

Waterborne polyurethane (WPU) reinforced with silk fibroin powders has been reported.[107] To achieve a good dispersion of the silk fibroin powder in the WPU matrix, both the negative charges in the periphery and the small particle size were key factors. It was found that the dispersion of the silk fibroin powders in the WPU matrix was strongly dependent on the powder content. At high silk fibroin powder content, the silk fibroin/WPU dispersion exhibited lower negative zeta potentials and larger particle sizes, leading to poorer mechanical properties of the resulting composite film. On the other hand, at lower powder content, the added silk fibroin dispersed well in the WPU matrix and the silk fibroin/WPU composite film showed a remarkable increase in mechanical properties in terms of tensile strength and modulus.

In a recent study, silk fibroin powders and chicken eggshells were simultaneously used as bio-fillers to reinforce poly(styrene-*b*-ethylene/butylene-*b*-styrene) (SEBS), a member of the styrene tri-block copolymers, in order to prepare a composite for biodegradable applications.[108] The fibroin/eggshell/SEBS composite was fabricated *via* a melt processing method with the use of a co-rotating twin-screw extruder. The fibroin played a key role in improving the interfacial interaction between the reinforcements and the polymer matrix. Compared to the pure SEBS, the thermal stability of the resulting composite was found to increase.

11.3.4 Regenerated Silk as a Composite Matrix

Traditionally, when the regenerated silk fibroin was used as a matrix, the composites were prepared using two main procedures. In the first method, the silk proteins were processed into the desired morphology before the fillers/reinforcements were impregnated into the as-prepared regenerated fibroin. Another preparation technique involved mixing of the filler dispersion with the silk fibroin solution prior to fabricating the composite material into the favorable form. However, a spin-assisted layer-by-layer assembly has recently been proposed as a novel route to prepare regenerated silk-based nano-composite films.

Since the regenerated silk fibroin is one of the most promising biomaterials for bone tissue engineering applications because of its appropriate degradability, several research studies have tried to provide an osteogenic environment by the incorporation of biominerals that can be found in native bone.

In addition, although the regenerated silk fibroin has several favorable biological characteristics, including excellent biocompatibility with low inflammatory and immunological response, the important drawbacks of the silk proteins are poor mechanical properties and low dimensional stability, especially for highly porous structures. As a result, much effort has been expended to improve its physical performance by the fabrication of composite materials with the use of organic nanofillers (whiskers obtained from bio-polymers). Table 11.4 shows some examples of silk-based composites with regenerated silk as a composite matrix containing various types of fillers, their improved properties, and their proposed applications.

11.3.4.1 Regenerated Silk with Inorganic Fillers

Titanium dioxide (TiO_2) nanoparticles were embedded in a silk fibroin matrix with the use of either the sol/gel technique[109,110] or the freeze-drying method.[111] The TiO_2 nanoparticles dispersed well in the fibroin matrix without reagglomeration and it was also reported that the embedment of TiO_2 was able to induce the structural transition of silk fibroin from silk I to silk II. Compared to the fibroin film, the TiO_2/fibroin nanocomposite showed higher mechanical and thermal properties but lower water solubility. However, the excessive TiO_2 nanoparticles tended to disrupt the crystal structure in the nanocomposite. Since the fabrication of the nanocomposite *via* these techniques avoided the reagglomeration of the TiO_2 nanoparticles, it was proposed that the fabricated TiO_2/fibroin nanocomposite should find potential applications in the biomedical field.

Hydroxyapatite (HA), an important constituent component of bone, was incorporated into silk fibroin materials, including nonwoven net, sheet, film, and scaffold, with the use of various techniques, such as a wet-mechanochemical route,[112,113] a co-precipitation method,[114–119] an alternating lamination,[120] and an alternate soaking method.[121–123] Moreover, a thin film of HA/silk fibroin was successfully prepared by using matrix-assisted pulsed laser evaporation (MAPLE).[124] Since the presence of the calcium ion plays a key role in the formation of HA crystals on polymeric surfaces, the HA/silk fibroin composite was fabricated by simply soaking the fibroin film containing calcium chloride in simulated body fluid (SBF).[125]

It was reported that the thermal stabilities of the HA/silk fibroin composites were improved while their microstructures and tensile properties were dependent on the HA content in the composite.[117,126] Since HA is a brittle compound, an excess HA content can presumably make the resulting composite become more brittle. Another possible reason why the silk fibroin composite at high HA content exhibited a reduction of tensile properties was inorganic/organic phase separation due to low interfacial adhesion. As a result, a number of research studies focused on the enhancement of the interfacial adhesion between the HA and the fibroin. For example, the silk fibroin was pretreated with different kinds of chemical reagents, including an aqueous solution of inorganic salts,[127] an aqueous alkali solution,[128–130] and the enzyme

Table 11.4 Examples of silk-based composite with regenerated silk as composite matrix.

Silk-based composite[a]	Fabrication method[a]	Improved properties	Proposed applications	Ref.
Regenerated silk with inorganic fillers				
TiO$_2$/silk fibroin	Sol/gel method, freeze drying	Mechanical properties, thermal properties	Biomedical applications	109–111
HA/silk fibroin	Wet-mechan-ochemical route, co-precipitation, lamination, alternative soaking, MAPLE, soaking in SBF	Mechanical properties, thermal properties, osteoconductive properties	Bone tissue engineering applications	112–143
	Gelation	Enhanced wound healing properties	Wound dressing materials	144
Silica/silk fibroin	Protein fusion, deposition	Osteoconductive properties	Bone tissue engineering applications	145–147
Silicon/silk fibroin	Co-precipitation	Induction of HA formation	Bone tissue engineering applications	148
CaSiO$_3$/silk fibroin	Freeze drying	Mechanical properties, water absorption abilities, induction of HA formation	Tissue engineering applications	149
MMT/silk fibroin	Solution intercalation	Mechanical properties	Environmentally friendly materials from tissue engineering scaffolds to biodegradable structural materials	150
	LBL assembly	Mechanical properties	Biomedical applications	151
Ag/silk fibroin	LBL assembly	Mechanical properties, optical properties	Photonic and optical utilizations	152
Regenerated silk with organic fillers				
Chitin whiskers/ silk fibroin	Freeze drying	Dimensional stabilities, mechanical properties	Tissue engineering applications	37
Cellulose whiskers/ silk fibroin	Solution casting	Mechanical properties	Biomedical applications	152
Alginate beads/ silk fibroin, PLGA/ silk fibroin	Porogen leaching, freeze drying	Sustained drug release	Drug delivery applications	153,154

[a]HA = hydroxyapatite; MAPLE = matrix-assisted pulsed laser evaporation; SBF = simulated body fluid; MMT = montmorillonite; LBL = layer-by-layer; PLGA = poly(lactide-*co*-glycolide).

proteinase K,[130,131] to disentangle surface fibrils for an effective contact between the inorganic and the organic phases. The structural modification was also done to improve the interaction between the HA and the fibroin by grafting the silk fibroin with other polymers, namely poly(γ-methacryloxypropyltrimethox-ysilane) [poly(MPTS)],[132–134] poly{2-[O-(1'-methylpropylideneamino)carbox-yamino]ethyl methacrylate},[135] and poly(4-methacyloyloxyethyl trimellitate anhydride) [poly(4-META)].[136] An alternative strategy to improve the inter-facial adhesion was to use a blend of silk fibroin and various kinds of polymers, such as hyaluronic acid[137] and chitosan,[138] a deacetylation derivative of chitin, as an organic matrix for the HA. The HA/fibroin composites showed potential use as either bone in growth or implant fixation in bone tissue engineering because a number of cells, including human bone mesenchymal stromal cells (BMSCs),[119] osteoblasts (MC3T3-E1),[120] human bone marrow stem cells (hMSCs),[121] osteosarcoma SaOs2 cells,[124] and rat marrow mesenchymal cells (MMCs),[134] were able to attach, spread, and differentiate well on the composite surfaces, indicating the osteoconductivity. To promote the osteogenesis, the HA/silk fibroin composites were combined with bone morphogenetic protein-2 (BMP-2).[121,139] Another study reported that the HA/silk fibroin composite scaffold also promoted the healing of mandibular bone defects in canines.[140] It is important to note that although *B. mori* silk fibroin was widely used as a matrix for the HA composites, the HA was successfully embedded in the film of genetically engineered *Nephila clavipes* spider silk.[141] In addition, the HA was not only deposited on regenerated silk material but also on silk fabric, or silk cloth, by using either co-precipitation or soaking in order to prepare bioactive materials for orthopedic implants.[142,143] It was further reported that the deposition of the HA was able to be promoted by the immobilization of urease onto the surface of silk cloth.[143]

Although most research has mainly focused on the osteoconductive prop-erties of the HA/silk fibroin composite, one study showed that the addition of the HA to the silk fibroin gel enhanced epidermal recovery from full thickness porcine skin wounds.[144] The HA/fibroin composite promoted the wound healing process, re-epithelization, and matrix formation, indicating its potential use as effective wound dressing material.

Regardless of the HA, the incorporation of silica particles into either the silk fibroin film or the spider silk film also yields osteoinductive properties.[145–147] Furthermore, the growth of HA in the HA/silk fibroin composite was found to be accelerated in the presence of silicon, one of the components in bone, although the morphology of the resulting HA crystal depended on the silicon content.[148]

A composite scaffold of silk fibroin with bioactive wollastonite ($CaSiO_3$), a classic example of calcium silicate-based ceramics, was fabricated *via* freeze drying.[149] Compared to the neat silk fibroin scaffold, the composite scaffold showed superior properties in terms of compressive strength, compressive modulus, surface hydrophilicity, and water-uptake capacity. The wollastonite/fibroin composite scaffold possessed good bioactivity by inducing HA forma-tion on its surface after being soaked in the SBF. The composite scaffold was compatible with cell line L929 of mouse connective tissue, a fibroblast-like cell, indicating its utilization in tissue engineering applications.

MMT clay was used as nanoscale reinforcement for the fabrication of MMT/ silk fibroin nanocomposite *via* the solution-intercalation technique.[150] An aqueous fibroin solution was added to a clay suspension before being lyophilized to obtain a nanocomposite powder. It was reported that the clay dispersion in the fabricated nanocomposite was dependent on the solution pH. The MMT/silk fibroin nanocomposite prepared under acidic conditions, or at a pH below the p*I* of fibroin protein, exhibited better clay dispersion compared to that prepared at neutral pH or at a pH above the p*I*, due to a stronger clay–silk fibroin interaction. Regardless of the clay dispersion, the solution pH also influenced the conformation of the fibroin protein. Under acidic conditions, the incorporated clay acted as an efficient nucleator of β-sheet formation of silk fibroin. On the other hand, the silk fibroin in the nanocomposite prepared at neutral pH preserved the random coil conformation. Since both the clay dispersion and the fibroin conformation play an important role in the composite performance, the solution pH should be properly adjusted. It was further suggested that the MMT/silk fibroin nanocomposite should be an environmentally friendly material which was suitable for a large number of uses, ranging from a tissue engineering scaffold to biodegradable structural materials.

A recent study reported on the preparation of MMT/silk fibroin nanocomposite film by using the spin-assisted LBL assembly.[151] Compared to the pure fibroin, the composite film was highly transparent and possessed superior mechanical properties in terms of strength, modulus, and toughness. As the MMT was replaced by a Langmuir monolayer of densely packed silver, the resulting composite film was found to be highly reflective, or mirror-like, although the silver nanoplates improved the mechanical properties in the same trend as the MMT. It was also proposed that this LBL strategy should be a new route for the preparation of nanocomposite films with exceptional mechanical properties for biomedical applications, and that a mirror-like silk-based nanocomposite should possibly be beneficial for both photonic and optical utilizations.

11.3.4.2 Regenerated Silk with Organic Fillers

A silk fibroin film was reinforced with cellulose whiskers in order to improve its mechanical performance.[152] The composite film was fabricated by using the solution casting technique and its mechanical properties were determined in terms of tensile strength, ultimate strain, and tensile modulus. Compared to pure fibroin film and pure cellulose film, the cellulose whisker-reinforced fibroin film possessed much higher mechanical properties. It was also reported that a highly ordered structure of cellulose whiskers was able to induce the structural transition of silk fibroin from random coil to β-sheet structure at the whisker/matrix interface.

In the case of highly porous materials like sponges or scaffolds, the induction of β-sheet formation of the regenerated silk fibroin in order to prepare water-insoluble materials usually causes high shrinkage, so it is difficult to control their size and shape. To improve the dimensional stability of the silk fibroin sponge, there was an attempt to utilize chitin whiskers as a nanofiller.[37] The chitin whisker-reinforced silk fibroin nanocomposite sponge was fabricated by adding a fibroin solution to a chitin whisker suspension with mechanical stirring before being freeze dried. The fabricated nanocomposite sponge was then treated with an aqueous methanol solution to induce β-sheet formation of the fibroin. The chitin whiskers homogeneously embedded in the fibroin matrix and the incorporation of the chitin whiskers successfully reduced shrinkage of the nanocomposite sponge after methanol treatment. It was also reported that the added whiskers not only improved the dimensional stability of the nanocomposite sponge but also increased its compression strength. To further evaluate the feasibility of the nanocomposite sponge for tissue engineering applications, the cytocompatibility test was done against the L929 cell line. The test cell attached well to the surface of the nanocomposite sponge, and the incorporation of the chitin whiskers into the fibroin matrix significantly promoted cell spreading. As a result, the fabricated nanocomposite sponge was not cytotoxic and should have potential as a scaffolding material.

Various kinds of microparticles, including alginate beads and poly(lactide-*co*-glycolide) (PLGA), were embedded in silk fibroin scaffolds *via* either the porogen leaching or the freeze-drying method, not for reinforcement but for control-release purposes.[153,154] The microparticles were loaded with desired substances, such as insulin-like growth factor I (IGF-I), bovine serum albumin (BSA), or insulin, prior to the embedment. Compared to the free microparticles, those embedded in the fibroin scaffold provided a superior sustained release rate, indicating their possible use as effective vehicles for drug delivery applications.

11.3.5 Miscellaneous

13.3.5.1 *Silk-reinforced Silk Composites*

Silk-based composites are being developed to fabricate new types of biocomposites with unique characteristics. For example, all-silk-reinforced silk composites have good mechanical properties because of the high interfacial adhesion between the reinforcement and the matrix. Moreover, the high biocompatibility also makes these silk-based composites one of the most promising materials for biomedical applications.

An electrospun-knitted silk composite scaffold for ligament tissue engineering has been reported.[155] The raw silk fibres were fabricated into a knitted scaffold before being subjected to the degumming process. After that, silk fibroin nanofibres were electrospun onto the degummed knitted silk scaffold to cover its macropores. With the use of an inclined electrospinning technique, the electrospun silk fibroin nanofibres adhered better on the knitted silk scaffold. The potential of the electrospun-knitted silk composite for ligament tissue engineering applications was examined in a comparison with a PLGA scaffold (one of the leading biomaterials used as tissue engineered ligament replacement). The cell culture test was performed against fibroblast cells for 28 days and it was found that the fabricated silk composite showed a higher cell adhesion with good cell proliferation and higher collagen content but possessed a lower *in vivo* degradability.

A uniaxially aligned silk fibre was used to reinforce the silk fibroin matrix in order to prepare a fibre-reinforced composite with good interfacial adhesion.[156] The silk fibre was degummed and treated with an aqueous LiBr solution before being impregnated in the silk fibroin solution. The composite film was then fabricated using the solution casting technique. The LiBr pretreatment was found to improve the impregnation of the silk fibre in the fibroin matrix by creating loosened fibroin segments on the silk fibre surface which were able to interact with the fibroin matrix and acted as mechanical interlocking between the reinforcement and the polymer matrix. Thus, the interfacial adhesion was increased and the composite performance, in terms of mechanical and thermal properties, was consequently enhanced. It was suggested that the high-performance all-silk composite material is a promising candidate for various applications.

A silk fibroin sponge was reinforced with silk fibroin microparticles to fabricate a protein/protein composite scaffold with desirable mechanical properties for *in vitro* osteogenic tissue formation.[157,158] The composite scaffold was prepared by using the salt leaching technique and it was reported that the silk microparticles dramatically increased the compressive modulus of the fabricated scaffold. An increase in the scaffold stiffness better mimicked the mechanical properties of native bone and enhanced the osteogenic differentiation of hMSCs, indicating the feasibility of the composite scaffold for bone repair.

11.3.5.2 Silk-based Conductive Composites

A recent emergence of interactive and smart textiles has led to the development of silk-based conductive composites. Either conductive polymers or CNTs can be incorporated into the silk material in order to provide the electrical conductivity. The fabricated silk-based composites should be beneficial for a wide range of applications, from traditional clothing to highly innovative electronics and biomedical end-uses.

A number of electrical conductive polymers, including polypyrrole (PPy), polyaniline (PANI), and poly(3,4-ethylenedioxythiophene) (PEDOT), were used to coat silk materials, such as nonwoven web, fibres, and fabric, *via in situ* polymerization in order to fabricate conductive polymer/silk composites.[159–161] The conductive polymers entirely covered the surface of the silk materials without damaging their physical morphology, indicating the high affinity between conductive polymers and silk. The conductive polymer coating did not affect the mechanical properties of the silk materials, but remarkably increased their thermal stabilities and provided electrical properties. The application of the conductive polymer/silk composites should not be limited to only the clothing industry but should also include the fields of technical textiles, from interactive and smart textiles to bio-based conductive composites for biomedical applications.

Regardless of the conductive polymer coating, the silk-based conductive composites were fabricated by the incorporation of CNTs, either single-walled carbon nanotubes (SWNTs) or multi-walled carbon nanotubes (MWNTs), in silk fibroin nanofibres via the electrospinning technique.[162–164] The CNTs were successfully embedded and well-aligned in the fibroin nanofibres, resulting in drastically increased mechanical strength and electrical conductivity. Therefore, the electrospun nanocomposite showed good efficacy in bone tissue engineering in which load-bearing tissue requires electrical conductivity for cell growth.[165]

11.3.5.3   Silk-based Composites as Biosensors

In most research, the regenerated silk fibroin is used as an important component in biosensors to provide a biocompatible environment for the enzyme immobilization. Regardless of the regenerated silk, the biosensors also consist of inorganic components, such as metal nanoparticles and CNTs.

A sensitive and stable amperometric biosensor was developed via the immobilization of acetylcholinasterase (AChE) on a platinum electrode modified with gold nanoparticles and silk fibroin for the determination of carbamate and organophosphate pesticides, such as carbofuran, phoxim, and methyl paraoxon.[166] A colloidal mixture of gold nanoparticles and silk fibroin was used as a composite matrix for the immobilization of AChE, an important enzyme for the proper functioning of the central nervous system in humans which is quite sensitive to pesticides,[167] prior to deposition on the platinum electrode. It was reported that the gold nanoparticles improved the sensitivity and response of the biosensor while the fibroin not only stabilized the immobilized AChE but also prevented its leakiness from the platinum electrode. The fabricated biosensor showed good sensitivity to carbofuran, phoxim, and methyl paraoxon, with good reproducibility and acceptable stability, indicating its efficacy as a promising tool for pesticide analysis.

In another study,[168] a colloidal mixture of gold nanoparticles and silk fibroin was also used as a composite matrix for enzyme immobilization to fabricate a third-generation biosensor on the basis of direct electrochemistry of redox enzymes for hydrogen peroxide detection. Horseradish peroxidase (HRP), a heme-containing enzyme, was immobilized in the composite matrix before being used to modify a glassy carbon electrode (GCE). The gold/silk fibroin composite matrix was found to effectively maintain the bioactivity of the immobilized HRP. The fabricated biosensor exhibited fast amperometric response, low detection limit, and wide linear range to H_2O_2, with good reproducibility and long-term stability, suggesting its potential utility as an electrochemical biosensor for H_2O_2 detection.

The preparation of amperometric biosensors based on immobilization of tyrosinase on a MWNTs/cobalt phthalocyanine (CoPc)/silk fibroin composite modified GCE for determination of 2,2'-bis(4-hydroxyphenyl)propane (bisphenol A, BPA), a key component in both PC and epoxy resin which has been identified as a harmful pollutant, has been reported.[169] A suspension of MWNTs functionalized with CoPc was used to modify the surface of the GCE. After drying, a mixed solution of silk fibroin and tyrosinase was dropped on the surface of the MWNTs/CoPc/GCE in order to prepare the biosensor. The silk fibroin maintained the bioactivity of the tyrosinase while the MWNTs and the CoPc gave excellent inherent conductivity and good electrocatalytic activity to the composite, respectively.

Instead of enzyme immobilization, a subsequent study reported another potential procedure to fabricate an electrochemical BPA biosensor based on dendrimer immobilization.[170] It is known that dendrimers are highly branched and monodisperse macromolecules with a globular structure containing many terminal functional groups and cavities for hosting other molecules.[171] In that previous work,[170] poly(amidoamine) (PAMAM) was immobilized on a gold/silk fibroin/GC electrode and the results showed that the modified electrode displayed better electrochemical response to BPA, with a higher adsorption capacity of BPA, compared to the unimmobilized GCE. Regardless of its high sensitivity, the biosensor also possessed good reproducibility and acceptable stability, indicating its possible use for tracing BPA amounts either in plastic products or in the environment.

11.4 Conclusions

Although the silk-based composites can be prepared in a diversity of forms, including fibres, films, and sponges or scaffolds, with several interesting properties, most research still focuses on the utilization of the silk fibres produced by mulberry silkworm moths due to their large-scale production. Advances in biotechnology perhaps can enhance silk fibre production from alternative sources, such as non-mulberry silks, spider silks, and recombinant silks, thus resulting in further development of the silk-based composites with desirable properties. It is also believed that the silk-based composites are good candidates for replacing synthetic materials as well as conventional composites in the near future because of their environmentally friendly characteristics. In addition, the excellent biocompatibilities of the silk-based composites might lead to their end-uses in high value-added applications, especially in bio-related fields. Hence, silk-based composites can be considered as promising materials for both environmental engineering and bioengineering applications.

Acknowledgements

The authors would like to acknowledge the Ratchadapiseksomphot Endowment Fund, Chulalongkorn University, for providing a postdoctoral fellowship to O.P.

References

1. Q. T. H. Shubhra, A. K. M. M. Alam, M. A. Gafur, S.M. Shamsuddin, M.A. Khan, M. Saha, D. Saha, M. A. Quaiyyum, J. A. Khan and M. Ashaduzzaman, *Fibers Polym.*, 2010, **11**, 725.
2. C. Vepari and D. L. Kaplan, *Prog. Polym. Sci.*, 2007, **32**, 991.
3. J. L. Capinera, *Encyclopedia of Entomology*, Springer, Dordrecht, 2008.
4. J. G. Hardy and T. R. Scheibel, *Prog. Polym. Sci.*, 2010, **35**, 1093.
5. J. G. Hardy, L. M. Römer and T. R. Scheibel, *Polymer*, 2008, **49**, 4309.
6. C. Vendrely and T. Scheibel, *Macromol. Biosci.*, 2007, **7**, 401.
7. O. Hakimi, D. P. Knight, F. Vollrath and P. Vadgama, *Composites, Part B*, 2007, **38**, 324.
8. G. H. Altman, F. Diaz, C. Jakuba, T. Calabro, R. L. Horan, J. Chen, H. Lu, J. Richmond and D. L. Kaplan, *Biomaterials*, 2003, **24**, 401.
9. A. Glišović and F. Vollrath, in *Industrial Applications of Natural Fibres: Structure, Properties, and Technical Applications*, ed. J. Müssig, Wiley, Chichester, 2010, p. 237.
10. M. Schoeser, *Silk*, Yale University Press, New Haven, 2007.
11. B. Leng, L. Huang and Z. Shao, in *Advances in Chemical Engineering: Engineering Aspects of Self-organizing Materials*, ed. R. J. Koopmans, Elsevier, Amsterdam, 2009, p. 119.
12. S. C. Kundu, B. C. Dash, R. Dash and D. L. Kaplan, *Prog. Polym. Sci.*, 2008, **33**, 998.
13. H.-Y. Cheung, M.-P. Ho, K.-T. Lau, F. Cardona and D. Hui, *Composites, Part B*, 2009, **40**, 655.
14. D. L. Kaplan, C. M. Mello, S. Arcidiacono, S. Fossey, K. Senecal and W. Muller, in *Protein-Based Materials*, ed. K. McGrath and D. L. Kaplan, Birkhäuser, Boston, 1997, p. 103.
15. E. S. Sashina, A. M. Bochek, N. P. Novoselov and D. A. Kirichenko, *Russ. J. Appl. Chem.*, 2006, **79**, 869.
16. S. Inoue, K. Tanaka, F. Arisaka, S. Kimura, K. Ohtomo and S. Mizuno, *J. Biol. Chem.*, 2000, **275**, 40517.
17. K. Tanaka, S. Inoue and S. Mizuno, *Insect Biochem. Mol. Biol.*, 1999, **29**, 269.
18. F. Sehnal and M. Zurovec, *Biomacromolecules*, 2004, **5**, 666.
19. J. Magoshi and Y. Magoshi, *J. Polym. Sci., Part B: Polym. Phys.*, 1977, **15**, 1675.
20. J. Magoshi, M. Mizuide and Y. Magoshi, *J. Polym. Sci., Part B: Polym. Phys.*, 1979, **17**, 515.
21. H.-J. Jin and D. L. Kaplan, *Nature*, 2003, **424**, 1057.
22. K. A. Trabbic and P. Yager, *Macromolecules*, 1998, **31**, 462.
23. T. Gamo, T. Inokuchi and H. Laufer, *Insect Biochem. Mol. Biol.*, 1977, **7**, 285.
24. M. Tsukada and G. Bertholon, *Bull. Sci. Inst. Text. Fr.*, 1981, **10**, 141.
25. Y.-Q. Zhang, *Biotechnol. Adv.*, 2002, **20**, 91.

26. S. A. Wainwright, W. D. Biggs, J. D. Currey and J. M. Gosline, *Mechanical Design in Organisms*, Princeton University Press, Princeton, 1982.
27. T. Arai, G. Freddi, R. Innocenti and M. Tsukada, *J. Appl. Polym. Sci.*, 2004, **91**, 2383.
28. C. Li, C. Vepari, H.-J. Jin, H. J. Kim and D. L. Kaplan, *Biomaterials*, 2006, **27**, 3115.
29. X. Wang, X. Hu, A. Daley, O. Rabotyagova, P. Cebe and D. L. Kaplan, *J. Controlled Release*, 2007, **121**, 190.
30. X. Wang, E. Wenk, X. Hu, G. R. Castro, L. Meinel, X. Wang, C. Li, H. Merkle and D. L. Kaplan, *Biomaterials*, 2007, **28**, 4161.
31. G. L. Jones, A. Motta, M. J. Marshall, A. J. El Haj and S. H. Cartmell, *Biomaterials*, 2009, **30**, 5376.
32. D. N. Breslauer, S. J. Muller and L. P. Lee, *Biomacromolecules*, 2010, **11**, 643.
33. X. Wang, T. Yucel, Q. Lu, X. Hu and D. L. Kaplan, *Biomaterials*, 2010, **31**, 1025.
34. Q. Lv, C. Cao, Y. Zhang, X. Ma and H. Zhu, *J. Appl. Polym. Sci.*, 2005, **96**, 2168.
35. Y. Yang, F. Ding, J. Wu, W. Hu, W. Liu, J. Liu and X. Gu, *Biomaterials*, 2007, **28**, 5526.
36. K. H. Lee, D. H. Baek, C. S. Ki and Y. H. Park, *Int. J. Biol. Macromol.*, 2007, **41**, 168.
37. P. Wongpanit, N. Sanchavanakit, P. Pavasant, T. Bunaprasert, Y. Tabata and R. Rujiravanit, *Eur. Polym. J.*, 2007, **43**, 4123.
38. P. Wongpanit, Y. Tabata and R. Rujiravanit, *Macromol. Biosci.*, 2007, **7**, 1258.
39. C. Chen, C. Chuanbao, M. Xilan, T. Yin and Z. Hesun, *Polymer*, 2006, **47**, 6322.
40. C. S. Ki, J. W. Kim, H. J. Oh, K. H. Lee and Y. H. Park, *Int. J. Biol. Macromol.*, 2007, **41**, 346.
41. C. S. Ki, J. W. Kim, J. H. Hyun, K. H. Lee, M. Hattori, D. K. Rah and Y. H. Park, *J. Appl. Polym. Sci.*, 2007, **106**, 3922.
42. B. Bondar, S. Fuchs, A. Motta, C. Migliaresi and C. J. Kirkpatrick, *Biomaterials*, 2008, **29**, 561.
43. G. Chang, H.-J. Kim, D. Kaplan, G. Vunjak-Novakovic and R. A. Kandel, *Eur. Spine J.*, 2007, **16**, 1848.
44. J. R. Mauney, T. Nguyen, K. Gillen, C. Kirker-Head, J. M. Gimble and D. L. Kaplan, *Biomaterials*, 2007, **28**, 5280.
45. S. Hofmann, H. Hagenmüller, A. M. Koch, R. Müller, G. Vunjak-Novakovic, D. L. Kaplan, H. P. Merkle and L. Meinel, *Biomaterials*, 2007, **28**, 1152.
46. P. Wongpanit, H. Ueda, Y. Tabata and R. Rujiravanit, *J. Biomater. Sci., Polym. Ed.*, 2010, **21**, 1403.
47. C. Jiang, X. Wang, R. Gunawidjaja, Y.-H. Lin, M. K. Gupta, D. L. Kaplan, R. R. Naik and V. V. Tsukruk, *Adv. Funct. Mater.*, 2007, **17**, 2229.

48. M. Lovette, C. Cannizzaro, L. Daheron, B. Messmer, G. Vunjak-Novakovic and D. L. Kaplan, *Biomaterials*, 2007, **28**, 5271.
49. I. C. Um, H. Y. Kweon, K. G. Lee and Y. H. Park, *Int. J. Biol. Macromol.*, 2003, **33**, 203.
50. J.-H. Yeo, K.-G. Lee, Y.-W. Lee and S. Y. Kim, *Eur. Polym. J.*, 2003, **39**, 1195.
51. X. Wang, E. Wenk, A. Matsumoto, L. Meinel, C. Li and D. L. Kaplan, *J. Control. Release*, 2007, **117**, 360.
52. H.-B. Yan, Y.-Q. Zhang, Y.-L. Ma and L.-X. Zhou, *J. Nanopart. Res.*, 2009, **11**, 1937.
53. X. Wang, J. A. Kluge, G. G. Leisk and D. L. Kaplan, *Biomaterials*, 2008, **29**, 1054.
54. S. Sukigara, M. Gandhi, J. Ayutsede, M. Micklus and F. Ko, *Polymer*, 2003, **44**, 5721.
55. S. Zarkoob, R. K. Eby, D. H. Reneker, S. D. Hudson, D. Ertley and W. W. Adams, *Polymer*, 2004, **45**, 3973.
56. I. C. Um, H. Y. Kweon, K. G. Lee, D. W. Ihm, J.-H. Lee and Y. H. Park, *Int. J. Biol. Macromol.*, 2004, **34**, 89.
57. I. C. Um, C. S. Ki, H. Y. Kweon, K. G. Lee, D. W. Ihm and Y. H. Park, *Int. J. Biol. Macromol.*, 2004, **34**, 107.
58. L. E. Nielsen and R. F. Landel, *Mechanical Properties of Polymers and Composites*, Dekker, New York, 1994.
59. M. C. Gupta and A. P. Gupta, *Polymer Composite*, New Age International, New Delhi, 2005.
60. B. Fisa, in *Composite Materials Technology: Processes and Properties*, ed. P. K. Mallick and S. Newman, Hanser, New York, 1990, p. 265.
61. M. E. Ryan and M. R. Kamal, in *Composite Materials Technology: Processes and Properties*, ed. P. K. Mallick and S. Newman, Hanser, New York, 1990, p. 321.
62. P. K. Mallick, in *Composite Materials Technology: Processes and Properties*, ed. P. K. Mallick and S. Newman, Hanser, New York, 1990, p. 67.
63. F. J. O'Brien, B. A. Harley, I. V. Yannas and L. Gibson, *Biomaterials*, 2004, **25**, 1077.
64. D. Cho, S. G. Lee, W. H. Park and S. O. Han, *Polym. Sci. Technol.*, 2002, **13**, 460.
65. S. M. Lee, D. Cho, W. H. Park, S. G. Lee, S. O. Han and L. T. Drzal, *Compos. Sci. Technol.*, 2005, **65**, 647.
66. D. K. Setua and S. K. De, *J. Mater. Sci.*, 1984, **19**, 983.
67. S. O. Han, S. M. Lee, W. H. Park and D. Cho, *J. Appl. Polym. Sci.*, 2006, **100**, 4972.
68. S. O. Han, H. J. Ahn and D. Cho, *Composites, Part B*, 2010, **41**, 491.
69. H.-Y. Cheung, K.-T. Lau, X.-M. Tao and D. Hui, *Composites, Part B*, 2008, **39**, 1026.
70. H.-Y. Cheung, K.-T. Lau, Y.-F. Pow, Y.-Q. Zhao and D. Hui, *Composites, Part B*, 2010, **41**, 223.
71. Y.-Q. Zhao, H.-Y. Cheung, K.-T. Lau, C.-L. Xu, D.-D. Zhao and H.-L. Li, *Polym. Degrad. Stab.*, 2010, **95**, 1978.

72. W. Li, X.-Y. Qiao, K. Sun and X.-D. Chen, *J. Appl. Polym. Sci.*, 2008, **110**, 134.
73. X.-Y. Qiao, W. Li, K. Sun, S. Xu and X.-D. Chen, *J. Appl. Polym. Sci.*, 2009, **111**, 2908.
74. X.-Y. Qiao, W. Li, K. Sun, S. Xu and X.-D. Chen, *Polym. Int.*, 2009, **58**, 530.
75. W. Li, X.-Y. Qiao, K. Sun and X.-D. Chen, *J. Appl. Polym. Sci.*, 2009, **113**, 1063.
76. X.-Y. Qiao, W. Li, K. Sun and X.-D. Chen, *Polym. Int.*, 2010, **59**, 447.
77. H.-S. Kim, B. H. Park, J.-S. Yoon and H.-J. Jin, *Polym. Int.*, 2007, **56**, 1035.
78. J. P. Craven, R. Cripps and C. Viney, *Composites, Part A*, 2000, **31**, 653.
79. S. P. Priya, H. V. Ramakrishna, S. K. Rai and A. V. Rajulu, *J. Reinf. Plast. Compos.*, 2005, **24**, 643.
80. S. P. Priya and S. K. Rai, *J. Reinf. Plast. Compos.*, 2005, **24**, 1605.
81. S. P. Priya and S. K. Rai, *J. Reinf. Plast. Compos.*, 2006, **25**, 33.
82. S. K. Rai and S. P. Priya, *J. Reinf. Plast. Compos.*, 2006, **25**, 565.
83. S. P. Priya, H. V. Ramakrishna and S. K. Rai, *J. Reinf. Plast. Compos.*, 2006, **25**, 339.
84. P. N. Khanam, M. M. Reddy, K. Raghu, K. John and S. V. Naidu, *J. Reinf. Plast. Compos.*, 2007, **26**, 1065.
85. K. Raghu, P. N. Khanam and S. V. Naidu, *J. Reinf. Plast. Compos.*, 2010, **29**, 343.
86. P. N. Khamnam, G. R. Reddy, K. Raghu and S. V. Naidu, *J. Reinf. Plast. Compos.*, 2010, **29**, 2124.
87. K. V. Arun, R. D. Kamat and S. Basavarajappa, *J. Reinf. Plast. Compos.*, 2010, **29**, 254.
88. X. Chen, Y.-Y. Qi, L.-L. Wang, G.-L. Yin, X.-H. Zou and H.-W. Ouyang, *Biomaterials*, 2008, **29**, 3683.
89. J. L. Chen, Z. Yin, W. L. Shen, X. Chen, B. C. Heng, X. H. Zou and H. W. Ouyang, *Biomaterials*, 2010, **31**, 9438.
90. Y.-K. Seo, H.-H. Yoon, K.-Y. Song, S.-Y. Kwon, H.-S. Lee, Y.-S. Park and J.-K. Park, *J. Orthop. Res.*, 2009, **27**, 495.
91. Q. T. H. Shubhra, A. K. M. M. Alam and M. D. H. Beg, *Mater. Lett.*, 2011, **65**, 333.
92. Q. T. H. Shubhra, A. K. M. M. Alam, M. A. Khan, M. Saha, D. Saha and M. A. Gafur, *Composites, Part A*, 2010, **41**, 1587.
93. S. Li, B. Li, J. Cheng and J. Hu, *Cem. Concr. Compos.*, 2008, **30**, 347.
94. K. Finlay, M. D. Gawryla and D. A. Schiraldi, *Ind. Eng. Chem. Res.*, 2008, **47**, 615. 92.
95. S. Akhtar, P. P. De and S. K. De, *J. Appl. Polym. Sci.*, 1986, **32**, 5123.
96. P. Potiyaraj, P. Kumlangdudsana and S. T. Dubas, *Mater. Lett.*, 2007, **61**, 2464.
97. A. R. Abbasi and A. Morsali, *J. Inorg. Organomet. Polym. Mater.*, 2010, **20**, 825.
98. Q. Dong, H. L. Su and D. Zhang, *J. Phys. Chem. B*, 2005, **109**, 17429.

99. Q. Dong, H. Su, W. Cao, J. Han, D. Zhang and Q. Guo, *Mater. Chem. Phys.*, 2008, **110**, 160.

100. J. Han, H. Su, Q. Dong, D. Zhang, X. Ma and C. Zhang, *J. Nanopart. Res.*, 2010, **12**, 347.

101. M. Brust, J. Fink, D. Bethell, D. J. Schiffrin and C. Kiely, *J. Chem. Soc., Chem. Commun*, 1995, **16**, 1655.

102. E. L. Mayes, F. Vollrath and S. Mann, *Adv. Mater.*, 1998, **10**, 801.

103. D. J. Kang, D. Xu, Z. X. Zhang, K. Pal, D. S. Bang and J. K. Kim, *Macromol. Mater. Eng.*, 2009, **294**, 620.

104. K. Hu, Q. Lv, F. Z. Cui, L. Xu, Y. P. Jiao, Y. Wang, Q. L. Feng, H. L. Wang and L. Y. Huang, *J. Bioact. Compat. Polym.*, 2007, **22**, 395.

105. Q. Lv, K. Hu, Q. Feng, F. Cui and C. Cao, *Compos. Sci. Technol.*, 2007, **67**, 3023.

106. K. Kesenci, A. Motta, L. Fambri and C. Migliaresi, *J. Biomater. Sci., Polym. Ed.*, 2001, **12**, 337.

107. Y. Tao, Y. Yan and W. Xu, *J. Polym. Sci., Part B: Polym. Phys.*, 2010, **48**, 940.

108. D. J. Kang, K. Pal, S. J. Park, D. S. Bang and J. K. Kim, *Mater. Des.*, 2010, **31**, 2216.

109. X.-X. Feng, L.-L. Zhang, J.-Y. Chen, Y.-H. Guo, H.-P. Zhang and C.-I. Jia, *Int. J. Biol. Macromol.*, 2007, **40**, 105.

110. X.-X. Feng, L.-L. Zhang, J.-Y. Chen and H.-P. Zhang, *J. Biomed. Mater. Res., A*, 2008, **84A**, 761.

111. X.-X. Feng, L. Zhou, H.-L. Zhu and J.-Y. Chen, *J. Appl. Polym. Sci.*, 2010, **116**, 468.

112. R. Nemoto, S. Nakamura, T. Isobe and M. Senna, *J. Sol-Gel Sci. Technol.*, 2001, **21**, 7.

113. L. Wang, R. Nemoto and M. Senna, *J. Nanopart. Res.*, 2002, **4**, 535.

114. X. D. Kong, F. Z. Cui, X. M. Wang, M. Zhang and W. Zhang, *J. Cryst. Growth*, 2004, **270**, 197.

115. K. Xiangdong, S. Xiaodan, Q. Min and X. Ruisheng, *Front. Mater. Sci. China*, 2007, **1**, 243.

116. R. Yongjuan, S. Xiaodan, C. Fuzhai and K. Xiangdong, *Front. Mater. Sci. China*, 2007, **1**, 258.

117. C. Du, J. Jin, Y. Li, X. Kong, K. Wei and J. Yao, *Mater. Sci. Eng., C*, 2009, **29**, 62.

118. C. Fan, J. Li, G. Xu, H. He, X. Ye, Y. Chen, X. Sheng, J. Fu and D. He, *J. Mater. Sci.*, 2010, **45**, 5814.

119. Y. Zhang, C. Wu, T. Friis and Y. Xiao, *Biomaterials*, 2010, **31**, 2848.

120. R. Kino, T. Ikoma, S. Yunoki, N. Nagai, J. Tanaka, T. Asakura and M. Munekata, *J. Biosci. Bioeng.*, 2007, **103**, 514.

121. H. J. Kim, U.-J. Kim, H. S. Kim, C. Li, M. Wada, G. G. Leisk and D. L. Kaplan, *Bone*, 2008, **42**, 1226.

122. Y. Zhao, J. Chen, A. H. K. Chou, G. Li and R. Z. LeGeros, *J. Biomed. Mater. Res., A*, 2009, **91**, 1140.

123. H. Tungtasana, S. Shuangshoti, S. Shuangshoti, S. Kanokpanont, D. L. Kaplan, T. Bunaprasert and S. Damrongsakkul, *J. Mater. Sci.: Mater. Med.*, 2010, **21**, 3151.

124. F. M. Miroiu, G. Socol, A. Visan, N. Stefan, D. Craciun, V. Craciun, G. Dorcioman, I. N. Mihailescu, L. E. Sima, S. M. Petrescu, A. Andronie, I. Stamatin, S. Moga and C. Ducu, *Mater. Sci. Eng., B*, 2010, **169**, 151.

125. R. Kino, T. Ikoma, A. Monkawa, S. Yunoki, M. Munekata, J. Tanaka and T. Asakura, *J. Appl. Polym. Sci.*, 2006, **99**, 2822.

126. L. Wang, R. Nemoto and M. Senna, *J. Eur. Ceram. Soc.*, 2004, **24**, 2707.

127. R. Nemoto, L. Wang, T. Ikoma, J. Tanaka and M. Senna, *J. Nanopart. Res.*, 2004, **6**, 259.

128. L. Wang, R. Nemoto and M. Senna, *J. Mater. Sci.: Mater. Med.*, 2004, **15**, 261.

129. L. Wang, G.-L. Ning and M. Senna, *Colloids Surf., A*, 2005, **254**, 159.

130. L. Wang, G.-L. Ning and M. Senna, *J. Nanopart. Res.*, 2007, **9**, 919.

131. L. Wang, R. Nemoto and M. Senna, *J. Nanopart. Res.*, 2004, **6**, 91.

132. T. Furuzono, A. Kishida and J. Tanaka, *J. Mater. Sci.: Mater. Med.*, 2004, **15**, 19.

133. T. Furuzono, S. Yasuda, T. Kimura, S. Kyotani, J. Tanaka and A. Kishida, *J. Artif. Organs*, 2004, **7**, 137.

134. T. Tanaka, M. Hirose, N. Kotobuki, H. Ohgushi, T. Furuzono and J. Sato, *Mater. Sci. Eng., C*, 2007, **27**, 817.

135. A. Korematsu, T. Furuzono, S. Yasuda, J. Tanaka and A. Kishida, *J. Mater. Sci.*, 2004, **39**, 3221.

136. A. Korematsu, T. Furuzono, S. Yasuda, J. Tanaka and A. Kishida, *J. Mater. Sci.: Mater. Med.*, 2005, **16**, 67.

137. R. Nemoto, L. Wang, M. Aoshima and M. Senna, *J. Am. Ceram. Soc.*, 2004, **87**, 1014.

138. L. Wang and C. Li, *Carbohydr. Polym.*, 2007, **68**, 740.

139. X. Jiang, J. Zhao, S. Wang, X. Sun, X. Zhang, J. Chen, D. L. Kaplan and Z. Zhang, *Biomaterials*, 2009, **30**, 4522.

140. J. Zhao, Z. Zhang, S. Wang, X. Sun, X. Zhang, J. Chen, D. L. Kaplan and X. Jiang, *Bone*, 2009, **45**, 517.

141. J. Huang, C. Wong, A. George and D. L. Kaplan, *Biomaterials*, 2007, **28**, 2358.

142. T. Furuzono, T. Taguchi, A. Kishida, M. Akashi and Y. Tamada, *J. Biomed. Mater. Res.*, 2000, **50**, 344.

143. H. Unuma, M. Hiroya and A. Ito, *J. Mater. Sci.: Mater. Med.*, 2007, **18**, 987.

144. R. Okabayashi, M. Nakamura, T. Okabayashi, Y. Tanaka, A. Nagai and K. Yamashita, *J. Biomed. Mater Res., B*, 2009, **90B**, 641.

145. C. W. P. Foo, S. V. Patwardhan, D. J. Belton, B. Kitchel, D. Anastasiades, J. Huang, R. R. Naik, C. C. Perry and D. L. Kaplan, *Proc. Natl. Acad. Sci. U. S. A.*, 2006, **103**, 9428.

146. A. J. Mieszawska, N. Fourligas, I. Georgakoudi, N. M. Ouhib, D. J. Belton, C. C. Perry and D. L. Kaplan, *Biomaterials*, 2010, **31**, 8902.

147. A. J. Mieszawska, L. D. Nadkarni, C. C. Perry and D. L. Kaplan, *Chem. Mater.*, 2010, **22**, 5780.
148. L. Li, K.-M. Wei, F. Lin, X.-D. Kong and J.-M. Yao, *J. Mater. Sci.: Mater. Med.*, 2008, **19**, 577.
149. H. Zhu, J. Shen, X. Feng, H. Zhang, Y. Guo and J. Chen, *Mater. Sci. Eng., C*, 2010, **30**, 132.
150. Q. Dang, S. Lu, S. Yu, P. Sun and Z. Yuan, *Biomacromolecules*, 2010, **11**, 1796.
151. E. Kharlampieva, V. Kozlovskaya, R. Gunawidjaja, V. V. Shevchenko, R. Vaia, R. R. Naik, D. L. Kaplan and V. V. Tsukruk, *Adv. Funct. Mater.*, 2010, **20**, 840.
152. Y. Noishiki, Y. Nishiyama, M. Wada, S. Kuga and J. Magoshi, *J. Appl. Polym. Sci.*, 2002, **86**, 3425.
153. E. Wenk, A. J. Meinel, S. Wildy, H. P. Merkle and L. Meinel, *Biomaterials*, 2009, **30**, 2571.
154. B. B. Mandal and S. C. Kundu, *Biomaterials*, 2009, **30**, 5170.
155. R.-F. Peh, V. Suthikum, C.-H. Goh and S.-L. Toh, *IFMBE Proc.*, 2007, **14**, 3287.
156. Q. Yuan, J. Yao, X. Chen, L. Huang and Z. Shao, *Polymer*, 2010, **51**, 4843.
157. R. Rajkhowa, E. S. Gil, K. Numata, L. Wang and X. Wang, *Macromol. Biosci.*, 2010, **10**, 599.
158. D. N. Rockwood, E. S. Gil, S.-H. Park, J. A. Kluge, W. Grayson, S. Bhumiratana, R. Rajkhowa, X. Wang, S. J. Kim, G. Vunjak-Novakovic and D. L. Kaplan, *Acta Biomater.*, 2011, **7**, 144.
159. A. Boschi, C. Arosio, I. Cucchi, F. Bertini, M. Catellani and G. Freddi, *Fibers Polym.*, 2008, **9**, 698.
160. Y. Xia and Y. Lu, *Compos. Sci. Technol.*, 2008, **68**, 1471.
161. I. Cucchi, A. Boschi, C. Arosio, F. Bertini, G. Freddi and M. Catellani, *Synth. Met.*, 2009, **159**, 246.
162. J. Ayutsede, M. Gandhi, S. Sukigara, H. Ye, C.-M. Hsu, Y. Gogotsi and F. Ko, *Biomacromolecules*, 2006, **7**, 208.
163. M. Gandhi, H. Yang, L. Shor and F. Ko, *Polymer*, 2009, **50**, 1918.
164. M. Kang, P. Chen and H.-J. Jin, *Curr. Appl. Phys.*, 2009, **9**, 595.
165. J. C. Anderson and C. Eriksson, *Nature*, 1970, **227**, 491.
166. H. Yin, S. Ai, J. Xu, W. Shi and L. Zhu, *J. Electroanal. Chem.*, 2009, **637**, 21.
167. S. Chapalamadugu and G. R. Chaudhry, *Crit. Rev. Biotechnol.*, 1992, **12**, 357.
168. H. Yin, S. Ai, W. Shi and L. Zhu, *Sens. Actuators, B*, 2009, **137**, 747.
169. H. Yin, Y. Zhou, J. Xu, S. Ai, L. Cui and L. Zhu, *Anal. Chim. Acta*, 2010, **659**, 144.
170. H. Yin, Y. Zhou, S. Ai, R. Han, T. Tang and L. Zhu, *Microchim. Acta*, 2010, **170**, 99.
171. S. M. Grayson and J. M. J. Frechet, *Chem. Rev.*, 2001, **101**, 3819.

CHAPTER 12

Hybrid Composite Structures from Collagenous Wastes and Environmental Friendly Polymers: Preparation, Properties and Applications

M. ASHOKKUMAR, P. THANIKAIVELAN* AND
B. CHANDRASEKARAN

Advanced Materials Laboratory, Centre for Leather Apparel & Accessories
Development, Central Leather Research Institute (Council of Scientific &
Industrial Research), Adyar, Chennai 600020, India
*Email: thanik8@yahoo.com

12.1   Introduction

Biomass is a non-fossil biological material derived from living organisms such as plants, animals and microorganisms.[1–3] Different sources of biomass include wood,[4] crops,[5] animal wastes,[6] agriculture wastes,[7] food wastes,[8] industrial wastes,[9] *etc*. The effective utilization of biomass or its wastes as valuable hybrid biocomposite materials has received considerable attention due to environmental concerns and global warming. There is a dire need to source and develop appropriate technology to satisfy this requirement. Global livestock animal biomass meant for food, fibre and labor accounts for nearly 200 million

RSC Green Chemistry No. 16
Natural Polymers, Volume 1: Composites
Edited by Maya J John and Thomas Sabu
© The Royal Society of Chemistry 2012
Published by the Royal Society of Chemistry, www.rsc.org

tonnes (dry weight), with approximately 3 billion counts. Livestock animals such as cattle, goats, sheep, pigs, *etc.*, are slaughtered by humans primarily for their meat. The skins, hides and other organic matter such as bones and offal containing high level of proteins are typically disposed off as wastes from the slaughterhouse.[10] On the other hand, the leather industries utilize a huge amount of solid proteinaceous biomass wastes, skins and hides, for making leather.

The transformation of skin/hides into leather requires a series of chemical and mechanical operations and generates solid, liquid and gaseous wastes.[11,12] Figure 12.1 shows a schematic of conventional leather processing, associated solid wastes generation and some of their major applications. The unavoidable solid wastes generated from the leather industry can be broadly classified as untanned collagenous, tanned collagen–chromium complexes and non-proteinous wastes. The preliminary stage in the leather making process generates an enormous quantity of proteinaceous biomass wastes such as raw fleshings, hide or skin trimmings, limed split wastes and limed fleshings.[13] These wastes contain high-value proteinaceous material, namely collagen. The other form of proteinaceous solid wastes predominantly contain heavy metals, such as trivalent chromium, or vegetable tannins bound with collagen and include wet blue trimmings, chrome-tanned leather shavings, wet blue split wastes, vegetable tanned leather shavings, a combination tanned leather shavings, buffing dust, crust trimmings, finished leather trimmings, leather scrap wastes, *etc.* Traditionally, these wastes are either disposed off, dumped or burned in suburban fields, thereby producing toxic chromium compounds and resulting in a significant environmental impact as well as destroying an economically valuable and protein-rich biomass.[14–18] The deposition of huge quantities of tanned and untanned proteinaceous solid wastes in land or by burning in an open area is a potential danger to public health due to the possibility of oxidation of chromium(III) to toxic chromium(VI).[19] Recent environmental regulations and escalating landfill costs have encouraged both researchers and industrialists to seek appropriate solutions for the utilization of these wastes. Presently, the untanned protein wastes are commercially utilized for the production of gelatin,[20] partially hydrolyzed collagen or glue.[21] In some parts of the world, glue production from animal bones and hides has been a lucrative business.[22] The chromium-containing protein wastes are utilized for making different low-value products, such as the formation of paste board,[23–25] sheets,[26–29] insulators,[30,31] biogas generation,[32] inner soles for shoes, *etc.*[33,34]

Among the different technologies employed for the conversion of waste biomasses, the preparation of hybrid composite materials plays an important role. Recent advances in composite science and technology and natural fibre research and development offer significant opportunities for the preparation of new and improved biocomposite materials from renewable resources that are optionally recyclable, biocompatible and biodegradable, thereby enhancing global sustainability. The development of biocomposite materials is consistent with the principles of green chemistry and engineering, which pertain to the design, commercialization and use of processes and products that are technically and economically feasible while minimizing the generation of pollution.

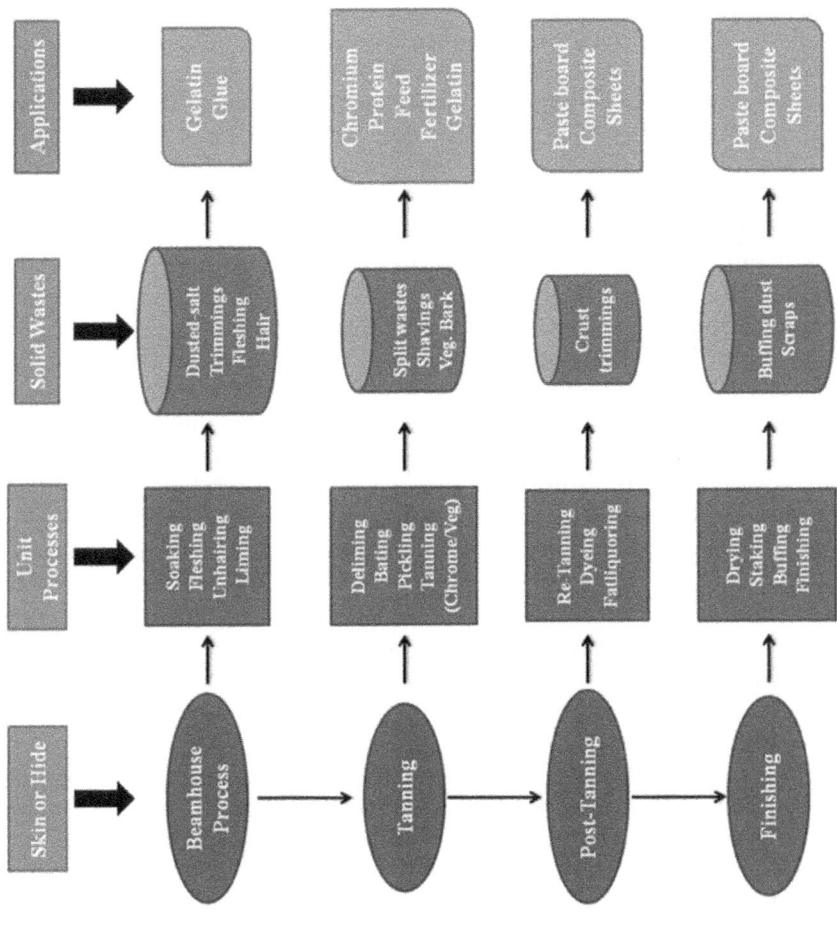

Figure 12.1 Schematic showing the generation of different types of solid wastes from a conventional leather processing chain and their present utilization areas.

This chapter provides detail on the recent research, development and future directions of hybrid biocomposites from natural proteinaceous biomass wastes such as skin trimmings and leather wastes employing environmentally friendly biopolymers.

12.1.1   Eco-benign Polymers

Environmentally friendly polymers have been defined as polymers that degrade in the environment by several mechanisms and culminate in complete biodegradation so that either no residue remains in the environment or only non-toxic residues are left. In recent years, there has been growing interest in the use of biodegradable polymers for different applications in order to reduce different forms of environmental pollution. Eco-benign polymers and their derivatives are diverse, abundant and important for life. These polymers are mostly produced from living organisms or through non-toxic precursors and reactions. They exhibit fascinating properties and are of increasing importance for different applications. These polymers are broadly classified into two main areas: renewable and non-renewable polymers. Essentially, renewable biodegradable polymers utilize a renewable resource (*i.e.*, animal or plant by-products)[35] in the development of the polymer, rather than a non-renewable (*i.e.*, petroleum-based) resource.[36] Obviously, long-term research and development focuses on the utilization of various renewable/biodegradable polymers for a variety of applications, including as composites. Some of the environmentally friendly polymers include cellulose, starch, chitin, chitosan, proteins, peptides, alginate, gelatin, poly(lactic acid), locust bean gum, pectin, plant exudate gums, poly(3-hydroxybutyrate), polyamide, *etc*. Originally, biopolymers were intended to be used in packaging,[37] farming[38] and other industries with low strength requirements. The performance limitations and high cost of biopolymers are major barriers for their widespread acceptance as substitutes for traditional non-biodegradable synthetic polymers. The high cost of some eco-benign polymers compared with traditional polymers is not due to the raw material costs for biopolymer synthesis but mainly to the low volume of production. New and advanced synthetic procedures for preparing biopolymers need to be developed for their high-value applications. The challenge for the development of biodegradable polymers lies in the fact that they should be stable during storage or use and again should degrade only when they are disposed of after their intended lifetime. Eco-benign polymers reinforced with natural biomass can produce novel biocomposites to replace and substitute different synthetic polymer fibre-reinforced composite materials for various applications.

12.1.2   Bio-based Materials *versus* Environment

Interest in eco-friendly or green polymeric materials is growing rapidly owing to concerns with the toxicity of degraded products and the limited availability of petroleum-based polymers. Biodegradable polymers from natural or biological resources have received much attention due to the fact that they are biocompatible and biodegradable. When a biodegradable material (neat polymer, blended product or composite) is obtained completely from renewable resources, it may be termed as a green polymeric material. The lifecycle of the renewable polymeric materials is a carbon-neutral process, and their use may reduce carbon dioxide accumulation in the atmosphere and dependence on valuable raw materials. Within the past few years, there has been a dramatic increase in the utilization of environmental wastes, including plant and animal biomasses. The animal biomass wastes, such as skin, hides, hairs, bones, *etc.*, and the plant wastes, such as leaves from flax, jute, hemp, pineapple leaf, sisal, *etc.*, have been utilized for making new types of environmentally friendly composite materials.[39] However, animal biomass waste, if not properly managed or utilized, can adversely impact the environmental stability by suppressing water quality through surface runoff and erosion, direct discharges to surface waters, spills and other dry-weather discharges, and leaching into soil and ground.[40] Natural biomass wastes are now emerging as viable alternatives to form multifunctional composites either alone or in combination with different polymers for various applications in automotive parts, building, structural and packaging materials. The advantages of using natural biomass waste are low cost, easy availability, sustainability, recyclability, biodegradability and waste minimization.[41] Another important application for the natural biomass waste is in the area of biomaterials. A biomaterial intended for biological applications must possess certain specific characteristics. The most fundamental requirement is being biocompatible, in other words, not having any adverse effect on the host. Therefore, in biomedical applications, the traditional composite structures with a non-biocompatible matrix or reinforcement are being substituted by engineered biocomposites.[42] Hence, it is imperative to consider the environment and lifecycle aspects of a material, including bio-based materials or composites, before attempting their development and application.

12.1.3   What are Biocomposites?

Biocomposites are made from a biodegradable polymer as a matrix material and natural or biodegradable fibres as reinforcing elements. They generally have low cost, low density, high toughness, significant strength, thermal properties, enhanced energy recovery and biodegradability. Biocomposites are capable of undergoing decomposition, primarily through the enzymatic action of microorganisms, to carbon dioxide, methane, inorganic compounds or biomass in a specified period of time. Matrix biodegradable polymers may be obtained from renewable resources, especially from plants and animals, and have gained much importance over petroleum-based biodegradable polymers in recent years. Biodegradable reinforcement fibres comprise renewable agricultural, animal and forestry feedstocks (biomass), including wood, agricultural waste, grasses and natural plant fibres composed of carbohydrates such as sugars, starch, lignin and cellulose,[43] as well as animal protein fibres. Biocomposites prepared by the combination of biomass and biopolymers provide the necessary performance either alone or in combination with petroleum-based polymers and offer a sustainable path to achieve eco-friendly materials. However, the need to produce 100% bio-based composites as substitutes for synthetic materials is not immediate. Here, different proteinaceous animal biomass wastes generated from the leather industry can be utilized as a matrix polymer or reinforcement fibre for making hybrid biocomposites that can substitute and compete with traditional synthetic polymer-based composites, thereby reducing the environmental impact and increasing their economic viability.

12.1.4 Value Added Products from Skin and Leather Wastes: Current Trends

A great deal of attention from various researchers is being paid to the valorization of chromium-free and chromium-containing proteinaceous wastes from the leather industry. The chromium-free solid proteinaceous wastes such as skin trimmings, fleshings and limed trimmings are predominantly used for the preparation of gelatin,[20] poultry feed,[44] biodiesel[45,46] and fertilizer.[47,48] Gelatin is a proteinaceous material obtained by the partial hydrolysis of collagen sources. Most of the methods of gelatin manufacture involve the utilization of bovine hide, cattle bones, pig skin, fish skins, tendons, *etc.* Hence, skin trimming wastes form an important source for gelatin manufacture. Gelatin possesses a wide range of applications due to its functional properties such as biodegradability, biocompatibility and physiological functions.[49] Fleshings are mainly used for vermicomposting to form nutrient-enriched organic manure[50] as well as biodiesel production.[44,51] Except gelatin, all the other products obtained from proteinaceous wastes of skins are found to be less proficient, having low-value applications. Hence, there is a great opportunity to convert the skin wastes into new innovative products for high-value applications. Along this line, several initiatives have been made in the recent past. Isolation of valuable protein products from untanned leather wastes by thermal and enzymatic treatments has been attempted for food applications.[52] Enzyme-based hydrolysis is becoming popular to extract collagen or its polypeptides from skin or limed trimmings.[44,53–55] Although these initiatives open up new avenues for the utilization of untanned proteinaceous wastes, there is still scope for high-value applications such as biomaterials, pharma products, *etc.* A major concern is the purity of collagen for biomedical applications; however, it has been shown that the collagen extracted from bovine hide wastes is pure and has increased biological and physiological properties.[55,56]

Another form of leather industry waste, chromium-containing proteinaceous wastes generated during the process of transformation of hides and skins into leather, is also being investigated by several groups for their utilization.[57,58] Most methods aim at recovery and isolation of the protein fraction, and are generally based on alkaline (using CaO, MgO or NaOH),[59–62] acidic[63] or enzymatic hydrolysis.[64,65] The chromium-containing proteinaceous wastes are also utilized for the formation of composites such as pasteboards,[23–25] sheets,[26–29] glue,[22,66] *etc.* Among various products reported for the utilization of chromium-containing proteinaceous wastes, the formation of composite sheets plays a distinctive role due to their ease of processability and improved mechanical and thermal properties. Moreover, the developed composite sheets find wide applications in leather goods, footwear, automotives, furnishings, *etc.* In the past, some researchers used chrome-tanned leather shaving wastes to prepare a powder, after multistep disintegration, which was used as a filler for butadiene–acrylonitrile rubber.[67] Several patents were filed for converting the leather wastes into semi-continuous sheet or boards for a variety of applications.[23–29] Chrome-tanned collagen wastes were used, beside silica, as a filler for rubber mixes containing synthetic *cis*-1,4-polyisoprene rubber and as a dispersing agent.[68] The polymerization of poly(methyl methacrylate) onto leather wastes to form composite sheets has been reported and the modification is claimed to improve the finishing, dyeing and stability of the formed composite sheets.[28,69,70] Composite sheets have also been prepared by the addition of synthetic polymers like poly(vinyl chloride) (PVC) to leather waste.[71,72] Leather wastes were also used as a filler in thermoplastic polymer composites to increase their mechanical and thermal properties.[73] Chemically modified short leather fibres were compression molded into a plasticized PVC matrix[74], and dioctyl phthalate plastisols[75,76] were further added to form composites with improved properties. Efforts have been made to form polymer composites by using chromium-containing wastes with butadiene rubber[77] and PVC, which were found to have improved properties.[78] Most of these methods employ expensive and toxic chemicals and involve noxious reactions, leading to unsatisfactory composite materials. To overcome these deficiencies, eco-friendly polymers and materials can be employed to form multifunctional composite sheets from chromium-containing protein wastes. In this chapter, we have discussed briefly the development of hybrid composite materials with multifunctional properties by utilizing collagen-containing protein wastes generated from the leather industry and environmentally benign polymers.

12.2 Biomaterials from Chromium-free Proteinaceous Wastes

Collagen, a widely used biopolymer, has been found to possess numerous biomedical applications. Collagen is extracted from different forms of living organisms and is classified accordingly in different types. Type I collagen is the most abundant protein present in animals. It has outstanding mechanical properties and is present in virtually every extracellular tissue. Some of the major applications of collagen include wound healing,[79] drug delivery, controlled transfer of therapeutic agents,[80] substrate culture of living cells,[81] gene delivery[82] and replacement/substitutes for human skin,[83] owing to its excellent biocompatibility and biodegradability. Collagen is a natural substrate for cellular attachment, growth, differentiation and promotes cellular proliferation.[84,85] However, the fast biodegrading rate and the low mechanical strength of the untreated collagen scaffold are crucial problems that limit the further use of this material. Glutaraldehyde-based cross-linking of the collagen-based scaffolds is a widely used method to modify the biodegradation rate and to improve the mechanical properties.[86] Alternatively, different cross-linking agents, such as formaldehyde, chromium, acyl azide and dimethyl suberimidate,[87–89] as well as synthetic polymers such as polyurethane,[90] poly(ethylene oxide)[91] and poly(glycolic acid),[92] have been used to obtain optimized hybrid biomaterials based on collagen. However, these biomaterials have some limitations regarding their stability, cytotoxicity and degradation properties. The cross-linking agents or synthetic polymers have an exacerbating effect on the calcification of prosthesis materials[93] and are also cytotoxic due to post-implantation depolymerization and monomer release.[94] To overcome these limitations, it has been shown that the collagen can be blended with suitable natural or biopolymers instead of synthetic polymers or cross-linking agents to improve the biomaterial stability and biological properties. Incorporation of human growth hormone or curcumin into collagen films has been shown to enhance the wound reduction property and cell proliferation at the dermal wound site.[80] Dermal skin substitutes (membranes) made of collagen and glycosaminoglycan were found to be suitable substrates for the culture of human epidermal keratinocytes.[81] The scaffold formed by blending collagen with chitosan is a potential candidate for dermal applications with enhanced biostability and biocompatibility.[95] Composite materials were formed by blending collagen with polycaprolactone[96] and poly(L-lactic acid)[97] for cell proliferation and their applications in tissue repair. The chromium-free proteinaceous wastes, such as skin trimmings, fleshings and limed trimmings, contain a valuable protein, collagen. There is great potential to extract collagen from these proteinaceous waste materials and utilize it as a biomaterial for a number of biomedical applications. This calls for the development of hybrid biomaterials for biomedical applications utilizing the collagen extracted from proteinaceous wastes and natural polymers without employing toxic cross-linking agents.

12.2.1 Formation of Hybrid Films Using Collagen and Natural Polymers

The process for the formation of hybrid films is shown schematically in Figure 12.2. The hide trimming wastes collected from a local tannery at Chennai were soaked, limed, dehaired, relimed, fleshed and delimed completely following a conventional leather processing procedure to remove unwanted proteins and materials.[98] The delimed hide pieces were subjected to gradient dehydration using acetone and methanol to completely remove the moisture. Finally, the hide pieces were thoroughly dried in a vacuum drier and made into fine powder using a Wiley mill of mesh size 2 mm. A known weight of hide powder was dissolved in 0.5 N glacial acetic acid by heating at $45 \pm 3\,^\circ$C to obtain a known concentration of collagen solution. The formed collagen solutions (C) were blended with different biopolymers, such as 2-hydroxyethyl cellulose (HEC),[99] starch (ST) or soy protein (SP),[100] to obtain viscous homogeneous solutions. These homogeneous solutions were poured into Petri dishes and air-dried at room temperature to obtain the desired biopolymer/ collagen hybrid films. All the developed hybrid films had an average thickness of $30 \pm 10 \, \mu m$.

12.2.2 Characteristics of Collagen/Biopolymer Hybrid Films

The prepared collagen/biopolymer hybrid films have been characterized for their mechanical, biological and thermal properties. As seen from Figure 12.3a, the C/HEC hybrid films exhibit a linear increase in the tensile stress as the HEC concentration increases when compared to collagen/starch/soy composite films. The percentage elongation or strain of the C/HEC hybrid films decreases slightly as the concentration of the HEC increases. Similar trends were observed for C/HEC hybrid films prepared from different concentrations of collagen, such as 10, 6.7 and 5 mg mL^{-1}.[99] The tensile stress values of the C/ST/ SP hybrid films increase with an increase in the concentration of starch (up to 75 wt%) but decrease with an increase in the concentration of soy protein (see Figure 12.3b). C/ST/SP hybrid films display an inverse relationship between tensile stress and strain, depending on the type of biopolymer.[100]

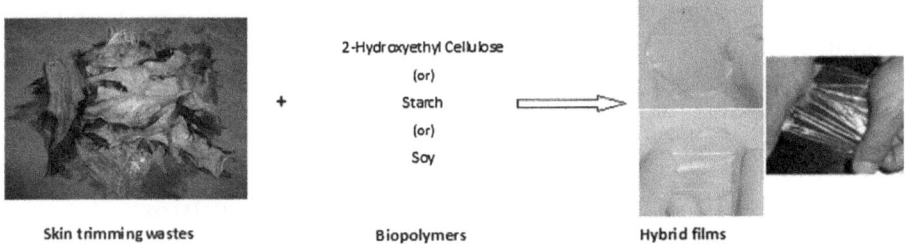

Figure 12.2   Schematic of formation of hybrid biocomposite films.[99,100]

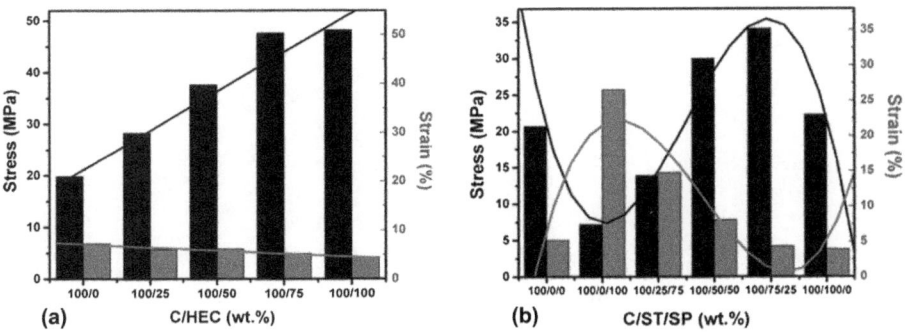

Figure 12.3   Stress–strain plots of the developed collagen/biopolymer hybrid films: (a) C/HEC hybrid films;[99] (b) C/ST/SP hybrid films.[100]

The elongation or strain values of C/ST/SP hybrid films increase with the increase in the concentration of soy protein and decrease with the increase in the concentration of starch. In other words, an increase in the concentration of soy protein provides more extendibility, while an increase in the concentration of starch provides more strength in the hybrid films when compared to pure collagen film.

The thermal stability of all the collagen/biopolymer hybrid films significantly improved as a function of polymer concentration, as evidenced from the increase in their degradation temperature. The swelling and the *in vitro* biodegradation of the formed hybrid biomaterials reduced significantly as the concentration of biopolymer (HEC/SP/ST) increased. Further, it has also been shown using mouse embryofibroblast cells and MTT [3-(4,5-dimethylthiazol-2-yl)-2,5-diphenyltetrazolium bromide] assay that the addition of biopolymers (HEC or starch or soy protein) along with collagen assists cell growth and proliferation.[99,100] The morphology of the formed hybrid films as visualized through scanning electron microscopy is shown in Figure 12.4. The cross-section of the hybrid films formed using a higher proportion of biopolymer shows the formation of tiny micro-pores and increased roughness, in comparison to the pure collagen film. The presence of pores and rough structure are expected to assist in the adhesion and growth of the cells, thereby increasing the cell proliferation and tissue generation.

Hence, the developed hybrid films exhibit superior mechanical, structural and thermal properties and are also biocompatible without employing toxic cross-linking agents when compared to the pure collagen films. Thus, the collagen sourced from skin trimming wastes is shown to be pure and not cytotoxic and hence suitable for various biomedical applications.

12.3   Flexible Composite Sheets from Chromium-containing Proteinaceous Wastes

Increasing environmental regulations and escalating landfill costs have forced researchers to look for alternative methodologies for converting

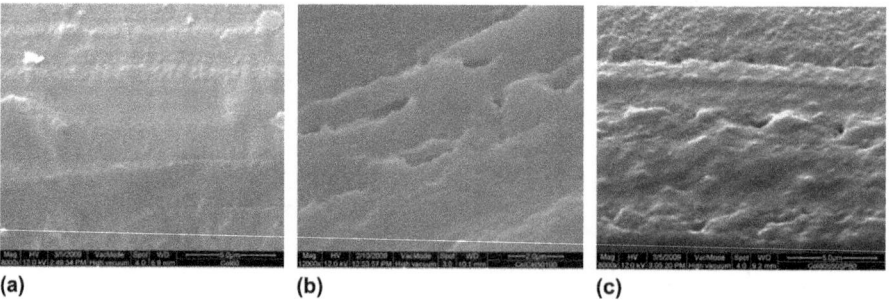

(a)   **(b)**   **(c)**

Figure 12.4   Scanning electron micrographs showing the cross-sectional images of the developed hybrid films: (a) pure collagen film; (b) 100/100 wt% C/HEC hybrid film; (c) 100/50/50 wt% C/ST/SP hybrid film.

chromium-containing proteinaceous wastes into useful products. Although several application avenues have been reported for the valorization of chromium-containing proteinaceous wastes, the formation of composite sheets plays a distinctive role due to their easy of processability and improved mechanical and thermal properties. Numerous articles and patents can be found on the conversion of these leather wastes into semi-continuous sheets or boards for a variety of applications. However, previous methods of sheet formation could not provide sheets with the desired characteristics. Hence, our group focused on the formation of flexible composite sheets with multifunctional properties using chrome-tanned leather wastes and a variety of environmentally friendly polymers such as 2-hydroxyethyl cellulose,[101] poly(dimethylsiloxane) (PDMS)[102] and poly(vinylpyrrolidone) (PVP).[103] All the chosen environmentally friendly polymers are water soluble and non-toxic. HEC is a derivative product of cellulose, a natural organic polymer from plants, widely used in cosmetics, pharmaceutical and household products.[104] PDMS is a non-toxic organosilicon compound and its applications range from cosmetics and medical devices to food additives.[105] Similar to PDMS, PVP is also a non-toxic polymer with applications ranging from cosmetics, medical devices and food additives to pharmaceutical and personal care products.[106] The amount of polymer has been optimized for achieving desired properties on the derived sheets. The formed composite sheets have been studied for their structural, physical and thermal properties.

12.3.1 Flexible Composite Sheet Formation

Figure 12.5 shows a schematic for the formation of flexible composite sheets by utilizing chromium-containing proteinaceous wastes and environmentally friendly polymers. The chrome-tanned leather wastes were initially ground into a fine powder of 2 ± 0.5 mm size in a Wiley mill and stored at room temperature. The powdered chrome-tanned leather wastes were partially hydrolyzed using dilute sulfuric acid followed by microwave heating to form a homogenous mass. The homogenous mass was mixed with environmentally friendly polymers such as HEC,[101] PDMS[102] or PVP[103] in different weight ratios ranging from 0 to 40 wt%, based on the weight of chrome-tanned leather shaving wastes. In the

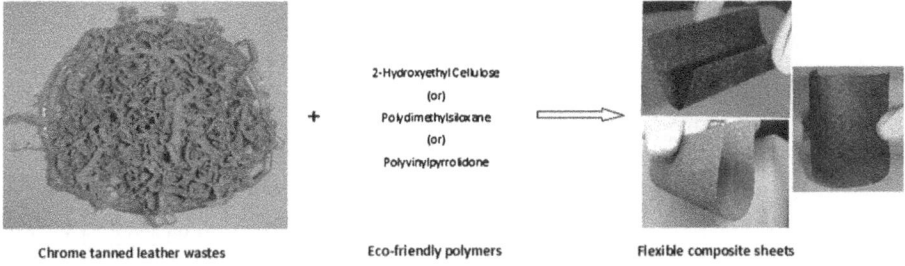

Chrome tanned leather wastes   Eco-friendly polymers   Flexible composite sheets

Figure 12.5 Schematic of flexible composite sheet formation.

case of PVP, pure collagenous paste (C), prepared by dissolving the skin powder using 0.5 N acetic acid, was used in conjunction with chrome shaving mass (CS) in a ratio of 1:10 (on raw weight basis) to form a homogenous mixture (CCS). The pure collagenous paste was used as a binder for the effective binding of PVP with chrome shaving mass. The corresponding mixtures were heated at $50 \pm 2\,°C$ for 10 min with stirring and transferred to a glass plate and spread uniformly or hot pressed. The dried composite sheets were removed from the plates and stored for further analysis.

12.3.2 Properties of the Flexible Composite Sheets

The mechanical properties of all the developed composite sheets show satisfactory improvement as the composition of the polymer increases, as shown in Table 12.1. It is seen that the tensile stress values of PVP- and HEC-incorporated composite sheets increase as the concentration of the PVP or HEC increases.[101,103] On the other hand, PDMS incorporation in the composite sheets leads to a mixed trend, with an initial increase but a sudden decrease of tensile stress after 20 wt.% polymer concentration.[102] In general, PVP-incorporated sheets were found to have improved mechanical properties when compared to that of cellulose- or siloxane-incorporated composite sheets. The order of mechanical properties of the developed composite sheets can be arranged as a function of the chosen environmentally friendly polymers as PVP > HEC > PDMS. The greater mechanical property of PVP-incorporated composite sheets compared to HEC or PDMS may be due to the strong binding between the collagen–chromium complex and oxygen functionalities of PVP. Akin to this, the thermal properties of the composite sheets exhibit a similar performance, as analyzed through thermogravimetric analysis. In general, a moderate improvement in the thermal properties of the composite sheets has been observed as the composition of the polymer increases. The softness of all the composite sheets is in inverse relation to the tensile property of the developed composite sheets, as evidenced from Table 12.1. Hence, it is possible to

Table 12.1 Tensile and softness properties of the developed composite sheets.[101–103]

Composition of CS/CCS/polymer composite sheets (wt%)	Tensile stress (MPa)			Softness (mm)		
	PVP	HEC	PDMS	PVP	HEC	PDMS
100/0	7.37 ± 2.2	–	0.52 ± 0.4	2.13 ± 0.2	–	6.5 ± 0.0
100/2.5	9.28 ± 1.8	0.17 ± 0.08	–	3.51 ± 0.1	6.73 ± 0.0	–
100/5.0	17.47 ± 0.7	0.73 ± 0.28	0.63 ± 0.2	1.95 ± 0.1	6.73 ± 0.0	6.5 ± 0.2
100/7.5	18.90 ± 1.2	1.30 ± 0.21	0.93 ± 0.8	1.69 ± 0.0	6.63 ± 0.1	6.5 ± 0.1
100/10	21.31 ± 5.9	2.10 ± 0.30	1.14 ± 0.5	1.52 ± 0.0	5.40 ± 0.2	5.1 ± 0.1
100/20	–	3.14 ± 0.45	1.72 ± 0.9	–	3.80 ± 0.2	4.1 ± 0.0
100/30	–	–	0.58 ± 0.01	–	–	6.5 ± 0.1
100/40	–	–	0.54 ± 0.07	–	–	6.5 ± 0.2

develop multifunctional composite sheets with tunable properties from chromium-containing proteinaceous wastes using environmentally friendly polymers.

Scanning electron micrographs of the cross-section of fractured surfaces after tensile strength analysis of selected composite sheets are shown in Figure 12.6. The CCS/PVP and CS/HEC composite sheets show a reduction of fibre pull-out (Figure 12.6a–c) or the presence of pores (Figure 12.6d–f) as the concentration of PVP or HEC increases, indicating better binding and formation of the hybrid composite sheets. The CS/PDMS composite sheets

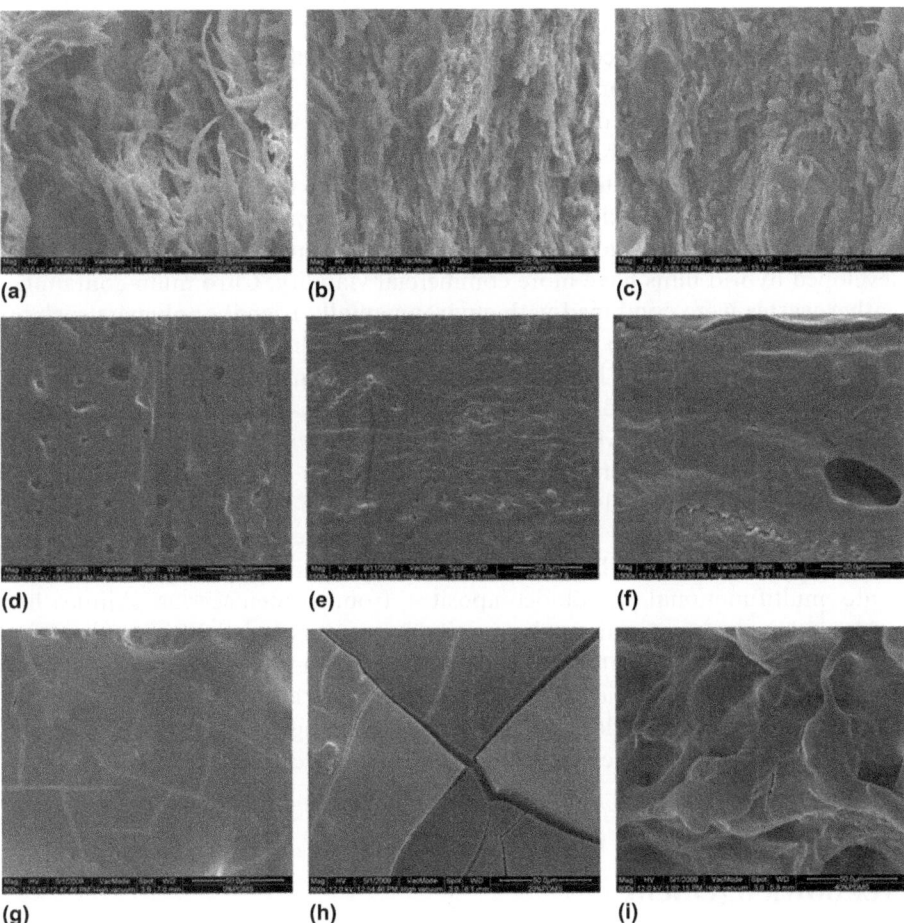

Figure 12.6 Scanning electron micrographs of the cross-section (d–f) or tensile fractured surfaces (a–c; g–i) of select composite sheets: (a) 100/2.5 wt% CCS/PVP; (b) 100/7.5 wt% CCS/PVP; (c) 100/10 wt% CCS/PVP; (d) 100/2.5 wt% CS/HEC; (e) 100/7.5 wt% CS/HEC; (f) 100/20 wt% CS/HEC; (g) 100/5 wt% CS/PDMS; (h) 100/20 wt% CS/PDMS; (i) 100/40 wt% CS/PDMS.[101–103]

(Figure 12.6g–h) have increased cracks and roughness on their surface as the concentration of the PDMS increases up to 20 wt%, which may be due to strong binding of PDMS with the chrome-tanned leather wastes. On the other hand, the presence of a very high concentration of PDMS (40 wt%) resulted in a collapsed surface without any cracks (Figure 12.6i). These results are in agreement with the mechanical properties and softness data.

12.4   Conclusion

In this chapter, we have discussed a possible solution for the valorization of different forms of proteinaceous solid biomasses arising from leather industries as wastes. The chromium-free proteinaceous wastes were blended with different natural polymers such as cellulose, soy and starch without employing any toxic cross-linking agents to form hybrid biomaterials. It has been shown that soy protein provides elastic properties to the hybrid films while the incorporation of starch provides more strength when compared to pure collagen film. The developed hybrid biomaterials were found to have promising mechanical, thermal, degradation, swelling and biocompatibility properties. Moreover, one of the biopolymers, collagen, is sourced from collagenous wastes and hence the developed hybrid films have more commercial viability. Chromium-containing leather wastes were combined with environmentally friendly polymers such as HEC, PVP and PDMS to form flexible composite sheets by a simple, efficient and scalable technique. It has been found that the tensile strength of PVP- and HEC-incorporated composite sheets increases gradually as a function of concentration, while the PDMS sheets show a mixed trend. For all the composite sheets, the softness property had an inverse relation with the tensile strength. The presence of pores and fibre pull-outs is found to decrease gradually as the concentration of polymer increases, indicating better binding and sheet formation, as evidenced by SEM analysis. Attempts are also being made to fabricate multifunctional nanobiocomposites from proteinaceous skin/leather wastes using nanomaterials such as carbon nanotubes, graphitic nanocarbon materials derived from leather industry wastes and inorganic nanoparticles and other environmentally friendly polymers for a variety of applications. The emergence of nanotechnology as well as biotechnology will provide viable and profitable solutions for the valorization of these protein-rich wastes in the near future.

Acknowledgement

The authors gratefully thank the Council of Scientific and Industrial Research (CSIR), India, for providing financial assistance under the YSA project scheme. The authors also wish to thank Dr A. B. Mandal, Director, Central Leather Research Institute, India, for his encouragement. The authors also thank Mr R. Murali and Ms A. Anumary for their valuable help in carrying out some of the experiments.

References

1. A. K. Bledzki and J. Gassan, *Prog. Polym. Sci.*, 1999, **24**, 221.
2. A. K. Mohanty, M. Misra and L. T. Drzal, *J. Polym. Environ.*, 2002, **10**, 19.
3. A. N. Netravali and S. Chabba, *Mater. Today*, 2003, **6**, 22.
4. B. P. Singh, M. R. Panigrahi and H. S. Ray, *Energy Sources*, 2000, **22**, 649.
5. B. K. Gullett and A. Touati, *Atmos. Environ.*, 2003, **37**, 4893.
6. M. Acaroglu, A. S. Aksoy and H. Ogut, *Energy Sources, Part A*, 1999, **21**, 339.
7. A. Demirbas, *Energy Policy*, 2008, **36**, 834.
8. S. K. Han and H. S. Shin, *Int. J. Hydrogen Energy*, 2004, **29**, 569.
9. E. Valdman and S. G. F. Leite, *Bioprocess Eng.*, 2000, **22**, 171.
10. K. Jayathilakan, K. Sultana, K. Radhakrishna and A. S. Bawa, *J. Food. Sci. Technol.*, 2012, **49**, 278. (DOI: 10.1007/s13197-011-0290-7).
11. A. Cassano, E. Drioli and R. Molinari, *Desalination*, 1997, **113**, 251.
12. A. Cassano, E. Drioli and R. Molinari, *Water Res.*, 1999, **4**, 443.
13. L. F. Cabeza, A. J. Mcaloon, W. C. Yee, M. M. Taylor, E. M. Brown and W. N. Marmer, *J. Am. Leather Chem. Assoc.*, 1998, **93**, 2990.
14. A. D. Dayan and A. J. Paine, *Hum. Exp. Toxicol.*, 2001, **20**, 439.
15. S. Tahiri, A. Albizane, A. Messaoudi, M. Azzi, J. Bennazha, S. A. Younssi and M. Bouhria, *Waste Manage.*, 2007, **27**, 89.
16. V. Sarin and K. K. Pant, *Bioresour. Technol.*, 2006, **97**, 15.
17. K. J. Sreeram, S. Saravanabhavan, J. R. Rao and B. U. Nair, *Ind. Eng. Chem. Res.*, 2004, **43**, 5310.
18. N. F. Fahim, B. N. Barsoum, A. E. Eid and M. S. Khalil, *J. Hazard. Mater.*, 2006, **136**, 303.
19. B. R. James and R. J. Bartlett, *J. Environ. Qual.*, 1983, **12**, 169.
20. A. Gammoun, S. Tahiri, N. Saffaj, A. Albizane, A. S. Younssi, M. Azzi and M. D. L. Guardia, *Phys. Chem. News*, 2008, **41**, 89.
21. S. Tahiri and M. D. L. Guardia, *J. Am. Leather Chem. Assoc.*, 2009, **104**, 52.
22. J. E. Hill and N. C. Hill, *U.S. Pat.* 2 517 487, 1950.
23. A. N. Mathieu, *U.S. Pat.* 20 020, 1868.
24. A. Adams and A. R. Brooklyn, *U.S. Pat.* 1 188 600, 1916.
25. A. W Case, *U.S. Pat.* 878 485, 1908.
26. N. S. Coulson, H. B. Kinsley and J. K. Nunn, *U.S. Pat. Appl.* US2007/0184742A1, 2007.
27. H. Pelzer, *U.S. Pat.* 5 624 619, 1997.
28. A. Klasek, J. Simonikova and F. Pavelka, *J. Appl. Polym. Sci.*, 1986, **31**, 2007.
29. E. W. Henke, *U.S. Pat.* 4 497 871, 1985.
30. O. A. Mohamed, N. H. E. Sayed and A. A. Abdelhakim, *J. Appl. Polym. Sci.*, 2010, **118**, 446.
31. T. Basegio, C. Haas, A. Pokorny, A. M. Bernardes and C. P. Bergmann, *J. Hazard. Mater.*, 2006, **137**, 1156.

32. A. Thangamani, S. Rajakumar and R. A. Ramanujam, *Clean Technol. Environ. Policy*, 2010, **12**, 517.
33. A. R. Horowitz, *U.S. Pat.* 2 148 904, 1939.
34. I. Newhall, *U.S. Pat.* 52 737, 1866.
35. C. K. Williams and M. A. Hillmyer, *Polym. Rev.*, 2008, **48**, 1.
36. B. Kamm and M. Kamm, *Appl. Microbiol. Biotechnol.*, 2004, **64**, 137.
37. R. N. Tharanathan, *Trends Food Sci. Technol.*, 2003, **14**, 71.
38. T. Gerngross, S. Slater and R. A. Gross, *Science*, 2003, **299**, 822.
39. T. Nishino, in *Green Composites*, ed. C. Baillie, Woodhead, Cambridge, 2004, p. 50.
40. E. Ceotto, *Bioresour. Technol.*, 2005, **96**, 191.
41. P. Sudha, H. I. Somashekhar, S. Rao and N. H. Ravindranath, *Biomass Bioenergy*, 2003, **25**, 501.
42. Z. Tang, Y. Wang, P. Podsiadlo and N. A. Kotov, *Adv. Mater.*, 2006, **18**, 3203.
43. M. Wollerdorfer and H. Bader, *Ind. Crops Prod.*, 1998, **8**, 105.
44. V. J. Sundar, A. Gnanamani, C. Muralidharan, N. K. Chandrababu and A. B. Mandal, *Rev. Environ. Sci. Biotechnol.*, 2011, **10**, 151.
45. H. Ozgunay, S. Colak, G. Zengin, O. Sari, H. Sarikahya and L. Yuceer, *Waste Manage.*, 2007, **27**, 1897.
46. K. Karel, B. Michaela and F. Tomas, *J. Am. Leather Chem. Assoc.*, 2009, **104**, 177.
47. A. R. Serrano, V. M. Maldonado and K. Kosters, *J. Am. Leather Chem. Assoc.*, 2003, **98**, 43.
48. J. P. Fontenot, L. W. Smith and A. L. Sutton, *J. Anim. Sci.*, 1983, **57**, 221.
49. S. Young, M. Wong, Y. Tabata and A. G. Mikos, *J. Control. Release*, 2005, **109**, 256.
50. B. Ravindran, S. L. Dinesh, J. Kennedy and G. Sekaran, *Appl. Biochem. Biotechnol.*, 2008, **151**, 480.
51. C. Olak, S. Zengin, G. Ozgunay, H. Sarıkahya, H. Sari and O. Yuceer, *J. Am. Leather Chem. Assoc.*, 2005, **100**, 137.
52. Z. Bajza and V. Vrcek, *Waste Manage.*, 2001, **21**, 79.
53. L. Yanchun, Z. Deyi, J. Liqiang, H. Lijie and S. Leyao, *J. Am. Leather Chem. Assoc.*, 2005, **89**, 103.
54. F. F. Hervas, P. Celma, I. Punti, J. Cisa, J. Cot, A. Marsal and A. Manich, *J. Am. Leather Chem. Assoc.*, 2007, **102**, 1.
55. E. V. Tonkova, M. Nustorova and A. Gushterova, *Curr. Opin. Microbiol.*, 2007, **54**, 54.
56. G. Y. Li, S. Fukunaga, K. Takenouchi and K. F. Nakamura, *J. Soc. Leather Technol. Chem.*, 2004, **88**, 66.
57. V. A. Lipsett, *J. Am. Leather Chem. Assoc.*, 1982, **77**, 291.
58. J. R. Rao, P. Thanikaivelan, K. J. Sreeram and B. U. Nair, *Environ. Sci. Technol.*, 2002, **36**, 1372.
59. D. F. Holloway, *U.S. Pat.* 4 100 154, 1978.
60. G. Guardini. *U.S. Pat.* 4 483 829, 1983.

61. S. Tahiri, M. Bouhria, A. Albizane, A. Messaoudi, M. Azzi, S. Y. Alami and J. Mabrour, *J. Am. Leather Chem. Assoc.*, 2004, **99**, 16.
62. A. Galatik, J. Duda and L. Minarik, *Czech. Pat.* 252 382, 1988.
63. G. Guardini, *U.S. Pat.* 4 483 829, 1984.
64. M. M. Taylor, E. J. Diefendorf, G. C. Na and W. N. Marmer, *U.S. Pat.* 5 094 946, 1992.
65. M. Sivaparvathi, K. Suseela and S. C. Nanda, *Leather Sci.*, 1986, **33**, 8.
66. L. S. Simeonova and P. G. Dalev, *Waste Manage.*, 1996, **16**, 765.
67. A. Przepiorkowska, K. Chronska and M. Zaborski, *J. Hazard. Mater.*, 2007, **141**, 252.
68. A. Przepiorkowska, M. Prochon and M. Zaborski, *J. Soc. Leather Technol. Chem.*, 2004, **88**, 223.
69. T. J. M. Santana and F. V. Moreno, *Polym. Bull.*, 1999, **42**, 329.
70. E. F. Jordan and S. F. Feairheller, *J. Appl. Polym. Sci.*, 1980, **25**, 2755.
71. T. J. M. Santana, A. C. Torres and A. M. Lucero, *Polym. Compos.*, 1998, **19**, 431.
72. T. J. M. Santana, M. J. A. Vega, A. M. Lucero and F. V. Moreno, *Polym. Compos.*, 2002, **23**, 49.
73. B. Ramaraj, *J. Appl. Polym. Sci.*, 2006, **101**, 3062.
74. T. J. M. Santana, M. J. A. Vega, A. Marquez, F. V. Moreno, M. O. W. Richardson and J. L. C. Machin, *Polym. Compos.*, 2002, **23**, 991.
75. A. G. Andreopoulos and P. A. Tarantili, *J. Macromol. Sci., Part A*, 2000, **37**, 1353.
76. K. Babanas, P. A. Tarantili and A. G Andreopoulos, *J. Elastomers Plast.*, 2001, **33**, 72.
77. K. Chronska and A. Przepiorkowska, *J. Hazard. Mater.*, 2008, **151**, 348.
78. J. Rajaram, B. Rajinikanth and A. Gnanamani, *J. Polym. Environ.*, 2009, **17**, 181.
79. M. Maeda, K. Kadota, M. Kajihara, A. Sano and K. Fujioka, *J. Control. Release*, 2001, **77**, 261.
80. D. Gopinath, M. R. Ahmed, K. Gomathi, K. Chitra, P. K. Sehgal and R. Jayakumar, *Biomaterials*, 2004, **25**, 1911.
81. S. T. Boyce, D. J. Christanson and J. F. Hsbrough, *J. Biomed. Mater. Res., B*, 1988, **22**, 939.
82. K. Y. Lee, I. C. Kwon, Y. H. Kim, W. H. Jo and S. Y. Jeong, *J. Control. Release*, 1998, **51**, 213.
83. F. A. Auger, M. Rouabhia, F. Goulet, F. Berthod, V. Moulin and L. Germain, *Med. Biol. Eng. Comput.*, 1998, **36**, 801.
84. F. J. O'Brien, B. A. Harley, I. V. Yannas and L. J. Gibson, *Biomaterials*, 2005, **26**, 433.
85. Z. Ruszczak, *Adv. Drug Delivery Rev.*, 2003, **55**, 1595.
86. L. H. H. Oldedamink, P. J. Dijkstra, M. J. A. Vanluyn, P. B. Vanwachem, P. Nieuwenhuis and J. Feijen, *J. Mater. Sci.: Mater. Med.*, 1995, **6**, 460.
87. R. J. Levy, F. J. Schoen, F. S. Sherman, J. Nichols, M. A. Hawley and S. A. Lund, *Am. J. Pathol.*, 1986, **122**, 71.

88. N. Barbani, P. Giusti, L. Lazzeri, G. Polacco and G. Pizzirani, *J. Biomater. Sci., Polym. Ed.*, 1996, **7**, 461.
89. V. Charulatha and A. Rajaram, *Biomaterials*, 2003, **24**, 759.
90. L. L. H. Huang, P. C. Lee, L. W. Chen and K. H. Hsieh, *J. Biomed. Mater. Res., B*, 1998, **39**, 630.
91. N. F. M. Nasir, M. G. Raha, N. A. Kadri, S. I. Sahidan, M. Rampado and C. A Azlan, *Am. J. Biochem. Biotechnol.*, 2006, **2**, 175.
92. T. Toba, T. Nakamura, Y. Shimizu, K. Matsumoto, K. Ohnishi, S. Fukuda, M. Yoshitani, H. Ueda, Y. Hori and K. Endo, *J. Biomed. Mater. Res., B.*, 2001, **58**, 622.
93. G. Golomb, F. J. Schoen, M. S. Smith, J. Linden, M. Dixon and R. J. Levy, *Am. J. Pathol.*, 1987, **127**, 122.
94. M. J. A. V. Luyn, P. B. V. Wachem, L. H. H. O. Damink, P. J. Dijkstra, J. Feijen and P. Nieuwenhuis, *Biomaterials*, 1992, **13**, 1017.
95. L. Ma, C. Gao, Z. Mao, J. Zhou, J. Shen, X. Hu and C. Han, *Biomaterials*, 2003, **24**, 4833.
96. N. T. Dai, M. R. Williamson, N. Khammo, E. F. Adams and A. G. A. Coombes, *Biomaterials*, 2004, **25**, 4263.
97. X. Yang, M. Yuan, W. Li and G. Zhang, *J. Appl. Polym. Sci.*, 2004, **94**, 1670.
98. P. Thanikaivelan, J. R. Rao, B. U. Nair and T. Ramasami, *Environ. Sci. Technol.*, 2002, **36**, 4187.
99. A. Anumary, P. Thanikaivelan, M. Ashokkumar, R. Kumar, P. K. Sehgal and B. Chandrasekaran, *Soft Mater.*, in press.
100. R. Murali, A. Anumary, M. Ashokkumar, P. Thanikaivelan and B. Chandrasekaran, *Waste Biomass Valor.*, 2011, **2**, 323.
101. M. Ashokkumar, P. Thanikaivelan, K. Krishnaraj and B. Chandrasekaran, *Polym. Compos.*, 2011, **32**, 1009.
102. M. Ashokkumar, P. Thanikaivelan and B. Chandrasekaran, *Polym. Polym. Compos.*, 2011, **19**, 497.
103. M. Ashokkumar, P. Thanikaivelan, R. Murali and B. Chandrasekaran, *Waste Biomass Valor.*, 2010, **1**, 347.
104. A. Sannino, C. Demitri and M. Madaghiele, *Materials*, 2009, **2**, 353.
105. S. J. Kim, J. K. Lee, J. W. Kim, J. W. Jung, K. Seo, S. B. Park, K. H. Roh, S. R. Lee, Y. H. Hong, S. J. Kim and K. S. Kang, *J. Mater. Sci.: Mater. Med.*, 2008, **19**, 2953.
106. F. Bian, L. Jia, W. Yu and M. Liu, *Carbohyd. Polym.*, 2009, **76**, 454.

CHAPTER 13

Spider Silk: The Toughest Natural Polymer

GANGQIN XU,[a] GUOYANG WILLIAM TOH,[a] NING DU[a,c] AND XIANG YANG LIU*[a,b]

[a] Department of Physics, National University of Singapore, Singapore 117542; [b] Department of Chemistry, National University of Singapore, Singapore 117542; [c] Singapore-MIT Alliance for Research and Technology, S16-07, 3 Science Drive 2, Singapore 117543
*Email: phyliuxy@nus.edu.sg

13.1 Introduction to Natural Spider Silk Fibres

In order to catch prey, many spiders possess different silk types to construct disparately shaped webs, among which orb-webs are the best studied. In general, there are four different kinds of silk in the construction of orb-webs: dragline silk [also termed as major ampullate (MA) spidroin silk], minor ampullate (MI) spidroin silk, flagelliform silk and piriform silk. Dragline silk is used in the building of the framework (frame and radii) and lifeline, as shown in Figure 13.1, which demonstrates a schematic illustration and real photograph of an established orb-web of the spider *Nephila pilipes*. MI silk is utilized to form a temporary auxiliary spiral to stabilize the web structure and to perform as a template for the succeeding capture spiral, which is formed by flagelliform silk. The gluey piriform silk is used as "attachment cement" to interconnect the different structures in an orb-web.[1,2] Dragline silk is the most extensively characterized spider silk among the miscellaneous types.

RSC Green Chemistry No. 16
Natural Polymers, Volume 1: Composites
Edited by Maya J John and Thomas Sabu
© The Royal Society of Chemistry 2012
Published by the Royal Society of Chemistry, www.rsc.org

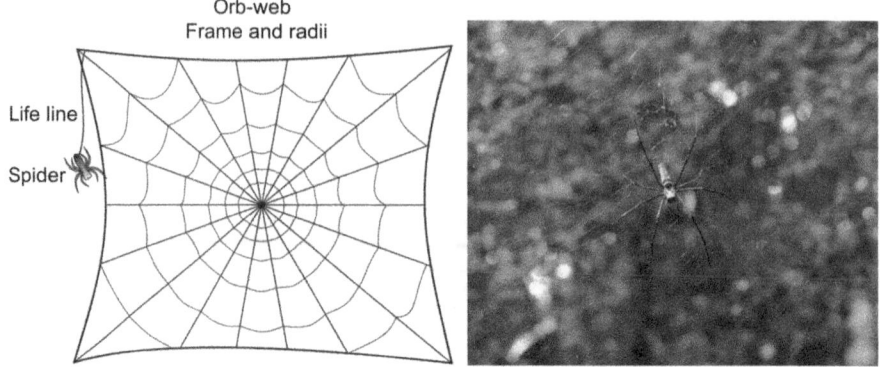

Figure 13.1   Schematic (*left*) and photograph (*right*) of the orb-web of the spider *Nephila pilipes*.

Spider dragline silk, used both as the web framework and safety line, is one of the most intriguing natural material marvels, owing to its exceptional strength and extensibility as well as light weight and biodegradability.[3–6] In contrast to current petrochemical-based synthetic polymers, the spider spins its ultra-strong, tough and totally recyclable fibres at ambient temperatures, low pressures and with water as solvent.[7,8] Spider silk provides a "green" alternative to high-performance synthetic fibres because of its environmental and human-friendly nature. Owing to the extraordinary properties of silk fibres, material scientists have generated increasing interest in designing and developing nature-inspired synthetic polymers with novel functions, as well as silk-based biomaterials.

13.2   General Structure of Silks

Spider dragline silk is a semicrystalline polymer, similar to silkworm silk, consisting of crystalline and non-crystalline (amorphous) regions.[9] The anti-parallel β-pleated sheet crystals, acting as the interlocks, are inter-dispersed in the amorphous region containing a range of conformations such as "random coil", β-turns and α-helices or the more compact and left-handed 3(1)-helices[10–16] to form a type of molecular network. The primary structures of silk proteins in both spider and silkworm are composed of repetitive sequences that can be separated into small blocks.[17] The amino acid sequence in the crystalline region of *Bombyx mori* silk fibroin is considered to be $(GAGAGS)_n$ (G, glycine; A, alanine; S, serine) and the sequence in the amorphous region contains Tyr-rich domains.[17] On the other hand, *N. pilipes* spider dragline silk is composed mainly of two proteins, spidroin I and II (SP I and SP II).[18] These proteins can be divided into block copolymers with $GAGA(A)_n$ blocks and non-crystalline glycine-rich domains.[19]

Table 13.1 A comparison between the structures of the spider *N. pilipes* dragline silk and the silkworm *B. mori* cocoon silk.[25]

Sample name	Overall content of β-sheets (%)	Crystal-linity (%)	Content of intramole-cular β-sheets (%)	Crystallite size (nm)			Inter-crystallite distance (nm)	
				a	*b*	*c*	Meridional direction	Equatorial direction
Spider dragline (10 mm s⁻¹)	51 (100%)	22 (43%)	29 (57%)	2.1	2.7	6.5	17.8	13.5
Silkworm cocoon silk	49 (100%)	40 (82%)	9 (18%)	2.3	4.1	10.3	4.8	7.2

Two types of β-structures can be identified in silks: the intramolecular β-sheets (non-crystalline β-sheets)[13] and intermolecular β-sheets (β-crystallites). The β-crystallites (orthogonal unit cell, $a = 1.03$, $b = 0.944$, $c = 0.695$ nm for the spider *Nephila* dragline silk,[20] and $a = 0.938$, $b = 0.949$, $c = 0.698$ nm for the silkworm *B. mori* silk[21]), constructed by several neighboring silk protein molecules, crosslink individual silk protein molecules[22] so as to form the molecular network.[13] The intramolecular β-sheets, on the other hand, are merely normal β-sheets in individual silk protein molecules. As shown in Table 13.1, for the dragline filaments of *N. pilipes* spiders, 57% of the total β-sheets are in the amorphous region, as characterized by Fourier transform infrared spectroscopy (FTIR). The remaining 43% of the β-sheets are β-crystal-linites,[13,23–25] as measured by X-ray diffraction (XRD). In contrast, the amount of intramolecular β-sheets in *B. mori* silkworm silk is much less. A comparison of the primary and crystalline structures between silkworm and spider silk proteins may explain the difference. *B. mori* silkworm silk proteins have significantly longer repetitive β-structural motifs than *N. pilipes* spider silk proteins. The β-structure sequence is GAGAGSGAAS(GAGAGS)$_n$, where $n = 1–11$ for *B. mori* silkworm silk, and GAGA(A)$_n$, where $n = 4–6$ for *N. pilipes* spider silk. This is in accordance with the larger crystallinity and crystallite size along the *c* direction (*i.e.* backbone direction) given by XRD measurements (Table 13.1). Although silkworm silk has larger crystallites, their inter-crystallite distance is much shorter than that of spider dragline silk, indicating that the crystallites in silkworm silk are packed much closer compared to spider draglines. The larger crystallites along the *c* direction and the shorter inter-crystallite distances may provide the clue why the silkworm silk contains much less intramolecular β-sheets than spider silk. Since protein chains containing larger portions of regular β-structures [polyA or poly(GA)] have a higher possibility to interact with each other to form intermolecular β-crystallites when the adjacent molecular chains are parallel aligned during silk spinning, this will suppress to a large extent the occurrence of intramolecular β-sheets.

13.3   The *in vivo* Formation of Silks

The unique mechanical properties of spider silk threads and an inability to domesticate spiders have induced numerous attempts to artificially manufacture spider dragline silks from recombinant/reconstituted silk protein[26-36] for industrial and medical applications.[37] However, the mechanical properties of the native spider silks have not been reproduced with either reconstituted or recombinant spider silk proteins,[31,38] mainly due to the difficulty in duplicating a spider's unique protein dope and spinning process in its natural way.

The difference of rheological behavior between native silk dope and reprocessed silk protein solution exhibits different structures of the spinning gel formed *in vivo*, in contrast to the one formed *in vitro*.[39,40] Holland *et al.*[40] also showed remarkable similarity between the rheologies for native spider dragline and silkworm cocoon silk, despite their independent evolution and substantial differences in protein structure.

Figure 13.2 shows spider silk processing in the spinneret in its natural way. Spider dragline silk proteins are stored at concentrations up to 50% (w/v) known as the silk spinning dope, a highly viscous liquid crystalline solution, in the MA gland until they are processed into fibres. Rheometry enables us to examine, *in vitro*, the spinning forces exerted on the dope *in vivo* along the spinning duct, thus imparting an analytical window into the silk spinning process. It also impresses us with the ability of both spider and silkworm to store silk proteins over long periods and stand by to spin a fibre instantly.

As displayed in Figure 13.2, during the natural spinning process the proteins move distally through the tapering duct which is divided into three limbs, where

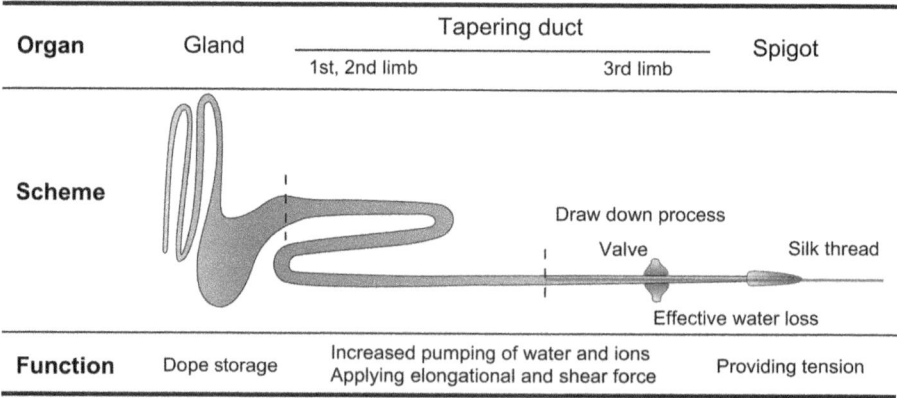

Figure 13.2   Schematic illustration of the spinning process of spider silk in its natural state. Silk dope is stored in the gland, then moved through the tapering duct with three limbs, encountering elongational flow, shear force and changes in their biochemical environment (ion exchange, decrease in pH and increased concentration). The preliminary fibre is reached after an internal drawdown process in the third limb and active evaporation of water after the valve. Finally, the silk thread exits the spider spigot with post-spin drawing and further evaporation of water.

they encounter elongational flow, shear force[2] and, simultaneously, changes in their biochemical environment, including ion exchange, decrease in pH and increased concentration.[41,42] The pH reduction improves the shear sensitivity of the dope to induce its gelation.[43] Thus, simultaneous acidification together with elongational flow are essential factors influencing the transformation from dope solution to fibres.[44] A simple protein denaturation model raised by Porter and Vollrath proposes that the spin stress and pH gradient as factors to unstable the amide–water hydrogen bond, and hence facilitate the replacement of stronger amide–amide interaction.[45] In addition, metallic cation changes also contribute to promote the formation of silk fibre.[46] For example, the K^+ content is considered to induce the formation of β-sheet and nanofibril structure in spider silk gland dope.[43,47] The liquid-crystalline spinning technology of spiders significantly reduces the viscosity of the silk dope. Hence, it enables the alignment of the protein molecules finely in a highly concentrated spinning dope ,with high energy efficiency.[3,42] The preliminary fibre is reached after an internal drawdown process in the last limb of the duct, with active transport of water after the valve.[41,42] The biochemical and physical changes are accompanied by a liquid–liquid phase separation followed by a liquid–solid phase transition that results in a preliminary silk fibre. Finally, the silk thread exits the spider through the spigot with flexible lips, which places the fibre under tension for the post-spin drawing, where the remaining solvent evaporates in air.[48]

Nature has created and refined protein structures through billions of years of evolution for various specific purposes. Amino acid sequences and their folding patterns make concerted efforts in creating elastic, rigid or tough materials, making nature's intricately designed products challenging examples for materials scientists to develop bio-inspired materials. To mimic spider dragline silk successfully, we need to duplicate the crucial features of feedstock dope as well as the spinning process. The key issues include how the spider stores protein dope in a highly concentrated liquid crystalline state and how it controls the processes in the spinning duct to form a supremely tough thread. Molecular biology helps us to extract, synthesize and assemble artificial genes to provide feedstocks for silk production,[27,49–51] whereas classical morphological studies discover the details of the spider's extrusion system.[48,52–55] Combining these two biological disciplines with process engineering can envision the design of advanced and benign fibre extrusion technology, and eventually allow commercialization.

13.4 Mechanical Properties

One of spider silk's advantages that attracts the increasing interest of scientists is its outstanding mechanical properties. Natural silks typically have a larger extension and higher toughness, although lower strength, compared to synthetic fibres. Strength and extension respectively are the stress and strain at which failure occurs. Toughness is calculated from the area under a stress–strain curve, representing the sum of the energy absorbed by the material before failure.[9] As shown in Table 13.2, most synthetic fibres have restrained toughness, although with high modulus and strength. In contrast, spider dragline

Table 13.2 Tensile mechanical properties of spider silk and other well-known natural and synthetic fibres.

Materials	Stiffness (GPa)	Strength (GPa)	Extensibility (%)	Toughness (MJ m^{-3})	Ref.
Araneus diadematus MA silk	10	1.1	27	160	9
Nephila edulis MA silk[a]	7.9	1.15	39	165	58
B. mori cocoon silk	7	0.6	18	70	9
Nylon 6.6	5	0.95	18	80	9
Kevlar 49	130	3.6	2.7	50	9
Carbon fibre	300	4	1.3	25	9
High-tensile steel	200	1.5	0.8	6	9

[a]Drawn at 20 mm s^{-1}.[25]

silks with moderate modulus show toughness more than three times higher than that of Kevlar 49 fibre, a type of synthetic fibre for bullet-proof vests. This unrivalled toughness benefits from a combination of strength and extensibility. Apart from its classical mechanical properties, dragline silk has the ability to undergo supercontraction. This phenomenon will be elucidated in Section 13.6. Furthermore, spider silk also has a torsional shape memory, which allows the spider dragline thread, after being twisted, to oscillate only slightly, and by this means to totally recover its initial form.[56,57] This unique property allows spiders to rapidly descend using dragline silk as a lifeline in case of danger.

13.4.1   Tensile Properties of Natural Silks

The mechanical properties of silks are determined from their response to tensile deformations, which are conducted in machines such as the Instron Micro-Tester,[15,25] deriving stress–strain curves. As shown in Figure 13.3, typical stress–strain curves show that both spider and silkworm silk fibres exhibit a linear stress–strain relation until a yield point. The linear section of the curve is the elastic region and the slope is defined as Young's modulus.[59] After the yield point, the spider dragline initially exhibits a characteristic J-shaped behavior of "strain hardening" (the slope of the stress–strain curves increases with strain) until point H, followed by so-called "strain weakening" behavior (the slope of stress–strain curves decreases with strain) until the ultimate break. Other species of spider draglines demonstrate similar stress–strain profiles,[60] indicating that strain hardening could be a general feature of spider draglines. On the other hand, the silkworm silk acts only in the "strain weakening" mode in its nonlinear portion of the stress–strain curve.

In addition to its capability of absorbing an enormous amount of impact energy for a soft dragline filament, the impact force, on the other hand, increases in a very slow manner owing to its marvelous extendibility and gradually hardening mode. In contrast, Kevlar comprises very rigid molecules

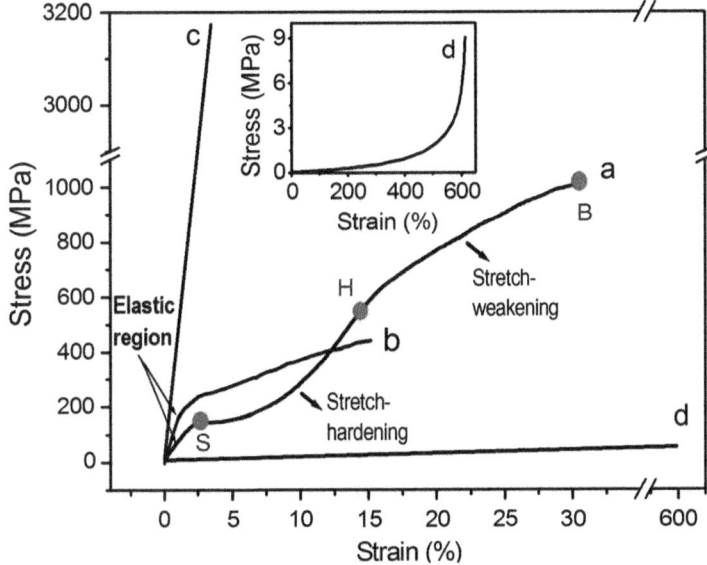

Figure 13.3 A comparison of typical stress–strain curves of silk fibres from the spider *N. pilipes* (by forced silking at 10 mm s^{-1}, curve a) dragline silk and silkworm *B. mori* cocoon silk (curve b), Kevlar 49 (curve c) and rubber (curve d). *Inset*: rescaled stress–strain curve of rubber. The stress–strain curves of silk fibres (curve a and b) are divided into two parts: linear (elastic) and non-linear regions. The mechanical characterizations were performed at 22 °C and the humidity was kept at 50–55%. (Reproduced with permission of Wiley-VCH from Du *et al.*[25])

which incline to form mostly planar sheet-like structures[61] whose filaments are very rigid and brittle.[62] If Kevlar 49 filaments become a substitute for the draglines in a spider web, the collision of a flying prey into it would result in a tremendous shock. As a consequence, the spider colonizing the web would have been knocked off upon impact.

13.4.2 Factors Influencing the Mechanical Properties of Natural Silks

The mechanical properties of silks can be utilized to probe the structures of these materials. However, the mechanical testing results of silks are influenced by several factors, such as forced reeling conditions, testing conditions, vapor or solvent treatment, to different extents. Such factors are regarded as valuable experimental and analytical tools to understand the structures of silks, instead of simply as impediments to obtain reproducible results.

Most natural spider silks used in research were obtained by forced silking technology, which is drawing the fibre from the spinneret of the immobilized spider and simultaneously winding it on a computer-controlled spindle. This method is capable of controlling the structure and thus mechanical properties of silks by tuning several key extrusion parameters, such as reeling speed, force, temperature and reeling medium (*e.g.* into air or solvents such as water[63]). Firstly, the spider dragline silks collected by forced silking at increased spinning speeds would possess higher breaking stress and initial modulus, but lower breaking strain.[15,64] It is found that this can be attributed to the varied crystalline alignment in the silks reeled at different speeds, as observed by X-ray diffraction.[15,65] Besides, the silking speed has also been reported to influence the size of the β-crystallites, which can also impact the mechanical performance.[15,66] In addition, the tensile performance of forcibly reeled silks can also be correlated to the reeling force, because the silks exhibit distinct mechanical properties when the reeling forces are different with all other parameters such as reeling speed kept the same.[63,67,68] Furthermore, the draw-down medium and the length of the aqueous post-draw duct also play important roles in modifying the mechanical properties of a silk. For instance, spider dragline silk spun into water displays greater strength, stiffness and resilience compared with silk drawn into air, owing to the better alignment of the polypeptide backbones, as discovered by Raman measurements. The resilience is defined as the ratio between energy recovered and energy input in a loading–unloading cycle. The difference in structure can be ascribed to the higher mobility and aligning time that the silk proteins gain in an aqueous environment.[63] Finally, the body temperature would affect the mechanical properties of forcibly reeled spider silks. An elevated silking temperature leads to an enhanced breaking strain and almost constant breaking stress, hence an improvement in the silk toughness.[58] This toughening effect may result from the temperature-induced changes in the viscosity of the spinning dope, thus inducing a more relaxed conformation in the amorphous regions when the spider silk forms.[69]

Testing conditions such as strain rate and measuring temperature also play important roles in influencing the mechanical properties. Normally, when tested under higher strain rates, viscoelastic synthetic polymers such as nylon are inclined to display increased stiffness and strength but decreased extensibility and toughness. It is interesting to observe that the stiffness, strength, extensibility and toughness of spider silk all increase with increasing strain rate.[9] Even measured at an extreme strain rate reaching $20–50 \text{ mm s}^{-1}$, *Araneus diadematus* MA silks exhibit extensibility at 20–50% with the initial modulus at 25–40 GPa and breaking stress at 2.0–4.0 GPa, thus obtaining toughness as high as $500–1000 \text{ MJ m}^{-3}$. According to the time–temperature equivalency principle, increasing the strain rate and decreasing the temperature play similar roles in affecting the mechanical properties of viscoelastic materials, such as silks. The conditions of dynamic mechanical temperature analysis on MA dragline silk of *N. edulis* spider at –60 °C enhance stiffness and strength, and breaking elongation, compared with that measured at room temperature.[70] Fractographic studies show that spider silk behaves as a ductile material at temperatures as low as liquid nitrogen temperature.[70] When changing the measuring temperature in the opposite direction, increasing from room temperature to 150 °C, the stiffness of the dragline silk decreases while the strength and extensibility almost keep the same. Higher temperatures induce spider dragline silks to contract, reaching up to 20% of their original length, indicating that the internal hydrogen bonds are modified at higher temperatures.[71] Further increasing the temperature above 200 °C causes thermal denaturation of spider dragline silks and hence reduces the intensity of the crystal diffraction and irreversibly changes the lattice constants.[71]

Treating silks with polar solvents or vapor always greatly reduces the stiffness and improves the breaking strain significantly.[72–74] Polar solvents with small molecules seem to outperform those with large molecules in influencing the mechanical properties of silks. This is because the small molecules can penetrate the silk more easily and also interact with polar groups of polypeptide chains inside the silk more strongly, owing to their high polarity. Therefore, water is regarded as a most active softener, followed by less active methanol, while ethanol and butanol show almost no softening effect on spider dragline silks which are immersed in them.[75]

Lee *et al.* revealed that metals can be successfully infiltrated into internal protein structures of biomaterials such as spider silk through multiple-pulsed vapor-phase infiltration (MPI) performed with equipment that is conventionally used for atomic layer deposition (ALD). In their research, zinc, titanium or aluminum, combined with water from corresponding ALD precursors, were infiltrated intentionally into spider dragline silks, with the toughness of the resulting silks greatly improved. The energy dispersive X-ray (EDX) and nuclear magnetic resonance (NMR) spectra were used to confirm the presence of the infiltrated metals such as titanium or aluminum inside the treated silks.[76]

In conclusion, factors such as speed, force, medium and temperature of the forced reeling, strain rate, temperature of testing and vapor/polar solvent treatment, can all influence mechanical properties of silks to different extents. However, by tuning these factors, we can change the structure of the material, leading to a difference in mechanical properties. In next section, we will discuss the property–structure relationship to reflect a general view on how to control the mechanical properties of silks by intrinsically tuning their structure.

13.5   Structure–Property Relationship

A deeper insight into the underlying structure–function correlation of silk proteins is fundamentally important for the potential biomimetic production of spider silk from recombinant and reconstituted silk protein with controlled mechanical properties. Such knowledge also benefits novel synthetic polymer-based materials, as spider silk organization can serve as a bio-inspired model for designing sophisticated composite materials.

13.5.1   Structure Related to Mechanical Properties

Generally, the structural origin of the extraordinary extensibility of silk has not been fully explained. Some ascribe the extensibility of the silk thread to the ability of the amorphous regions to stretch and move from their random coil/other conformation into a more extended helix or β-strand configuration.[9] Another model proposes the nanofibrillar morphology of the thread (*i.e.* in a helical or a zigzag manner) as a major feature contributing to the extensibility of silkworm silk.[77] The morphology in a helical manner could contribute because these helical fibrils would straighten parallel to the fibre as the original silk threads extend and break. The morphology in a zigzag manner would play a role because the zigzag fibrils would compress like a spring when the silk fibre undergoes buckling or shear failure. Both effects could help to reduce failure under a sharp bend while the former tensile effect could contribute to the well-known large tensile strains of spider silks at breaking points.

Strength has normally been attributed to the presence of β-sheets packed into crystalline regions and hydrophobic interactions within those crystalline regions, which bind molecules together in the silk fibre.[78,79] Crystalline regions consisting of poly-Ala β-sheets are found to form more hydrophobic interactions, and thus have a higher binding energy, than poly-Gly-Ala β-sheets.[78] Given similar numbers and lengths of repeats, silks with poly-Gly-Ala repeats with fewer interactions seem to have lower tensile strengths relative to those with poly-Ala repeats. This is in accord with the lower tensile strength of MI silk (with poly-Gly-Ala) compared with MA silk (with poly-Ala). It also agrees with the lower tensile strength of silkworm silk (rich in poly-Gly-Ala blocks) in contrast to spider dragline silk (rich in poly-Ala blocks). The higher crystallinity of silkworm silk compared to spider dragline silk may also account for its brittle feature.

It appears that the overall toughness of silk threads is connected to its structure at all hierarchical levels, including the secondary structure of polypeptide chain folding as well as the higher order structure of the protein constituents. For instance, recent studies have proposed that spider silk filaments can possess a complex "coat–skin–core" organization[80,81] that is likely to impart energy-dispersive properties.[82] This structural complexity with microfibrils[83–85] enclosing fine channels[80] is likely to contribute significantly to the toughness of spider silk. Du *et al.* have illustrated how the intramolecular β-sheet-induced strain-hardening contributes to the breaking energy, which will be described in detail in the following section, covering many models raised to address the issue of the structure–property relationship.[25]

13.5.2   Models

Many models have been proposed to interpret the mechanical properties of natural silks from the perspective of structure. In Termonia's model,[12] the dragline silk is described as many small stiff crystallites embedded in amorphous regions, which are made of rubber-like chains. The crystallites serve as multifunctional interlock sites and create a thin layer of high modulus in the amorphous region. The simulated properties of dragline silk are in accord with the experimental results, while assuming the modulus of β-sheet crystals as for fully extended crystals, *i.e.* 160 GPa. β-Spiral secondary structures formed by the motif GPGXX in MA and flagelliform silk are suggested to act as springs, being responsible for the extensibility of the silk.[78] The proline residue would be the focal point for the retraction energy after stretching. By forcing the proline bonds to torque in response to extension, a large force can be generated for retraction. It is proved by the evidence that the GPGXX motif only exists in MA and flagelliform silks, which are also the stretchiest of spider silks. A further support for the GPGXX motif providing the elasticity modulus is the correspondence between the number of tandemly arrayed GPGXX repeats and the different extensibilities of the two silks. MA silk, with up to 35% extension,[86] has at most nine β-turns in a row before interruption by another motif.[87] Flagelliform silk with 200% extensibility[86] has at least 43 contiguously linked β-turns in its spring-like spirals.[88]

In addition, mean field theory for polymers has been applied to predict structure–property relationships in terms of chemical composition and morphological structure, as well as the dynamic mechanical properties of silk, derived in terms of the ordered fraction.[89] This model predicts the range of silk tensile properties in quantitative agreement with experimental results, and suggests a numerical relation between the ordered fraction and in-water shrinking capacity of spider silk. Zhou *et al.* has proposed a hierarchical chain model to explain the nonlinear force–extension response of spider silks, and the prediction well agrees with the experimental observations.[90,91] In this model, a polymer consists of many structural motifs, organized into structural modules and supra-modules in a hierarchical manner, with their own characteristic force for each module. The repetitive patterns in the amino acid sequence of the flagelliform protein of spider capture silk give support to this model. A hairpin-turn model is proposed based on studies of the molecular structure in silk, which suggested that silk has a folded hairpin structure[13,78,92] with about six peptide segments per fold. In this model, three types of material in the silks are defined as a "string of beads" and their approximate fractions are suggested for silk nanofibrils, which determine the strength of natural silks.[39,93] Typical predicted stress–strain profiles at different reeling speeds for spider and silkworm silk are in good agreement with experimental measurements by implementing the "β-sheet splitting" mechanism, and varying the secondary structure of the protein macromolecules.[94] The work by Buehler's group[95] highlighted the important role of hydrogen bonds contributing to the mechanical property of silk. The work well explains why the nano-crystallites can achieve high strength.

The molecular origin of strain hardening of spider dragline silk, compared with rubber, Kevlar and silkworm silk, has been addressed by Du *et al.*[25] The occurrence of strain hardening originates from the unfolding of the intramolecular β-sheets inside spider silk, which perform as "molecular spindles" to store long molecular chains in space in a compact manner. With continuous unfolding and alignment of proteins, the protein backbones and nodes of the molecular network start to be stretched to support the load during fibre stretching. As a result, the dragline filaments become progressively hardened, which enables efficient energy buffering as an abseiling spider tries to escape from its predator through the safety line.

Specifically, during stretching, the intramolecular β-sheets inside the amorphous matrix will first unfold to release lengthy protein chains, with the β-crystallites being unaffected. This will lead to a high extension of the dragline silk, with the intermolecular linkages exempt from breakage. This process occurs as the stretching reaches the yield point S in Figure 13.3. At the beginning stage of protein unfolding, the modulus declines to nearly zero as the breaking of weak intramolecular hydrogen bonds makes the major contribution to fibre elongation. As the elongation continues, the progressive unfolding and alignment of protein chains result in stretching of the protein backbones and nodes of the molecular network to support the load. As a consequence, the dragline silk thread becomes stiffer owing to the contribution coming from enthalpic component.[96] These processes exhibit how strain hardening happens in spider dragline silk. Similar unfolding behavior in recombinant spider silk protein[97] and some other elastic proteins[98] are also shown by single protein molecule stretching of these proteins using atomic force microscopy (AFM). The further stretching exceeding the inflection point H will generate breaking of the β-crystallites, destroying the interlocks of the molecular networks with the silk threads, thus giving rise to the weakening of the silk filaments (also termed as strain weakening). The process of β-crystallite breakage in the spider dragline silk is verified by the XRD measurements, demonstrating that once a spider dragline silk is stretched beyond point H, the total β-crystallinity of the dragline will decrease correspondingly (Figure 13.4). In contrast, there are much fewer intramolecular β-sheets in the amorphous region of *B. mori* silkworm silk. Therefore, silkworm silk only exhibits strain weakening after the yield point, with inferior extensibility.

The response of spider dragline silk upon stretching is found to deviate greatly from what occurs in rubbers.[99] As revealed in Figure 13.4, instead of inducing crystallization like that in rubber, the tensile deformation of spider draglines results in breaking of the crystallites, giving rise to the decrease of crystallinity shown by the XRD measurements. The distinction between the behavior of silk and rubber can be credited to their disparity in molecular structure and intermolecular connections established upon the formation of the material. The nano-sized β-crystallites together with the orientated amorphous region[3,100] in spider silk are formed during the silk-spinning process in the animal's sophisticated spinning organ known as the spinneret.[3,101] During the silk-spinning process, substantial shearing is applied to the silk protein molecules to promote the alignment of the adjacent protein molecules.[3,101–103] The β-structures (polyA or polyGA regions) can thus be crystallized into nano-β-crystallites.[3,15] Once the silk fibre is formed, the protein chains inside are unlikely to form new β-crystallites by stretching. On the contrary, rubbers, consisting of entirely random-oriented molecular chains possessing the same subunit,[104] can smoothly form crystallites while the amorphous chains are aligned under stretching.[99]

The different structural characteristics in spider dragline silk and rubber also result in their remarkably different recovery dynamics from a stretch. Rubber

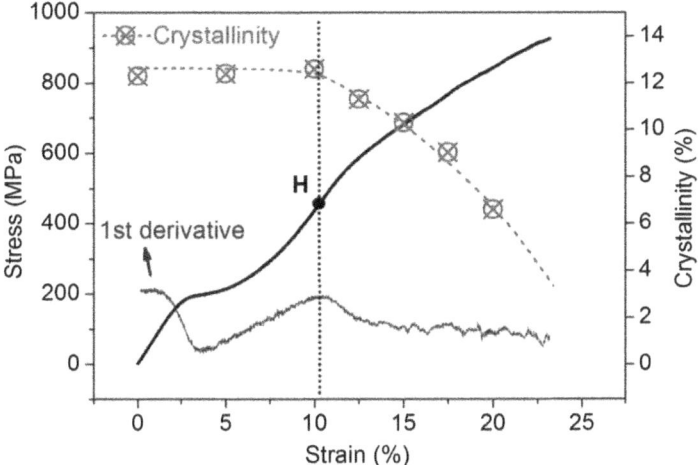

Figure 13.4 Effect of tensile deformations on the β-crystallinity of spider *N. pilipes* dragline fibres. The lower curve is the first derivative line of the stress–strain curve of the spider draglines (upper line, 12 samples averaged). (Reproduced with permission of Wiley-VCH from Du *et al.*[25])

will oscillate severely, with the amplitude declining to zero at a slow speed after being stretched.[25] Spider dragline fibre, on the other hand, decays rapidly to equilibrium, with small oscillations. This demonstrates that spider dragline silk rarely oscillates after a disturbance like elastomers such as rubbers, with similar behavior also observed in a torsional stability study of spider dragline silk.[56] The different recovery responses between spider dragline silk and rubber indicate an optimized viscoelasticity in silk protein molecules, which can be attributed to the special organization of molecular chains in spider dragline silks. The features encompass a more compact packing of molecules by folding into intramolecular β-sheets or crystallizing into β-crystallites, and a higher degree of molecular orientation in comparison to rubber. As a result, spider silk is capable of absorbing energy without generating a drastic oscillation like the way elastomers behave.

The dragline silk fibres extruded at faster reeling speeds have larger strain-hardening-induced strain, because they contain an increased content of the intramolecular β-sheets compared with those drawn at lower speeds.[25] The strain-hardening-induced strain is defined as the strain produced in the strain-hardening curve segment (the strain difference between points S and H). The strain-hardening strain ratio is defined as the strain-hardening induced strain over the total strain of the non-linear region. The strain-hardening strain ratio is found to grow substantially with the ratio of intramolecular β-sheets to the total content of β-sheets.[25] As shown in the experimental results, the crystallinity decreases from 0.22 to 0.11 when the reeling speed increases from 10 mm s^{-1} to 200 mm s^{-1}, as is indicated from the XRD measurements. The content of the intramolecular β-sheets rises from 0.29 to 0.45, while the overall content of the β-sheets keeps almost the same (from 0.51 to 0.56), as shown by FTIR spectroscopy.[25]

The underlying mechanism could be explained by the well-controlled formation process of spider dragline silk.[25] An escaped nematic liquid crystalline phase exists in the silk gland and in the first half of the duct of the spider, in a tight hexa-columnar packed arrangement of compactly coiled molecules.[23,105] The flow elongation and the exchange of ions along the duct are thought to encourage the molecules to partly unfold. The retention of some hairpin loops results from the incomplete extension of the molecules. It would help to provide an anti-parallel arrangement,[96] which could facilitate the formation of β-structures with extension flow. More of such hairpin loops are retained in the duct when the silk fibres are extruded from a spider at high speeds, as they do not have sufficient time to respond to the fast external drawing. As a consequence, the content of the intramolecular β-sheets rises, and thus the strain-hardening effect becomes more significant with the faster drawing of the fibre, while the overall content of the β-sheets keeps almost the same.[25]

13.5.3   Techniques to Study the Structure of Silk Fibres

In this section we summarize and compare several techniques that are widely used to characterize silk fibres, revealing the general morphological and structural characteristics in their static or dynamic process.

Firstly, the microscopic morphology can be observed by scanning electron microscopy (SEM). AFM can be utilized to scrutinize the topography of fibres such as *B. mori* silkworm silk and spider silk,[106] offering details on the nanometer scale. Analytical transmission electron microscopy (TEM) can demonstrate the diffraction patterns from organized structures in a fibre and assess the crystallite content in dragline silk.[107]

Secondly, advanced imaging techniques such as Raman spectroscopy, FTIR spectroscopy, XRD and NMR are important tools for analyzing the secondary and tertiary structure of silkworm and spider silk threads.

Raman and FTIR spectroscopy are non-invasive techniques used to investigate the dominant conformational content of the fibre, by monitoring the amide I, II and III frequencies values, which are indicative of α-helix, β-sheet, random coil and turns in the fibre.[75,108,109] The advantages of Raman spectroscopy over FTIR involves the low interference by water in the natural biofibres and the chance to acquire spectra of single fibres with a diameter of less than 10 μm.[110] Raman microspectroscopy was utilized to obtain the orientation of the carbonyl groups in silks *via* the amide I band by recording polarized spectra, therefore determining the orientation of the β-sheet crystallites.[110] Conformational and orientation changes in fibres under stress and strain can also be observed by Raman spectroscopy.[111–113] Raman spectroscopy on single silk threads during gradual stretching shows that the silk structure undergoes uniform stress during deformation,[72,111,114] suggesting the presence of micro-fibrillar structures within the silk thread that evenly distribute any applied stress.

XRD can also be applied to characterize the structure of silk threads during gradual stretching together with the structure in the static state. Wide-angle X-ray diffraction (WAXD) and small-angle X-ray scattering (SAXS), when combined with synchrotron radiation, can be used to characterize small samples such as single fibres. The interference patterns coming from diffraction of the radiation can be related to the distance between domains in the fibre. Therefore the patterns can reflect the relative arrangements of the domains. A length scale similar to the radiation wavelength ($\sim 1 \text{ Å}$) is probed by this technique, discovering information of fibre structure such as crystallinity, crystal size, inter-crystalline distance and orientation distribution. Varying alignments of crystalline structure, crystallinity and crystal size in the silks reeled at different speeds were observed by XRD studies.[15,115–118] XRD patterns obtained from single threads being spun[119] also revealed that the thread consists of small crystalline blocks in a matrix containing both oriented and disoriented amorphous material, with the crystalline fraction having a β-poly(L-alanine) structure.[117]

Solid-state NMR (SS-NMR) can characterize the dynamics and reveal the local conformation where specifically labeled amino acids are discovered. The structural information gained by SS-NMR depends on the measurement of chemical shifts, torsion angles and inter-nuclear distances. The packing arrangement and size are acquired through experiments which probe the dynamics and manifest information on the crystalline and non-crystalline content. For example, proton-driven ^{13}C two-dimensional NMR spin-diffusion experiments on dragline silk thread suggested that the poly-Ala segments adopt a highly ordered β-sheet structure[13,120] and the glycine-rich segments have 3(1) helical structures,[14,121] both oriented in line with the thread. However, it demands a greater amount of sample and isotopic labeling of the fibre in most situations. For a biological fibre, SS-NMR does not allow sequencing a specific amino acid assignment by itself, nor does it provide a complete structure.

In comparison, microscopy-related techniques are used to image fibre morphology, while Raman and FTIR spectroscopy can give a global portrait conformational content in the fibre. The XRD techniques display information in the crystalline regions of a material.

SS-NMR, as a powerful technique to study natural protein fibres, allows the study of the molecular structure as well as the dynamics of semi-crystalline and amorphous material.[120,122–130]

13.6 Supercontraction

Supercontraction, as its name suggests, is a phenomenon describing the large shrinkage in the length of unrestrained spider dragline silk fibre when it is immersed in water or exposed to high relative humidity environments at room temperature.[131–133] It is a distinctive feature of the MA and flagelliform silk fibres spun by orb web-building spiders.[3,134] When wetted in an axially unrestrained conditions, spider dragline silk retracts up to 50% of its original length, while increasing in overall volume, decreasing in stiffness by several orders of magnitude and causing a dramatic change in its mechanical behavior.[131,132] This process generates substantial stress in restrained fibres, owing to restricted supercontraction, in excess of 50 MPa[131,133,135] and amounts up to 4.7% of the breaking strength.[131,136] Wetting-induced supercontraction of spider dragline silk is presumed to be exploited to take up slack in webs and restores web shape and tension after deformation by prey capture, precipitation or wind as they become loaded with dew or rain.[131–133,135,137–139] This shape memory behavior of dragline spider silk is unique in that it is brought about totally in a benign condition at room temperature when wetted with moisture. Similar shape memory behavior can also be observed for some polymers, but only induced by external stimuli such as high temperature or high pressure or harsh solvent conditions.[100,140,141]

Dry fibres, either forcibly silked or naturally spun fibres, behave as glassy polymers: there is an initial large modulus followed by a linear-elastic regime up to a yielding point, with a subsequent decrease in the slope of the stress–strain curve to a minimum and, lastly, accompanied by a gradual increase in the slope until breaking point.[133] In contrast, the tensile behavior of supercontracted silk tested in water seems to correspond to an elastomeric material: an initial low modulus and large ultimate strain. This difference in the mechanical properties between dry and supercontracted spider silk fibres has been modeled through a double network of hydrogen bonds and protein chains.[142] The model assumes that supercontracted fibres correspond to a state where the hydrogen bond network is disrupted and the observed elastomeric behavior tensile properties are controlled by the alignment of the protein chains. The elastomeric behavior only appears in wet supercontracted fibres since drying freezes the conformation of the chains by re-establishing the hydrogen bond network.[12,142,143] These conformational changes are reversible when the humidity is increased and when chain reorientation is allowed.[139]

Research carried out by Gosline[10] revealed that supercontraction has profound implications on the mechanical behavior of spider silk, even hypothesizing a relationship between supercontraction and tensile properties.[12,142] It is found that the whole range of variability of the tensile behavior of spider silk, either naturally spun or forcibly silked, can be predictably and reproducibly tailored through a combination of supercontraction and stretching, suggesting that the variability corresponds to different degrees of alignment of the protein chains.[133] Controlled supercontraction may thus be exploited as a mechanism to tailor the properties of spider silk, allowing it to be implemented in the operation of spinning artificial silk fibres in large quantities for different biomimetic applications, as well as improving the reproducibility of the fibre tensile properties.[133,144]

Some authors found that supercontraction stress was transient and that silk fibres lose tension within 200 s when the humidity remains high, suggesting that supercontraction is ephemeral and supercontraction stress could not maintain tension in wet webs.[135] They argued that this stress relaxation is a major impediment to technological applications of spider silk that are based on the protein and network structure of spider dragline silk. Subsequently, Savage *et al.*[145] replied that the stress relaxation was largely an artifact of Bell *et al.* using forcibly silked material, rather than naturally spun silk, since spiders impart a large frictional force on forcibly drawn silk through their internal friction brake.[67] In contrast, others demonstrated no signs of stress relaxation in forcibly silked dragline silk, despite it belonging to both different and similar species.[133,145,146] They showed that supercontraction tension can be sustained for long periods of time, suggesting a permanent change in the molecular structure within the silk fibre. Therefore the hypothesis that supercontraction maintains tension in the web after deformation by water restoring its shape is clearly supported in these studies. However, the supercontraction stress generated varied widely between studies by an order of magnitude (10–140 MPa[133,136,145–147] and 300–400 MPa[135]), perhaps due to the different species sampled,[64,136,148–150] possibly related to the differences in content of the critical amino acids such as proline,[22,59] spinning effects resulting from variation in forces applied to the fibre by the spider during artificial silking,[67] spinning conditions on fibre orientation[100] and external factors. Agnarsson *et al.* demonstrated that the observed phenomenon in the variability of supercontraction force depends upon the rate of increase in relative humidity[147] to the threshold level of humidity causing supercontraction. Rapid increases in humidity ($41\% \, s^{-1}$), over a few seconds, generated supercontraction stresses exceeding 100 MPa, whereas slower increases in humidity, over several minutes, led to relatively low supercontraction stresses not exceeding 80 MPa in supercontracted forcibly silked dragline fibres.

It is known that spider dragline silk can be modeled as a semicrystalline material, consisting of multiple fibroins joined by polyalanine nano-β-crystallites embedded in a non-crystalline amorphous network.[10,15,16,78,86] Spider dragline silk is composed mainly of two proteins, spidroin I and II, which can be divided into block copolymers with $GAGA(A)_n$ blocks and glycine-rich domains.[18] The amorphous network consists of relatively ordered glycine-rich linker domains and proline-containing random coils. The glycine-rich linker domains that retain fibroin orientation before supercontraction mostly occur in the major ampullate spidroin I (MaSp I), whereas the random coils are produced by the greater abundance of proline in MaSp II.[151] These two proteins are partially phase-separated during fibre formation, such that they are dispersed heterogeneously throughout the silk, with MaSp II occurring largely in the interiors of threads and MaSp I dispersed throughout.[152] Infiltration of water during supercontraction is hypothesized to plasticize the silk fibres by disrupting hydrogen bonding within the amorphous region of the proteins, thereby increasing molecular mobility, allowing a subsequent entropy-driven reorientation and coiling of silk molecules into a less organized configuration.[10,73,121,133,145,153,154] Water molecules exert a similar plasticizing effect on silkworm silk.[143] Recent studies have focused on the importance of disrupting secondary structure in the glycine-rich blocks for mobilization of the proteins within the amorphous network.[22,92,149] Hydrogen bonding within the random coils is relatively weak and disrupted by small amounts of water, while the stronger hydrogen bonds of the glycine-rich linker regions are partially disrupted at high humidity.[133,145,146] This disruption in secondary structure is sufficient to reconfigure the silk protein network, driving the reorientation of the silk molecules and allowing the silk proteins to move rapidly to more disordered and higher entropy configurations. This causes a sudden contraction in the entire unrestrained silk fibre while expanding in overall volume, behaving like a filled rubber with a relatively low modulus, or generates substantial stresses in the restrained silk.[146]

Dragline silk exhibits a cyclic contraction phenomenon distinct from supercontraction: it extends and contracts repeatedly like a muscle, by cyclically raising and lowering the relative humidity (Figure 13.5).[146] Cyclic contraction is repeatable both before and after supercontraction. This cyclic response produces high forces, exceeding supercontraction forces, which are completely reversible and that can be precisely controlled through humidity alone. Agnarsson suggested that spider silk emerges as an attractive model for biomimetic muscle fibres as it generates a power density 50 times higher than biological muscle.[155] Water binds to silk permanently during supercontraction, resulting in a change in its structure, whereas cyclic contraction results from a reversible loss of water during drying. This cyclic phenomenon is not explained by molecular models developed to explain supercontraction.[10,22,154,156] Initial exposure to humidity results in infiltration of water molecules into the silk fibre. The water molecule first binds itself to the random coil within the amorphous

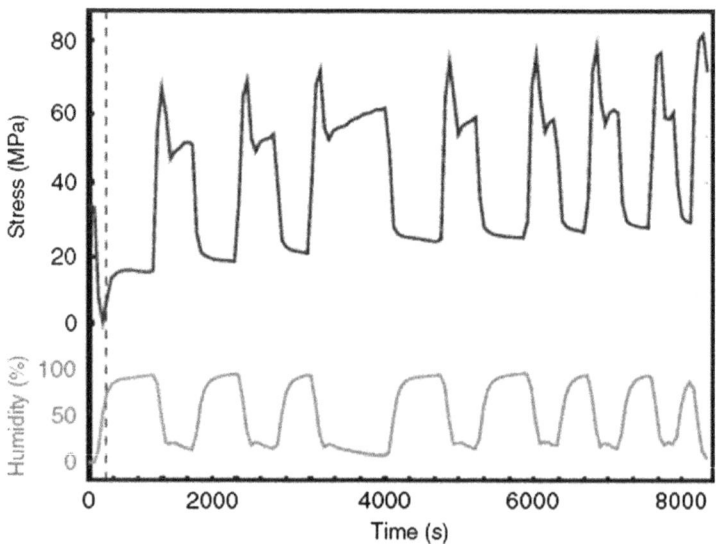

Figure 13.5 Cyclic contraction of spider dragline silk. (Reproduced with permission of the Company of Biologists from Blackledge *et al.*[146])

region, disrupting the relatively weak hydrogen bonding. This relaxes the silk and explains the initial dip in the stress–strain curve when silk is first expose to humidity just before the critical threshold of 70–75%. When the humidity reaches the critical threshold level, most of the hydrogen bonds within the random coil have been disrupted. Now the water molecules start to bind to the oriented glycine-rich linkers within the amorphous region, disrupting the relatively strong hydrogen bonding. This disruption in hydrogen bonding within oriented glycine-rich linker regions increases the local mobility of molecules, allowing them to reconfigure to a less organized state that shrinks the fibre and traps the water molecules in the silk structure, causing super-contraction.[59,121] Upon drying immediately after supercontraction, the water molecules are lost only from the random coil region, rebuilding hydrogen bonds in the process that immobilize the silk molecules, contracting the fibre. Subsequent wetting results in water molecules binding to the random coil region, breaking the hydrogen bonds and causing the silk fibre to relax. In short, high humidity in cyclic contraction causes the silk to relax if the mobilization of silk molecules enabled by moisture is limited to the random coils region within the silk, by breaking the relatively weak hydrogen bonds. However, high humidity in supercontraction causes the silk to supercontract and hence manifest itself as an increase in tension in restrained silk fibre, if the local mobility of silk molecules enabled by moisture is limited to the oriented glycine-rich region within the silk, by breaking the relatively strong hydrogen bonds.[146] Thus supercontraction is a two-step process that depends on the wetting-induced mobilization in two different regions within the amorphous region: oriented glycine-rich linkers and random coil. Cyclic contraction is also observed in silkworm silk, as well as natural fibres such as cotton and hair, whereas it is not observed in hydrophobic synthetic fibres such as polyethylene or acrylic.[155] This suggests that cyclic contraction depends on the hydrophilic nature of the constituent amino acids. The contraction stress generated is 16 MPa in silkworm silk,[157] 0.6 MPa in cotton, 6.1 MPa in hair and an estimated 60 MPa in dragline silk.[155]

Supercontraction and cyclic contraction generate substantial forces, providing a possibility that these phenomena could be exploited to perform work. Contrary to popular belief that supercontraction is a one-time response and hence a limit in its application to perform work, Blacklege *et al.*[146] have demonstrated that supercontraction has a multiple occurrence within the same dragline silk if the fibre is re-extended to its original length before drying after each unrestrained supercontraction. This happens because physical energy provides energy to the silk, reconfiguring the silk proteins within the amorphous region to a more ordered state. Once dried, the reinstating of the hydrogen bonds maintains the structural organization of the silk proteins. Subsequent wetting increases the mobility of silk proteins in different regions, reconfiguring the silk proteins to a higher entropy state, driving the super-contraction of the fibre.[146]

In summary, spider silk makes use of supercontraction to maintain tension in the web and restore web shape after deformation by prey capture, precipitation or wind as they become wetted with morning dew or rain. The powerful cyclic contraction exhibited by spider silk also results in a green and energy efficient mimicry of biological muscles. The multiple occurrences of supercontraction and fatigueless cyclic contraction offer possibilities for performing work in industry and in clinical sciences.

13.7   Silk Protein-based Novel Biomaterials

One appealing application of spider silk is to mimic the disparate material functions of the proteins to design novel biomaterials. Efforts have been continually focused on studying the assembly and processing of spider silk proteins into diverse material forms to expand applications to medical, electronic and optical fields. Owing to its biocompatiblity and controllable degradability, the silk protein material can be promisingly anticipated as use in sutures, drug delivery systems, fibre-based tissue, flexible electronics for physiological recording and functions, and optical systems for diagnosis and treatment.[158–161]

For example, scientists are investigating spider silk as "artificial skin" to replace the skin from a body for a graft. The spider silks are biocompatible and controllably biodegradable, meeting the requirements of a graft. Compared with other materials, they are also strong and tough enough to resist all the rigors that ordinary skin would experience. Wendt *et al.* successfully cultivated skin cells on spider silk meshes to replace lost skin tissues, forming equivalents of the epidermis, the outermost layer of the skin, and the dermis, the living tissue below the epidermis containing blood capillaries, nerve endings and other structures.[158]

Apart from pure silk proteins for materials, protein block copolymers and chimeric proteins will become more available. The most extensively studied recombinant spider silks encompass those from *N. clavipes* (MaSp I and MaSp II), *A. diadematus* (dragline ADF-3 and ADF-4) and *N. clavipes* flagelliform.[26–28,30,32,37,162–164] Much effort has been exerted to explore a range of different host systems for large-scale expression of gene coding spider silk, containing yeast, insect cells, bovine mammary epithelial alveolar cells, baby hamster kidney cells, transgenic plants such as *Arabidopsis*, soybean, potato and tobacco, transgenic mammals such as mouse and goat, and transgenic silkworms.[26,28,32,37,162–169]

Native spider silks and recombinant silk proteins can be solubilized with solvents, including formic acid, hexafluoroisopropanol (HFIP), calcium nitrate, lithium salts or ionic liquids, then processed into different structures covering fibres, films, gels, porous sponges and other related systems.[170–172] Techniques utilized to form fibres involve solvent extrusion and microfluidic and electrospinning methods. Solvent extrusion refers to drawing the thread through a coagulation bath in a controlled way. Microfluidic methods modulate the geometry and chemistry of the outcome fibre by contracting channels and diverse solvent inputs. Electrospinning processes integrate strong voltage gradients with syringe pump extrusion to fabricate either random or aligned fibre deposition.[173]

Spider silk fibres of 10–60 μm diameter have been fabricated through the electrospinning of genetically engineered silk spin dopes in urea containing Tris buffer and salts. This aqueous method has generated fibres similar to natural spider silk in terms of molecular orientation and mechanical properties.[174] Smaller fibre diameters up to 40 μm can be formed through electrospinning of spin dopes derived from ADF-3 recombinant spider-silk protein. This method generates fibres with similar toughness and modulus, but with lower tenacity compared to native *A. diadematus* dragline silk.[37] Fibre diameters down to 300 nm can be generated through electrospinning of *N. clavipes* MaSp I silk dope prepared in HFIP. Thus fibre properties can be potentially improved through the control of conformation and molecular alignment.[175]

Water-insoluble, transparent films with thickness in the range 0.5–1.5 μm can be obtained by casting solutions that contain recombinant forms of the dragline spider silk proteins ADF-3 and ADF-4 through the use of potassium phosphate or methanol. This treatment promotes the conversion of the protein's secondary structure from α-helix to β-sheet, creating a level of chemical stability in certain denaturants that is comparable to that of native dragline silk. Various cloning techniques and the approach of modifying film surfaces with functional molecules envision these materials as promising biomedical and clinical applications, such as wound dressings and scaffolds.[176,177]

Hydrogels with excellent superior mechanical response and stability over weeks are utilized in the field of tissue engineering to form porous but stable tissue scaffolds.[178] Silk nanofibres of diameter ~3 nm and lengths less than 1 μm appear after the addition of methanol to recombinant forms of the dragline silk protein ADF-4, and then transform into a hydrogel fibre network over several days. These hydrogels demonstrate a non-linear viscoelastic material response with low stiffness and strength, which can be improved by a cross-linking process that will promote a hydrogel with a fairly linear material response, a much greater modulus and strength response. Cross-linking can be induced by visible light and treatment with ammonium peroxodisulfate and tris(2,2′-bipyridyl)dichlororuthenium(II).

Microcapsules with high mechanical stability and controlled properties of a constrained degradation response to tissue-specific enzymes can be developed by controlling the silk self-assembly at an oil–liquid interface of the recombinant form of ADF-4 spider silks. Therefore, these materials show promising applications, ranging from drug delivery to microreactor design.[179]

Surface functionalization has been applied to silk surface chemistry to influence cell and tissue development through targeting carboxylic acid groups on the amino acids in the protein.[170] Spider silk has a significantly higher number of carboxyl side groups on the amino acids compared to silkworm silk, which provides higher degrees of cell-modifying functional groups.[180] EDC [1-ethyl-3-(3-dimethylaminopropyl)carbodiimide] has been used to functionalize ADF-3 and ADF-4 garden spider *A. diadematus* dragline silks by coupling films formed from the proteins with fluorescein and β-galactosidase using carbodiimide chemistry.[176] This approach could be used for biosensors and reactive media and to influence cell and tissue functions by increasing high levels of surface decoration with enzymes.

Additional benefits of providing designed functionalities encoded during gene construction can be provided from genetically engineered spider silks. Incorporation of cell binding domains, such as RGD (Arg-Gly-Asp),[109] or domains that interact selectively with inorganic components, are some strategies employed for *N. clavipes* spider silks to generate novel organic–inorganic composite material systems.[181,182] Such incorporation allows the highly tailored hybrid or chimeric spider silks to selectively undergo mineralization with silica[181] and hydroxyapatite.[12] By controlling the processing conditions, the morphology and structure of the composite could be manipulated. A biomimetic approach toward tougher and stiffer silk composite materials can be achieved by controlling the size of the inorganic domains, as well as the molecular-level interactions between the organic and inorganic domains in these composites. These results suggest that chimeric spider-silk proteins might provide new uses for spider silk in biomedical and other specialty materials, such as in hard tissue remodeling or in adhesive fillers. The ability to process and assemble these novel composite materials in aqueous ambient processing conditions is another benefit of this approach.

13.8 Summary

In summary, we have reviewed the inherent properties of spider dragline silk as well as the technological advances on the possible applications of dragline silk in the biomedical and clinical fields.

Spider dragline silk exhibits excellent mechanical properties in terms of breaking energy, coupled with both high strength and strain at failure. The mechanical properties could be further tuned by varying external factors such as strain rate, temperature, reeling speed and force, the medium of the forcibly draw silk, and treatment with solvents and vapors. In addition, spider silk utilizes its inherent property, supercontraction, to maintain tension in the web and restore web shape after deformation by prey capture, precipitation or wind as they become wetted with morning dew or rain. The powerful cyclic contraction exhibited by spider silk also results in a green and energy efficient mimicry of biological muscles. The multiple occurrences of supercontraction and fatigueless cyclic contraction offer possibilities for performing work in industries and clinical sciences.

The outstanding mechanical properties of spider silk are determined by its unique structural characteristics formed through its incomparable dope storage and spinning process. Scientists have applied many techniques in order to study the structure of the dragline silk. SEM and AFM are employed to study the morphology and topography of the fibre, and WAXD and FTIR are used to assess the β-crystallite and total β-sheet content in dragline silk, respectively. Besides, NMR, Raman spectroscopy and SAXS are also important tools in analyzing the secondary and tertiary structures in dragline silk. Generally speaking, dragline silk can be modeled as a semicrystalline material, in which the polyalanine nano-β-crystallites are embedded in the non-crystalline amorphous region to form a network. In particular, the model by Du *et al.*[25] suggested that intramolecular β-sheets acting as "molecular spindles" addressed the molecular origin of strain hardening of spider silk filaments.

Last but not least, reconstituted and recombinant silk proteins can be used to fabricate novel biomaterials for application in the biomedical and clinical fields. The proteins are processed into different structures ranging from fibres, films, gels, hydrogels and porous sponges to microcapsules. Surface functionalization through targeting carboxylic acid groups on the amino acids in the protein could be used for biosensors and influence cell and tissue functions by increasing high levels of surface decoration with enzymes.

References

1. *Fibrous Proteins*, ed. T. Scheibel, Landis Bioscience, Austin, TX, 2008.
2. M. Heim, D. Keerl and T. Scheibel, *Angew. Chem. Int. Ed.*, 2009, **48**, 3584–3596.
3. F. Vollrath and D. P. Knight, *Nature*, 2001, **410**, 541–548.
4. Z. Shao and F. Vollrath, *Nature*, 2002, **418**, 741–741.
5. R. W. Work, *Text. Res. J.*, 1976, **46**, 485.
6. W. W. Adams, D. L. Kaplan, B. Farmer, C. Viney, in *Silk Polymers: Materials Science and Biotechnology*, ed. D. L. Kaplan, W. W. Adams, B. Farmer and C. Viney, American Chemical Society, Washington, 1994, ch. 1.
7. H. Heslot, *Biochimie*, 1998, **80**, 19–31.
8. T. Asakura and D. L. Kaplan, in *Encyclopedia of Agricultural Science*, ed. C. J. Arntzen and E. M. Ritter, Academic Press, New York, 1994, vol. 4, pp. 1–11.
9. J. Gosline, P. Guerette, C. Ortlepp and K. Savage, *J. Exp. Biol.*, 1999, **202**, 3295–3303.
10. J. M. Gosline, M. W. Denny and M. E. DeMont, *Nature*, 1984, **309**, 551–552.
11. J. M. Gosline, in *Silk Polymers: Materials Science and Biotechnology*, ed. D. L. Kaplan, W. W. Adams, B. Farmer and C. Viney, American Chemical Society, Washington, 1994, pp. 328–341.
12. Y. Termonia, *Macromolecules*, 1994, **27**, 7378–7381.
13. A. H. Simmons, C. A. Michal and L. W. Jelinski, *Science*, 1996, **271**, 84–87.
14. J. Kummerlen, J. D. van Beek, F. Vollrath and B. H. Meier, *Macromolecules*, 1996, **29**, 2920–2928.
15. N. Du, X. Y. Liu, J. Narayanan, L. Li, M. L. M. Lim and D. Li, *Biophys. J.*, 2006, **91**, 4528–4535.

16. A. Glisovic, T. Vehoff, R. J. Davies and T. Salditt, *Macromolecules*, 2008, **41**, 390–398.

17. Fibroin heavy chain precursor (Fib-H) (H-fibroin) [gij9087216], P05790, http://www.gl.iit.edu/frame/genbank.htm; dragline silk spidroin 1 [Nephila pilipes], [gi:55274134], AAV48947, http://www.gl.iit.edu/frame/genbank.htm.

18. M. Xu and R. V. Lewis, *Proc. Natl. Acad. Sci. U. S. A.*, 1990, **87**, 7120–7124.

19. M. S. Engster, *Cell Tissue Res.*, 1976, **169**, 77–92.

20. M. A. Becker, in *Silk Polymers: Materials Science and Biotechnology*, ed. D. L. Kaplan, W. W. Adams, B. Farmer and C. Viney, American Chemical Society, Washington, 1994, ch. 17.

21. Y. Takahashi, in *Silk Polymers: Materials Science and Biotechnology*ed, ed. D. L. Kaplan, W. W. Adams, B. Farmer and C. Viney, American Chemical Society, Washington, 1994, ch. 15.

22. K. N. Savage and J. M. Gosline, *J. Exp. Biol.*, 2008, **211**, 1937–1947.

23. T. Lefèvre, M.-E. Rousseau and M. Pézolet, *Biophys. J.*, 2007, **92**, 2885–2895.

24. O. Rathore and D. Y. Sogah, *J. Am. Chem. Soc.*, 2001, **123**, 5231–5239.

25. N. Du, Z. Yang, X. Y. Liu, Y. Li and H. Y. Xu, *Adv. Funct. Mater.*, 2011, **21**, 772–778.

26. S. R. Fahnestock and S. L. Irwin, *Appl. Microbiol. Biotechnol.*, 1997, **47**, 23–32.

27. J. T. Prince, K. P. McGrath, C. M. Digirolamo and D. L. Kaplan, *Biochemistry*, 1995, **34**, 10879–10885.

28. S. R. Fahnestock and L. A. Bedzyk, *Appl. Microbiol. Biotechnol.*, 1997, **47**, 33–39.

29. R. V. Lewis, M. Hinman, S. Kothakota and M. J. Fournier, *Protein Express. Purif.*, 1996, **7**, 400–406.

30. S. Arcidiacono, C. Mello, D. Kaplan, S. Cheley and H. Bayley, *Appl. Microbiol. Biotechnol.*, 1998, **49**, 31–38.

31. S. R. Fahnestock, Z. Yao and L. A. Bedzyk, *Rev. Mol. Biotechnol.*, 2000, **74**, 105–119.

32. J. Scheller, K. H. Guhrs, F. Grosse and U. Conrad, *Nat. Biotechnol.*, 2001, **19**, 573–577.

33. O. Liivak, A. Blye, N. Shah and L. W. Jelinski, *Macromolecules*, 1998, **31**, 2947–2951.

34. K. A. Trabbic and P. Yager, *Macromolecules*, 1998, **31**, 462–471.

35. A. Seidel, O. Liivak and L. W. Jelinski, *Macromolecules*, 1998, **31**, 6733–6736.

36. A. Seidel, O. Liivak, S. Calve, J. Adaska, G. D. Ji, Z. T. Yang, D. Grubb, D. B. Zax and L. W. Jelinski, *Macromolecules*, 2000, **33**, 775–780.

37. A. Lazaris, S. Arcidiacono, Y. Huang, J.-F. Zhou, F. Duguay, N. Chretien, E. A. Welsh, J. W. Soares and C. N. Karatzas, *Science*, 2002, **295**, 472–476.

38. J. P. O'Brien, S. R. Fahnestock, Y. Termonia and K. H. Gardner, *Adv. Mater.*, 1998, **10**, 1185–1195.

39. D. Porter and F. Vollrath, *Adv. Mater.*, 2009, **21**, 487–492.
40. C. Holland, A. E. Terry, D. Porter and F. Vollrath, *Nat. Mater.*, 2006, **5**, 870–874.
41. O. Hakimi, D. P. Knight, F. Vollrath and P. Vadgama, *Composites, Part B*, 2007, **38**, 324–337.
42. F. Vollrath and D. P. Knight in *Biopolymers*, ed. S. R. Fahnestock, and A. Steinbuchel, WILEY-VCH, Weinheim, 2003, pp. 25–44.
43. X. Chen, D. P. Knight and F. Vollrath, *Biomacromolecules*, 2002, **3**, 644–648.
44. S. Rammensee, U. Slotta, T. Scheibel and A. R. Bausch, *Proc. Natl. Acad. Sci. U. S. A.*, 2008, **105**, 6590–6595.
45. D. Porter and F. Vollrath, *Soft Matter*, 2008, **4**, 328–336.
46. D. Knight and F. Vollrath, *Naturwissenschaften*, 2001, **88**, 179–182.
47. C. Dicko, J. M. Kenney, D. Knight and F. Vollrath, *Biochemistry*, 2004, **43**, 14080–14087.
48. F. K. Vollrath and D. P. Knight, *Int. J. Biol. Macromol.*, 1998, **24**, 243–249.
49. S. J. Lombardi and D. L. Kaplan, *J. Arachnol*, 1990, **18**, 297–306.
50. S. K. Winkler and D. L. Kaplan, *Rev. Mol. Biotechnol.*, 2000, **74**, 85–93.
51. P. Guerette, D. Ginzinger, B. Weber and J. Gosline, *Science*, 1996, **272**, 112–115.
52. J. Kovoor, *Ann. Biol.*, 1977, **16**, 97–171.
53. J. Kovoor and L. Zylberberg, *Tissue Cell*, 1982, **14**, 519–530.
54. J. Kovoor, in *Ecophysiology of Spiders*, ed. W. Nentwig, Springer, Berlin, 1987, pp. 160–186.
55. F. Vollrath, X. Wen and Hu and D. P. Knight, *Proc. R. Soc. London, Ser. B*, 1998, **263**, 817–820.
56. O. Emile, A. L. Floch and F. Vollrath, *Nature*, 2006, **440**, 621–621.
57. O. Emile, A. Le Floch and F. Vollrath, *Phys. Rev. Lett.*, 2007, **98**, 167402-1–167402-4.
58. F. Vollrath, B. Madsen and Z. Z. Shao, *Proc. R. Soc. London, Ser. B*, 2001, **268**, 2339–2346.
59. M. Denny, *J. Exp. Biol.*, 1976, **65**, 483–506.
60. Y. Liu, A. Sponner, D. Porter and F. Vollrath, *Biomacromolecules*, 2008, **9**, 116–121.
61. M. G. Dobb, D. J. Johnson and B. P. Saville, *Philos. Trans. R. Soc. London, Ser. A*, 1980, **294**, 483–485.
62. A. R. Bunsell, *J. Mater. Sci.*, 1975, **10**, 1300–1308.
63. Y. Liu, Z. Shao and F. Vollrath, *Chem. Commun.*, 2005, 2489–2491.
64. B. Madsen, Z. Z. Shao and F. Vollrath, *Int. J. Biol. Macromol.*, 1999, **24**, 301–306.
65. C. Riekel, M. Muller and F. Vollrath, *Macromolecules*, 1999, **32**, 4464–4466.
66. B. L. Thiel and C. Viney, *J. Microsc. (Oxford, U. K.)*, 1997, **185**, 179–187.
67. C. S. Ortlepp and J. M. Gosline, *Biomacromolecules*, 2004, **5**, 727–731.

68. J. Perez-Rigueiro, M. Elices, G. Plaza, J. I. Real and G. V. Guinea, *J. Exp. Biol.*, 2005, **208**, 2633–2639.
69. X. Chen, Z. Shao and F. Vollrath, *Soft Matter*, 2006, **2**, 448–451.
70. Y. Yang, X. Chen, Z. Z. Shao, P. Zhou, D. Porter, D. P. Knight and F. Vollrath, *Adv. Mater.*, 2005, **17**, 84–88.
71. A. Glisovic and T. Salditt, *Appl. Phys. A: Mater. Sci. Process.*, 2007, **87**, 63–69.
72. Z. Shao and F. Vollrath, *Polymer*, 1999, **40**, 1799–1806.
73. A. Schafer, T. Vehoff, A. Glisovic and T. Salditt, *Eur. Biophys. J.*, 2008, **37**, 197–204.
74. T. Vehoff, A. Gliaovi, H. Schollmeyer, A. Zippelius and T. Salditt, *Biophys. J.*, 2007, **93**, 4425–4432.
75. Z. Shao, R. J. Young and F. Vollrath, *Int. J. Biol. Macromol.*, 1999, **24**, 295–300.
76. S.-M. Lee, E. Pippel, U. Gosele, C. Dresbach, Y. Qin, C. V. Chandran, T. Brauniger, G. Hause and M. Knez, *Science*, 2009, **324**, 488–492.
77. S. Putthanarat, N. Stribeck, S. A. Fossey, R. K. Eby and W. W. Adams, *Polymer*, 2000, **41**, 7735–7747.
78. C. Y. Hayashi, N. H. Shipley and R. V. Lewis, *Int. J. Biol. Macromol.*, 1999, **24**, 271–275.
79. O. Liivak, A. Flores, R. Lewis and L. W. Jelinski, *Macromolecules*, 1997, **30**, 7127–7130.
80. S. Frische, A. B. Maunsbach and F. Vollrath, *J. Microsc. (Oxford, U. K.)*, 1998, **189**, 64–70.
81. R. W. Work, *Trans. Am. Microsc. Soc.*, 1984, **103**, 113–121.
82. Z. Shao, X. W. Hu, S. Frische and F. Vollrath, *Polymer*, 1999, **40**, 4709–4711.
83. D. V. Mahoney, D. L. Vezie, R. K. Eby, W. W. Adams and D. Kaplan, in *Silk Polymers: Materials Science and Biotechnology*, ed. D. L. Kaplan, W. W. Adams, B. Farmer and C. Viney, American Chemical Society, Washington, 1994, pp. 196–210.
84. C. Viney, A. E. Huber, D. L. Dunaway, K. Kerkam and S. T. Case, in *Silk Polymers: Materials Science and Biotechnology*, ed. D. L. Kaplan, W. W. Adams, B. Farmer and C. Viney, American Chemical Society, Washington, 1994, pp. 120–136.
85. F. Vollrath, T. Holtet, H. C. Thogersen and S. Frische, *Proc. R. Soc. London, Ser. B*, 1996, **263**, 147–151.
86. J. M. Gosline, M. E. DeMont and M. W. Denny, *Endeavour*, 1986, **10**, 37–43.
87. M. B. Hinman and R. V. Lewis, *J. Biol. Chem.*, 1992, **267**, 19320–19324.
88. C. Y. Hayashi and R. V. Lewis, *J. Mol. Biol.*, 1998, **275**, 773–784.
89. D. Porter, F. Vollrath and Z. Shao, *Eur. Phys. J. E*, 2005, **16**, 199–206.
90. N. Becker, E. Oroudjev, S. Mutz, J. P. Cleveland, P. K. Hansma, C. Y. Hayashi, D. E. Makarov and H. G. Hansma, *Nat. Mater.*, 2003, **2**, 278–283.
91. H. Zhou and Y. Zhang, *Phys. Rev. Lett.*, 2005, **94**, 028104.

92. J. D. van Beek, S. Hess, F. Vollrath and B. H. Meier, *Proc. Natl. Acad. Sci. U. S. A.*, 2002, **99**, 10266–10271.
93. F. Vollrath and D. Porter, *Soft Matter*, 2006, **2**, 377–385.
94. X. Wu, X.-Y. Liu, N. Du, G. Xu and B. Li, *Appl. Phys. Lett.*, 2009, **95**, 093703.
95. S. Keten, Z. Xu, B. Ihle and M. J. Buehler, *Nat. Mater.*, 2010, **9**, 359–367.
96. J. J. Gilman, *Eletronic Basis of the Strength of Materials*, Cambridge University Press, Cambridge, 2003, pp. 174–175.
97. E. Oroudjev, J. Soares, S. Arcdiacono, J. B. Thompson, S. A. Fossey and H. G. Hansma, *Proc. Natl. Acad. Sci. U. S. A.*, 2002, **99**, 6460–6465.
98. M. Rief, M. Gautel, F. Oesterhelt, J. M. Fernandez and H. E. Gaub, *Science*, 1997, **276**, 1109–1112.
99. S. Toki and B. S. Hsiao, *Macromolecules*, 2003, **36**, 5915–5917.
100. Y. Liu, Z. Shao and F. Vollrath, *Nat. Mater.*, 2005, **4**, 901–905.
101. T. Asakura, K. Umemura, Y. Nakazawa, H. Hirose, J. Higham and D. Knight, *Biomacromolecules*, 2007, **8**, 175–181.
102. F. Hagn, L. Eisoldt, J. G. Hardy, C. Vendrely, M. Coles, T. Scheibel and H. Kessler, *Nature*, 2010, **465**, 239–242.
103. G. Askarieh, M. Hedhammar, K. Nordling, A. Saenz, C. Casals, A. Rising, J. Johansson and S. D. Knight, *Nature*, 2010, **465**, 236–238.
104. J. Moore, *Br. J. Appl. Phys.*, 1950, **1**, 6.
105. D. P. Knight and F. Vollrath, *Philos. Trans. R. Soc. London, Ser. B*, 2002, **357**, 155–163.
106. L. D. Miller, S. Putthanarat, R. K. Eby and W. W. Adams, *Int. J. Biol. Macromol.*, 1999, **24**, 159–165.
107. B. L. Thiel, D. D. Kunkel and C. Viney, *Biopolymers*, 1994, **34**, 1089–1097.
108. D. B. Gillespie, C. Viney and P. Yager, in *Silk Polymers: Materials Science and Biotechnology*, ed. D. L. Kaplan, W. W. Adams, B. Farmer and C. Viney, American Chemical Society, Washington, 1994, pp. 155–167.
109. E. Bini, C. W. P. Foo, J. Huang, V. Karageorgiou, B. Kitchel and D. L. Kaplan, *Biomacromolecules*, 2006, **7**, 3139–3145.
110. M.-E. Rousseau, T. Lefèvre, L. Beaulieu, T. Asakura and M. Pézolet, *Biomacromolecules*, 2004, **5**, 2247–2257.
111. J. Sirichaisit, R. J. Young and F. Vollrath, *Polymer*, 2000, **41**, 1223–1227.
112. J. Sirichaisit, V. L. Brookes, R. J. Young and F. Vollrath, *Biomacromolecules*, 2003, **4**, 387–394.
113. M.-E. Rousseau, L. Beaulieu, T. Lefèvre, J. Paradis, T. Asakura and M. Pézolet, *Biomacromolecules*, 2006, **7**, 2512–2521.
114. W. Y. Yeh and R. J. Young, *Polymer*, 1999, **40**, 857–870.
115. B. L. Thiel, K. B. Guess and C. Viney, *Biopolymers*, 1997, **41**, 703–719.
116. Z. Yang, D. T. Grubb and L. W. Jelinski, *Macromolecules*, 1997, **30**, 8254–8261.
117. D. T. Grubb and L. W. Jelinski, *Macromolecules*, 1997, **30**, 2860–2867.

118. C. Riekel, B. Madsen, D. Knight and F. Vollrath, *Biomacromolecules*, 2000, **1**, 622–626.
119. C. Riekel, C. Branden, C. Craig, C. Ferrero, F. Heidelbach and M. Muller, *Int. J. Biol. Macromol.*, 1999, **24**, 179–186.
120. D. H. Hijirida, K. G. Do, C. Michal, S. Wong, D. Zax and L. W. Jelinski, *Biophys. J.*, 1996, **71**, 3442–3447.
121. J. D. van Beek, J. Kummerlen, F. Vollrath and B. H. Meier, *Int. J. Biol. Macromol.*, 1999, **24**, 173–178.
122. G. P. Holland, R. V. Lewis and J. L. Yarger, *J. Am. Chem. Soc.*, 2004, **126**, 5867–5872.
123. T. Asakura, M. Y. Yang, T. Kawase and Y. Nakazawa, *Macromolecules*, 2005, **38**, 3356–3363.
124. B. W. Hu, P. Zhou, I. Noda and G. Z. Zhao, *Anal. Chem.*, 2005, **77**, 7534–7538.
125. J. M. Yao, Y. Nakazawa and T. Asakura, *Biomacromolecules*, 2004, **5**, 680–688.
126. G. Estrada, E. Villegas and G. Corzo, *Nat. Prod. Rep.*, 2007, **24**, 145–161.
127. G. P. Holland, M. S. Creager, J. E. Jenkins, R. V. Lewis and J. L. Yarger, *J. Am. Chem. Soc.*, 2008, **130**, 9871–9877.
128. G. D. McLachlan, J. Slocik, R. Mantz, D. Kaplan, S. Cahill, M. Girvin and S. Greenbaum, *Protein Sci.*, 2009, **18**, 206–216.
129. T. Izdebski, P. Akhenblit, J. E. Jenkins, J. L. Yarger and G. P. Holland, *Biomacromolecules*, 2010, **11**, 168–174.
130. J. E. Jenkins, M. S. Creager, R. V. Lewis, G. P. Holland and J. L. Yarger, *Biomacromolecules*, 2010, **11**, 192–200.
131. R. W. Work, *Text. Res. J.*, 1977, **47**, 650–662.
132. R. W. Work, *J. Arachnol.*, 1981, **9**, 299–308.
133. G. V. Guinea, M. Elices, J. Perez-Rigueiro and G. Plaza, *Polymer*, 2003, **44**, 5785–5788.
134. G. V. Guinea, M. Cerdeira, G. R. Plaza, M. Elices and J. Perez-Rigueiro, *Biomacromolecules*, 2010, **11**, 1174–1179.
135. F. I. Bell, I. J. McEwen and C. Viney, *Nature*, 2002, **416**, 37–37.
136. R. W. Work, *J. Exp. Biol.*, 1985, **118**, 379–404.
137. R. V. Lewis, *Acc. Chem. Res.*, 1992, **25**, 392–398.
138. E. K. Tillinghast and M. Townley, *Ecophysiology of Spiders, Spinger*, Berlin, 1987, pp. 203–210.
139. M. Elices, J. Perez-Rigueiro, G. Plaza and G. V. Guinea, *J. App. Polym. Sci.*, 2004, **92**, 3537–3541.
140. A. Lendlein and S. Kelch, *Angew. Chem. Int. Ed.*, 2002, **41**, 2034–2057.
141. A. Lendlein and R. Langer, *Science*, 2002, **296**, 1673–1676.
142. Y. Termonia, in *Structural Biological Materials*, ed. M. Elices, Elsevier, Oxford, 2000, p. 335.
143. J. Pérez-Rigueiro, C. Viney, J. Llorca and M. Elices, *Polymer*, 2000, **41**, 8433–8439.
144. J. Perez-Rigueiro, M. Elices and G. V. Guinea, *Polymer*, 2003, **44**, 3733–3736.

145. K. N. Savage, P. A. Guerette and J. M. Gosline, *Biomacromolecules*, 2004, **5**, 675–679.

146. T. A. Blackledge, C. Boutry, S. C. Wong, A. Baji, A. Dhinojwala, V. Sahni and I. Agnarsson, *J. Exp. Biol.*, 2009, **212**, 1980–1988.

147. I. Agnarsson, C. Boutry, S. C. Wong, A. Baji, A. Dhinojwala, A. T. Sensenig and T. A. Blackledge, *Zoology*, 2009, **112**, 325–331.

148. T. A. Blackledge and J. M. Zevenbergen, *Anim. Behav.*, 2007, **73**, 855–864.

149. K. N. Savage and J. M. Gosline, *J. Exp. Biol.*, 2008, **211**, 1948–1957.

150. R. W. Work and N. Morosoff, *Text. Res. J.*, 1982, **52**, 349–356.

151. J. Gatesy, C. Hayashi, D. Motriuk, J. Woods and R. Lewis, *Science*, 2001, **291**, 2603–2605.

152. A. Sponner, E. Unger, F. Grosse and W. Klaus, *Nat. Mater.*, 2005, **4**, 772–775.

153. L. W. Jelinski, A. Blye, O. Liivak, C. Michal, G. LaVerde, A. Seidel, N. Shah and Z. T. Yang, *Int. J. Biol. Macromol.*, 1999, **24**, 197–201.

154. Z. T. Yang, O. Liivak, A. Seidel, G. LaVerde, D. B. Zax and L. W. Jelinski, *J. Am. Chem. Soc.*, 2000, **122**, 9019–9025.

155. I. Agnarsson, A. Dhinojwala, V. Sahni and T. A. Blackledge, *J. Exp. Biol.*, 2009, **212**, 1989–1993.

156. P. T. Eles and C. A. Michal, *Macromolecules*, 2004, **37**, 1342–1345.

157. G. Y. W. Toh and X. Y. Liu, unpublished results.

158. H. Wendt, A. Hillmer, K. Reimers, J. W. Kuhbier, F. Schafer-Nolte, C. Allmeling, C. Kasper and P. M. Vogt, *Plos One*, 2011, **6**, e21833-1– e21833-10.

159. F. G. Omenetto and D. L. KapLan, *Nat. Photonics*, 2008, **2**, 641–643.

160. D. H. Kim, J. Viventi, J. J. Amsden, J. L. Xiao, L. Vigeland, Y. S. Kim, J. A. Blanco, B. Panilaitis, E. S. Frechette, D. Contreras, D. L. Kaplan, F. G. Omenetto, Y. G. Huang, K. C. Hwang, M. R. Zakin, B. Litt and J. A. Rogers, *Nat. Mater.*, 2010, **9**, 511–517.

161. D. H. Kim, Y. S. Kim, J. Amsden, B. Panilaitis, D. L. Kaplan, F. G. Omenetto, M. R. Zakin and J. A. Rogers, *Appl. Phys. Lett.*, 2009, **95**.

162. R. Menassa, H. Zhu, C. N. Karatzas, A. Lazaris, A. Richman and J. Brandle, *Plant Biotechnol. J.*, 2004, **2**, 431–438.

163. C.-Z. Zhou, F. Confalonieri, M. Jacquet, R. Perasso, Z.-G. Li and J. Janin, *Proteins: Struct., Funct., Bioinf.*, 2001, **44**, 119–122.

164. J. Yang, L. A. Barr, S. R. Fahnestock and Z. B. Liu, *Transgenic Res.*, 2005, **14**, 313–324.

165. Y. Miao, Y. Zhang, K. Nakagaki, T. Zhao, A. Zhao, Y. Meng, M. Nakagaki, E. Park and K. Maenaka, *Appl. Microbiol. Biotechnol.*, 2006, **71**, 192–199.

166. D. Huemmerich, T. Scheibel, F. Vollrath, S. Cohen, U. Gat and S. Ittah, *Curr. Biol.*, 2004, **14**, 2070–2074.

167. C. N. Karatzas, in *Biopolymers. Polyamides and Complex Proteinaceous Materials II*, ed. S. R. Fahnestock and A. Steinbüchel, Wiley-VCH, Weinheim, 2003, pp. 500–510.

168. L. A. Barr, S. R. Fahnestock and J. Yang, Mol, *Breeding*, 2004, **13**, 345–356.

169. K. Lee, B. Kim, Y. Je, S. Woo, H. Sohn and B. Jin, *J. Biosci.*, 2007, **32**, 705–712.

170. C. Vepari and D. L. Kaplan, *Prog. Polym. Sci.*, 2007, **32**, 991–1007.

171. G. H. Altman, R. L. Horan, H. H. Lu, J. Moreau, I. Martin, J. C. Richmond and D. L. Kaplan, *Biomaterials*, 2002, **23**, 4131–4141.

172. D. Huemmerich, C. W. Helsen, S. Quedzuweit, J. Oschmann, R. Rudolph and T. Scheibel, *Biochemistry*, 2004, **43**, 13604–13612.

173. P. Corsini, J. Perez-Rigueiro, G. V. Guinea, G. R. Plaza, M. Elices, E. Marsano, M. M. Carnasciali and G. Freddi, *J. Polym. Sci., Part B: Polym. Phys.*, 2007, **45**, 2568–2579.

174. S. Arcidiacono, C. M. Mello, M. Butler, E. Welsh, J. W. Soares, A. Allen, D. Ziegler, T. Laue and S. Chase, *Macromolecules*, 2002, **35**, 1262–1266.

175. J. S. Stephens, S. R. Fahnestock, R. S. Farmer, K. L. Kiick, D. B. Chase and J. F. Rabolt, *Biomacromolecules*, 2005, **6**, 1405–1413.

176. D. Huemmerich, U. Slotta and T. Scheibel, *Appl. Phys. A: Mater. Sci. Process.*, 2006, **82**, 219–222.

177. F. Junghans, M. Morawietz, U. Conrad, T. Scheibel, A. Heilmann and U. Spohn, *Appl. Phys. A: Mater. Sci. Process.*, 2006, **82**, 253–260.

178. S. Rammensee, D. Huemmerich, K. D. Hermanson, T. Scheibel and A. R. Bausch, *Appl. Phys. A: Mater. Sci. Process.*, 2006, **82**, 261–264.

179. K. D. Hermanson, D. Huemmerich, T. Scheibel and A. R. Bausch, *Adv. Mater.*, 2007, **19**, 1810–1815.

180. K. McGrath and D. Kaplan, Protein-Based Materials, Birkhauser, Basel, 1997.

181. C. W. P. Foo, S. V. Patwardhan, D. J. Belton, B. Kitchel, D. Anastasiades, J. Huang, R. R. Naik, C. C. Perry and D. L. Kaplan, *Proc. Natl. Acad. Sci. U. S. A.*, 2006, **103**, 9428–9433.

182. J. Huang, C. Wong, A. George and D. L. Kaplan, *Biomaterials*, 2007, **28**, 2358–2367.

CHAPTER 14

Mussel Byssus Fibres: A Tough Biopolymer

F. G. TORRES,* O. P. TRONCOSO AND C. E. TORRES

Department of Mechanical Engineering, Catholic University of Peru,
Av. Universitaria 1801, Lima 32, Peru
*Email: fgtorres@pucp.edu.pe

14.1 Introduction

Byssus fibres produced by mussels are tough biopolymers composed mainly of proteins and water. These natural biopolymer fibres have been intensively studied owing to their mechanical and adhesive properties. The remarkable strength, unmatched toughness and extensibility of byssus fibres allow mussels to withstand the large and repetitive forces produced by crashing waves with velocities of over $10 \, \mathrm{m \, s^{-1}}$ and accelerations of approximately $400 \, \mathrm{m \, s^{-2}}$. They are composed of three collagenous proteins that make up the bulk of the thread core. Their toughness is considered six times greater than that of the human tendon collagen and comparable with that of Kevlar and carbon fibres.

In addition, the adhesive properties of byssus fibres allow mussels to be attached to hard and soft substrates in aquatic environments through their glue proteins located in their adhesion plaques. These plaques contain at least six different types of adhesive proteins that have attracted great interest for their potential applications.

In this chapter, we review the properties, characterization and applications of byssus fibres. First, we give a brief introduction to the natural history of mussels. Then, we discuss the structure at different levels of byssus fibres,

RSC Green Chemistry No. 16
Natural Polymers, Volume 1: Composites
Edited by Maya J John and Thomas Sabu

including their chemical composition and morphology. We then continue by describing the structural, mechanical, thermal and adhesive properties of byssus fibres. Finally, we present several applications of the mussel adhesive proteins found in the byssus fibres. The last section includes the analysis of adhesive properties for applications in medicine and biotechnology, such as cell culture, tissue engineering, medical, dental and biosensor applications. All these aspects are discussed in order to present the potential use of byssus fibres that suggest new models for the design of novel biomimetic polymers.

14.2 Natural History

14.2.1 Biological Functions

Mussels live in a variety of habitats, such as rocky intertidal, salt marshes and subtidal and hydrothermal vents. Byssal fibres provide mussels with a secure attachment to the rocks present in their aquatic habitats.[1] These fibres present a combination of strength and extensibility, together with a low resilience, that give them an unmatched toughness that allows mussels to survive the large and repetitive forces produced by waves.[2] With gentle waves, the byssus fibres provide an elastic response, which allows mussels to move freely with the waves. With violent waves, byssal threads are able to dissipate energy and increase their attachment strength.[3] Without these properties, mussels would be dashed against rocks by the waves.

14.2.2 Mussel Anatomy

Mussels are aquatic invertebrate animals living in freshwater and marine habitats. They belong to the phylum "mollusca" under the class "bivalve". Their anatomy is characterized as having two shells or valves that are joined together by a ligament and are closed by strong internal muscles. Internally, mussels have a large organ called a foot that is used for mussel locomotion. Byssus fibres are secreted by this "foot" and are responsible for tethering the mussel to the rock surface *via* some adhesive plaques (Figures 14.1a and 14.1b).

Three main glands are located in the mussel foot: the stem forming gland, the thread forming gland and the plaque forming gland. The first gland forms the stem, an organic tissue that supports each byssus fibre serving as a connection between the foot and the byssus fibres (Figure 14.1). Byssus fibres are generated from the stem located in the mussel foot in an injection moulding-like process.[4] As shown in Figure 14.2, the byssal fibres are located all along the stem. The fibres attach to the stem with a hoop of thin tissue. An adhesive plaque is located at the end of each fibre. The total number of fibres along the stem varies from mussel to mussel, and increases with water turbulence.[5] Large numbers of threads allow mussels to attach to several surfaces.

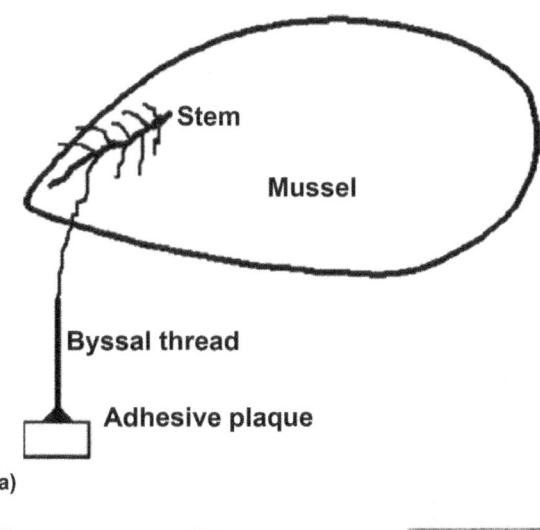

(a)

(b)

Figure 14.1   (a) Schematic view of a mussel showing the stem, byssal thread and adhesive plaque;[8] (b) photograph of byssus fibres from *Aulacomya ater* attached to various stones.

14.2.3   Mussel Species

Byssal threads of a broad range of mussel species from diverse environments have been studied. Intertidal and subtidal mussels in freshwater and marine habitats such as *Modiolus modiolus*, *Geukensia demissa*, *Bathymodiolus thermophilus*, *Dreissena polymorpha*, *Aulacomya ater* and *Perna canaliculus* are among these species.[1,6–8] Byssus from the genus *Mytilus*, such as *Mytilus edulis*,

Figure 14.2   SEM micrograph of a stem and its byssus fibres from the species
Aulacomya ater.

M. californianus, *M. galloprovinicialis*, *M. coruscus* and *M. trossulus*, has also
received considerable attention.[1,9–14] The results show that byssal fibres from
this genus have a similar composition and structure.

   However, the threads of byssus produced from different species of mussels
are not equivalent.[1] The differences found in the mechanical properties of
mussel byssus from different species are probably related to adaptations that
help mussels survive in different environments.[15] Flow, salinity and temperature are among the environmental factors studied for their influence on the
mechanical properties of mussel byssus.[16–18] As an example, the byssus fibres of
M. californianus are stronger than those of *M. edulis*.[19,20] The adhesive
attachment of *M. galloprovincialis* is stronger than that of *M. edulis*.[21,22] Even
the season and wave exposure are factors that affect the attachment strength
within specific species.[20,23,24]

14.3   Morphology and Structure

14.3.1   Chemical Composition

Table 14.1 shows the identified constituents of mussel byssus. They are composed mainly of proteins and water. In the case of *M. edulis*, freshly collected
byssal threads are about 70% water, by weight. Of the dry weight, protein is
about 96%.[25] Small amounts of inorganic residues such as ash, heavy metals
and transition metals have also been found in mussel byssus.[26]

Table 14.1 Compositional analysis of *M. edulis* byssal threads.

	Percentage (%)	*Ref.*
Water content (% total wt)	70 ± 3	46
Dry weight (% total wt)	30 ± 2	46
Protein (% dry wt)	95 ± 3	25
Ash (% dry wt)	4.8	46
Cu/Fe/Zn (% dry wt)	0.1	26
Other cations (% dry wt)	0.2	26

The byssus fibre is primary composed of collagenous proteins known as "pre-pepsinized collagens" (PreCol) that make up the bulk of the thread core.[25,27–29] The adhesive plaque located at the end of the fibre is composed of at least six different types of proteins known as "foot proteins" (fp). These proteins are labelled according to their specific mussel species. As an example, five distinctive types of foot proteins have been isolated from the adhesion plaque of *M. edulis* and are labelled from Mefp-1 to Mefp-5.[30–34]

14.3.2 Macro Structure of Byssal Threads

The formation of byssal threads begins when the mussel foot touches down on a surface that the mussel finds suitable for attachment.[4] Soluble thread precursors are secreted into the ventral groove of the foot, where muscular contractions mould them into functional threads. This process lasts around 5 min.[15] The grooves depicted in Figure 14.3 are a consequence of the injection moulding-like process that forms byssal threads as they are secreted in the liquid state and moulded against the walls of ventral grooves.[35]

The byssus fibres of most mussel species can be divided into three parts: proximal and distal regions and an adhesive plaque. The proximal region has a corrugated surface and represents approximately one third of the total thread length, whereas the distal region is smoother, narrower and approximately twice the length of the proximal region. These two regions have been identified in byssus from *M. califonianus*, *M. edulis*, *M. trossulus* and *M. galloprovincialis*.[3] Some mussel species show no differences between the proximal and distal region. As an example, byssal threads from the South American mussel *Aulacomya ater* show no morphological differences along the entire thread of the byssus. In addition, problems have been reported when trying to characterize the two regions of the thread of *Bathymodiolus thermophilus*, *Geukensia demissa* and *Dreissena polymorpha*.[3]

Byssal threads have a fibrous core covered by a thin protective cuticle or sheath. The thickness of this cuticle is in the range 2–4 μm.[36] The cuticular proteins show highly repetitive sequences, dominated by proline/hydroxyproline, lysine and 3,4-dihydroxyphenyl-L-alanine (DOPA), and mussel foot protein-1 (mfp-1).[37,38] The function of such a cuticle is to protect the underlying collagen fibres from abrasion and microbial attack.[36]

Figure 14.3   SEM micrograph of the external morphology of mussel byssus from
Aulacomya ater.

14.3.3   Collagens Present in Mussel Byssus

Collagen is the major component of mussel byssus. The type of collagen that is
predominant in the distal region of byssus is known as Col-D. Col-D is a
homotrimer with a chain mass of 60 kDa. Col-P is present in the proximal
region of the thread and has an apparent mass of 55 kDa.[28]

Mussel foot extracts have been used to identify precursors of Col-P and
Col-D. The precursors are named preCol-P, preCol-D and preCol-NG. These
precursors have three major domains: a tough central collagen domain, flanked
on either side by structural domains. These structural domains resemble elastin
in preCol-P, spider dragline silk in preCol-D, and Gly-rich cell-wall proteins in
preCol-NG.[15] PreCol-P is abundant in the elastic proximal region, while pre-
Col-D predominates in the stiffer distal portion. By contrast, PreCol-NG is
evenly distributed.

There are histidine-rich sequences in the terminal regions of the preCols in
mussel byssus. Histidines are proposed to be reversible ligands for coordination
with transition metal ions such as zinc(II) and copper(II). Such reversibly
broken coordination complexes are stronger than non-covalent bonds, but
possess only half the strength of a covalent bond.[39] In addition, histidines are
suspected to play an integral role as sacrificial bonds in yield and self-healing in
the distal region of byssal threads.[40]

14.4 Characterization

14.4.1 Structural Characterization

The byssus fibre is composed of a fibrous core surrounded by a protective cuticle. The structural morphology varies from species to species. The cuticle of *P. canaliculus* is uniform and homogeneous. In contrast, the cuticle of *M. galloprovincialis* has been described as a composite-like material having granules immersed in a homogeneous protein matrix. These granules have a diameter of around 0.8 μm and comprise about 50% of the cuticle volume.

In the case of *M. galloprovincialis*, the granular cuticle exhibits hardness and stiffness approximately one order of magnitude greater than those of the fibrous core. Holten-Andersen *et al.*[41] have shown that this cuticle is extensible (breaking strain ~ 70%). They found that during the stretching of byssal threads, microcracks were formed within the matrix of the cuticle. These cracks did not propagate through the entire cuticle owing to the fact that the granules arrest microcracks so that they cannot develop into microtears. Byssus cuticle is the only known coating that has both high compliance and hardness which co-exist without mutual detriment.[42]

Structural characterization of the adhesive plaque has been made in order to understand the underlying principles of the attachment of mussels to hard and soft surfaces. Byssal threads mediate attachment of the soft mushy body of the mussel to a very stiff and hard substratum through their adhesive plaques. When the adhesive plaque is joined to a hard substrate such as a rock, there is a significant mismatch in the stiffness of the rock and the stiffness of the mussel body. In order to prevent an excessive deformation of mussel body and mussel byssus, the adhesive proteins within the adhesive plaque are present as solid foam.[43,44]

The structure of the adhesive plaque from *M. californianus* is shown in Figure 14.4. It is a continuous, partially open cell network in which the pore size gradually increases from the interface towards the cuticle. The microfibrils descending from the thread extend into the adhesive plaque. The damage caused by contact deformation is mitigated by the pore size gradient within the adhesive plaque.

The external morphology of plaques of *A. ater* is shown in Figures 14.4a and 14.4b. The shape of the plaques is elliptical. The patterned surface of plaques plays an important role in the interaction with a variety of both organic and inorganic substrates in aqueous environments. Through these plaques, mussels secret adhesive proteins that allow them to attach themselves to hard substrates. The adhesive proteins are characterized by the presence of an unusual amino acid known as DOPA (3,4-dihydroxy-l-phenylalanine) that is thought to be responsible for the strong adhesion of byssal threads.[25,39,45]

The morphological characteristics of assembled PreCols have been explored by both transmission electron microscopy (TEM) and atomic force microscopy (AFM).[40,46] TEM of both the liquid crystalline polymer secretor granules and byssal threads reveal a smectic packing that consists of "6 + 1" bundles; that is,

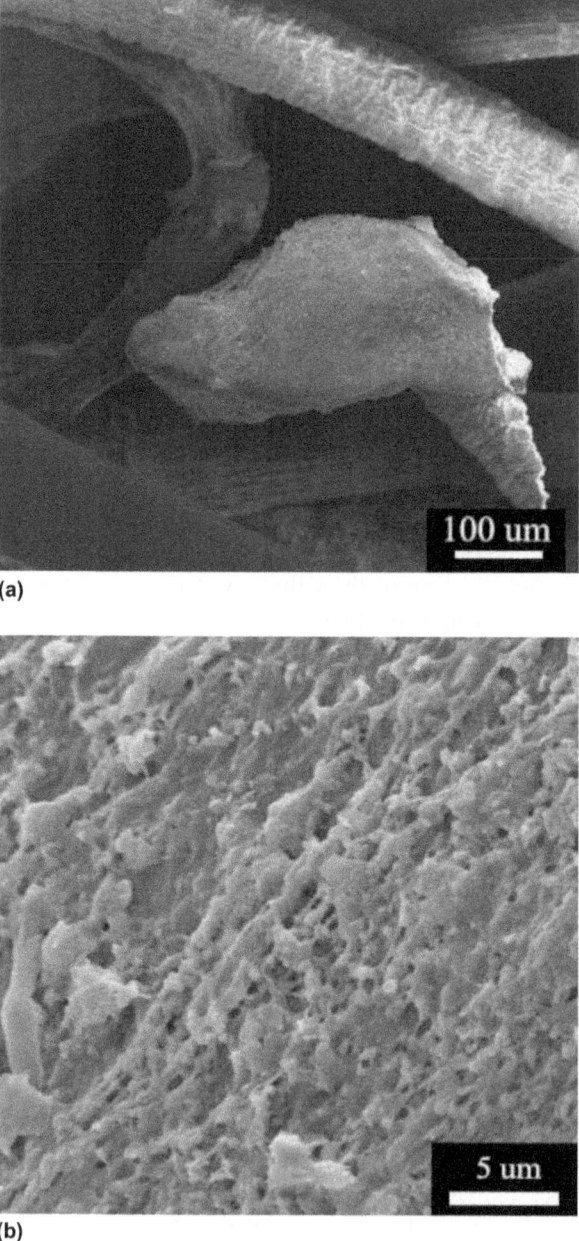

(a)

(b)

Figure 14.4   (a) SEM micrograph of an *Aulacomya ater* byssal adhesive plaque; (b) SEM micrograph showing a closer view of the morphology of an *A. ater* byssal adhesive plaque.

seven trimers per unit with six hexagonally arranged on the outside and one on the inside. These can be modelled in three dimensions as flower or banana shapes. AFM images have detected the banana variety. AFM studies have also confirmed the extensive smectic arrays of these banana or bent core bundles.[40]

14.4.2 Mechanical Properties

Byssus fibres need to meet specific mechanical requirements of stiffness, strength, toughness and extensibility in order to withstand water velocities of over $10\,\mathrm{m\,s^{-1}}$ and water acceleration of around $400\,\mathrm{m\,s^{-2}}$ in the marine inter-tidal zone.[47,48] In addition, the actual values of the mechanical properties, such as ultimate stress, ultimate strain and others, vary from species to species.[14] Figure 14.5 shows some representative stress–strain curves of different byssal fibres from several mussel species. The energy absorbed during the deformation process is represented by the area below the tensile curves depicted in Figure 14.5. The representative curve that corresponds to *A. ater* indicates that this material absorbs more energy during the deformation process than the other types of byssal fibres.

The mechanical properties of mussel byssus are also influenced by environmental conditions such as temperature, humidity, salinity, *etc.*[5,23,24,49] For instance, Moeser *et al.*[5] observed an increase in the strength and extensibility of

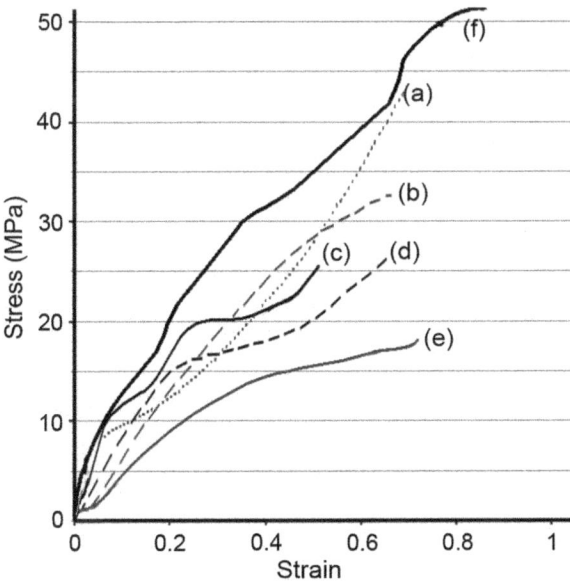

Figure 14.5 Representative stress–strain curves of byssus from (a) *Dreissena poly-morpha*,[1] (b) *Bathymodiolus thermophilus*,[1] (c) *Modiolus modiolus*,[1] (d) *Mytilus edulis*,[1] (e) *Geukensia demissa*[1] and (f) *Aulacomya ater*.

byssus fibres after fall and winter. They also report that the strongest attachment occurs during spring, while the weakest attachment takes place in fall.

The distal and proximal regions of byssal fibres from several mussel species have different mechanical properties among each other. In the case of *M. edulis*, Allen *et al.*[9] found that the distal region was stronger than the proximal region. Smeathers and Vincent[10] reported that the proximal region is more extensible than the distal region. Waite *et al.*[40,46] found that the distal region of the mussel byssus from *M. galloprovincialis* is about 10 times stiffer than the proximal region.

Bell and Gosline investigated the stress–strain behaviour of full threads and individual proximal and distal regions of several mussel species.[14] They showed that the byssal threads of edulis-like species have both similar compositions and mechanical behaviour. They found major differences between the distal and proximal regions. Typical stress–strain curves for isolated distal and proximal regions showed that the proximal region has a lower Young's modulus, a lower ultimate stress and a higher ultimate strain than the distal region. The distal region exhibits a striking yield behaviour (at around 35 MPa) and is considerably stronger than the proximal region.

In the case of entire byssal fibres, the same authors divided the stress–strain curve into three phases: the mechanical behaviour of the first phase is due to the stretching of the proximal region; the second phase of the curve is dominated by the yield of the distal region; whereas the third phase shows the stiffening of the distal region.[14]

Carrington and Gosline have reported cyclical loading tests carried out on byssal threads. The mechanical behaviour of mussel byssus depends on the loading history of the thread. In *Mytilus californianus*, whole threads cycled below the yield point were highly resilient and increased slightly in stiffness when loaded once again. By contrast, threads cycled beyond the yield point had much lower resilience and were dramatically less stiff when reloaded.[3]

Harrington and Waite have observed the behaviour of native seawater-treated distal threads when cycled multiple times, to a slightly higher strain each time. They noted a second yield point for cycles beyond the initial yield point. After each successive yield, the threads showed a reduction in the elastic modulus. The largest reduction in the stiffness occurs during the 20% strain cycle, with the initial modulus reduced by almost 40% from the first cycle.[15]

Troncoso *et al.* have studied the mechanical properties of mussel byssus from *A. ater*. They used true stress and strain to model the mechanical behaviour of byssal threads in hydrated and dry environments. Linear, power law-type and Mooney–Rivlin relationships were used to model the results obtained from their uniaxial tensile tests.[8]

The results of the tensile tests carried out by Troncoso *et al.* are summarized in Table 14.2.[8] The mechanical properties of byssus immersed in distilled and sea water are similar. By contrast, hydrated byssus is more extensible than dry byssus, as it reaches higher maximum strain values. Furthermore, differences between true and engineering ultimate stresses are similar for both dry and hydrated byssus.

Table 14.2   Values of mechanical properties of mussel byssus from *A. ater*
under different conditions.

Condition	Ultimate stress (MPa)	Ultimate strain
Dry	144.71 ± 33.58	0.27 ± 0.018
Immersed in distilled water	59.47 ± 25.02	0.84 ± 0.097
Immersed in sea water	70.16 ± 20.46	0.79 ± 0.275

Figure 14.6 shows representative stress–strain curves of byssal threads. Dry byssal thread (Figure 14.6a) behaves as an elastoplastic polymer while humid threads (Figure 14.6b) immersed in both distilled and sea water display elastomeric behaviour. Figure 14.6a shows that the curves of dry byssus are characterized by two clearly different regions. Initially, they follow a linear elastic behaviour at small strains until they reach a critical strain where plastic deformation begins. A power-law equation has been used to describe such plastic behaviour.[8] The suitability of this power-law to model the tensile behaviour of dried byssal threads is clear from Figure 14.6a, where a representative stress–strain curve for dry byssus is shown.

Figure 14.6b shows a representative stress–strain curve for hydrated byssus, demonstrating that the experimental data fit well with the chosen model. In contrast to dry byssus, lower stresses are needed to deform the hydrated byssal threads. Troncoso *et al.* used the Mooney–Rivlin relation of rubber elasticity theory to model the stress–strain relationships, owing to the fact that hydrated byssus displayed an elastomer-like tensile behaviour (theoretical points in Figure 14.6b). The Mooney–Rivlin model has been used in the past to model the deformation of rubber-like materials and has been applied to the study of soft tissues such as tendons and ligaments.[50–53]

14.4.3   Thermal Properties

Differential scanning calorimetry (DSC), thermomechanical analysis (TMA) and dynamic mechanical analysis (DMA) have been used to evaluate the thermal properties of byssus fibres. Tsukada *et al.* used DSC to examine thermal transitions of byssus fibres.[54] They found a minor endothermic peak at 230 °C and a major broad endothermic peak at 310 °C. The last broad endothermic peak was associated with the thermal decomposition of the collagenous byssus fibres. Figure 14.7a shows DSC curves of byssus fibres. Dried byssus fibres have a thermal transition at around 103 °C, whereas hydrated byssus fibres show a thermal transition at around 113 °C. Although the temperature of the thermal transitions of hydrated byssus fibres is different from the thermal transition of dried fibres, when the amount of water content rises, the thermal transition temperature does not change accordingly.

TMA tests showed that the byssus fibres slightly contracted at a constant rate during heating from room temperature to about 200 °C. After that, the contraction increased abruptly, reaching a maximum value at 240 °C. Beyond this

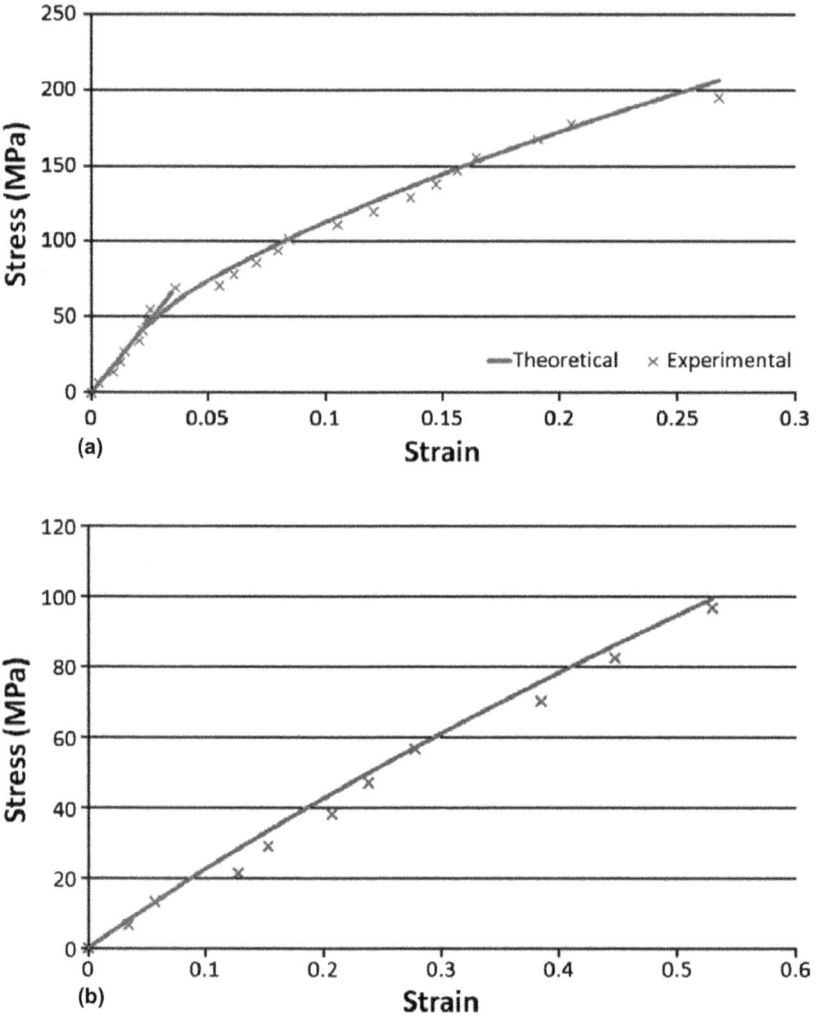

Figure 14.6 (a) Experimental (×) and theoretical (–) stress–strain curve for dry byssus from *A. ater*; (b) experimental (×) and theoretical (–) stress–strain curve for hydrated byssal threads from *A. ater* immersed in sea water.

temperature, the byssus fibres started to expand rapidly. The latter might be attributed to the fast thermal movement of the molecules, along with the decomposition and partial reforming of the intermolecular bonds. These results suggested that both the high degree of contraction registered by the TMA test above 200 °C and the minor endothermic peak (around 230 °C) are related to the occurrence of some changes in the byssus fibres at a molecular level.[54]

Aldred *et al.* have used DMA tests to study mussel byssus fibres (Figure 14.7b).[55] They found that the denaturation and thermal decomposition of

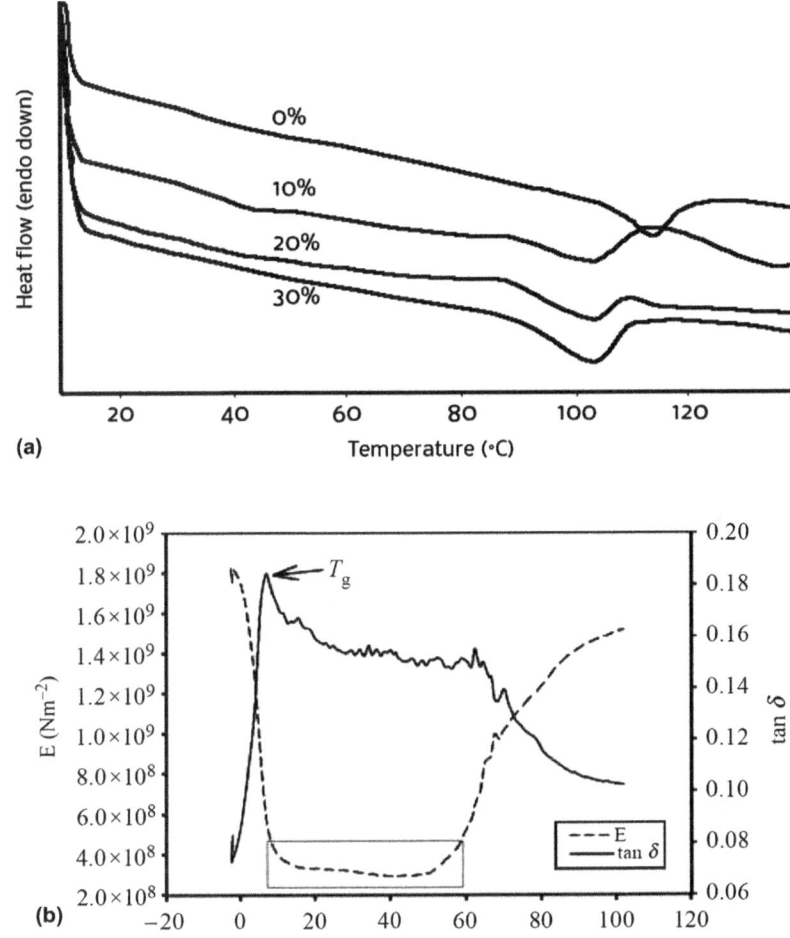

Figure 14.7 (a) Representative thermograms of mussel byssus with different water contents; (b) dynamical mechanical analysis of mussel byssus in dry conditions.[55]

byssus fibres start at 217 °C. They determined that the glass transition temperature (T_g) of dried byssus fibres takes place at around 6 °C.

14.4.4 IR Analysis

IR analyses of peptides and proteins are characterized by typical absorptions peaks such as amide bands. Figure 14.8 shows the IR spectra of byssus fibres from *A. ater*. These byssus fibres exhibit an amide I band showing a narrow peak at $\sim 1655\,cm^{-1}$. Amide I (~ 1600–$1700\,cm^{-1}$) is associated with the C=O

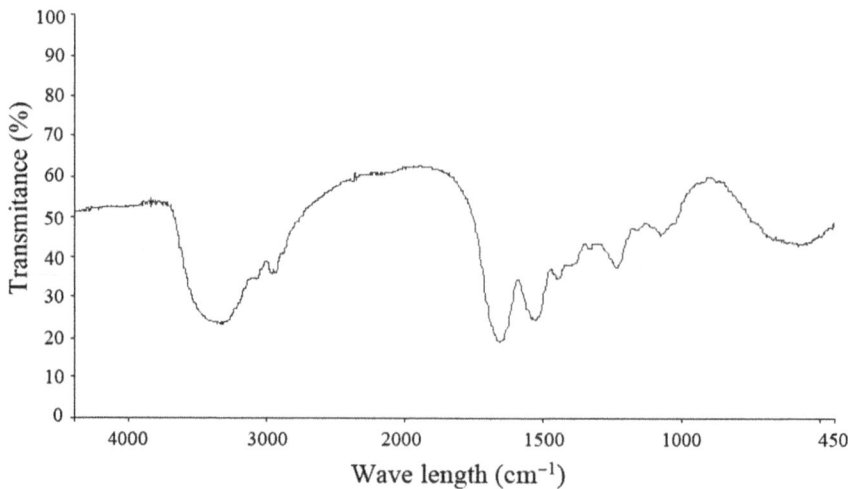

Figure 14.8 FTIR spectra of a byssal thread from *A. ater.*

stretching vibration (70–85%) and the C–N stretching vibration (10–20%).[56] Different species show similar results but not identical values. Hagenau *et al.* studied the species *M. galloprovincialis*, separating distal and proximal regions.[57] Their results showed that the amide I peak of the distal region occurs at 1633 cm^{-1}, whereas the amide I peak of the proximal region takes place at 1626 cm^{-1}. The differences on the exact position of this band are determined by the backbone conformation and the hydrogen bonding pattern.[56]

Byssus fibres also show amide II and III peaks at ~ 1555 cm^{-1} and ~ 1235 cm^{-1}, respectively (Figure 14.8). The amide II peak (~ 1520–1580 cm^{-1}) is associated with the N–H bending vibration (40–60%), C–N stretching vibration (18–40%) and the C–C ($\sim 10\%$) stretching vibrations, whereas the amide III peak (1229–1301 cm^{-1}) is related to the N–H bending and C–N stretching vibrations within the triple helical structure of collagen.[56] Hagenau *et al.* have reported the amide II peak for the distal and proximal regions at 1530 cm^{-1} and 1543 cm^{-1}, respectively.[57]

Figure 14.8 also shows a broad peak at 3300 cm^{-1}. This peak is associated with N–H stretching vibrations that give rise to amide A (3310–3270 cm^{-1}).[58] The narrow signals at around 2900 cm^{-1} correspond to C–H stretching vibrations; C–H deformation vibrations appear at around 1450 cm^{-1}. The IR signals between 1200 and 1300 cm^{-1} are typically assigned to collagen types I and II.[57]

Kong *et al.* have used FTIR curves to perform a quantitative estimation of the content of different secondary structures.[56] They used a curve fitting procedure to analyze the deconvoluted spectrum of the amide I band. They argue that any protein can be considered as a linear sum of a few fundamental secondary structural elements and that the percentage of each element is only related to the spectral intensity. In order to study the secondary structure of byssus fibres from *A. ater*, the amide I band was deconvoluted, and the

Table 14.3 Percentage of each secondary structure of byssus fibres from *A. ater.*

Secondary structure	Wavelength (cm^{-1})	Percentage (%)
α-Helix	1657	22.5
β-Sheet	1622, 1639	36.8
β-Turn	1676	17.8

deconvoluted spectrum was be fitted with Gaussian band shapes. The deconvoluted amide I band frequencies, assignments to secondary structure for byssus fibres of *A. ater*, as well as the percentage of their secondary structures, are shown in Table 14.3. The results confirm the presence of crystalline and amorphous regions in the secondary structure of byssus fibres with a high β-sheet ($\sim 37\%$), α-helix (23%) and β-turn (18%) structural content.

Hagenau *et al.*[57] studied the secondary structure of byssus fibres of *M. galloprovincialis*. They found that for byssus fibres which can be divided into two parts, distal and proximal regions, they present secondary structure in the following way. Proteins of the distal region are oriented along the thread axis and are well ordered with a high β structural content (70%) and collagen typical triple helices (22%). Proteins of the proximal part are not as well oriented and show less order, with primarily α-helical structure (47%) with less β conformations (15%) and triple helical structures (13%).

14.4.5 Adhesive Properties

The natural adhesive used by mussels is composed of glue proteins that are able to bind strongly to virtually all inorganic and organic surfaces in aqueous environments in which most adhesives work poorly.[39] It is known that byssal threads counteract the residual stresses that arise at the interface between the substratum and the adhesive plaque of mussels by the production of a structural adhesive that appears in the form of solid foam.[43,59–61] Several theories have been developed to explain the process by which this foam is made, but it is still unclear.

The adhesive plaque is composed of six distinct types of proteins known as foot proteins (fp) (Table 14.4).[30–34,62–64] Even though *M. edulis* has been the main mussel species used for identifying foot proteins, studies carried out on other mussel species, such as *M. galloprovincialis*, *M. coruscus* and *M. californianus*, confirm that they have similar amino acid sequences of adhesive proteins.[25,65,66]

Studies reveal that the mussel adhesive proteins of *M. edulis* contain a high level of the amino acid DOPA. DOPA is a catabolic amino acid that is produced by post-translational hydroxylation of tyrosine using polyphenol oxidase.[31,32,34] DOPA-containing adhesive proteins are also found in other mussels and even in other phyla as well. It was reported that mussel adhesive proteins which are closer to the adhesion interface have a higher proportion of

Table 14.4 Mass (kDa) and DOPA content (mol%) of mussel byssus foot proteins.

Proteins	Mass (kDa)	DOPA (mol%)	Ref.
Mefp-1	∼110	∼10–15	33,64
Mefp-2	∼42–47	∼2–3	73
Mefp-3	∼5–7	∼20–25	31,64,73
Mefp-4	∼79–80	∼4–5	31,73,74
Mefp-5	∼9	∼27–30	73,77
Mefp-6	∼12	∼4	77

DOPA residues, whereas mussel adhesive proteins with less or without DOPA show a reduced ability for adhesion.[31,33,34] In addition, DOPA residues allow mussel adhesive protein molecules to cross-link each other by oxidative conversion to DOPA-quinone. It has been suggested that the reactive quinone provides the water resistance characteristic of mussel adhesion.[67,68] Therefore, DOPA functions as a cross-linking agent and mediates adhesion to the substratum.

The first mussel adhesive protein to be identified was fp-1 from *M. edulis* (Mefp-1).[33,69] This protein is found on the thin protective cuticle that covers the entire structure of the adhesion plaque and the distal region of the thread. Mefp-1 has a high molecular mass as it is composed of about 80 repeats of a decapeptide (Ala-Lys-Pro-Ser-Tyr-Hyp-Hyp-Tyr-DOPA-Lys). The DOPA content is 10–15 mol% and has the highest molecular mass among mussel adhesive proteins (110 kDa). Mefp-1 is directly related to the durability of byssus fibres and its robust coating is 4–5 times stiffer and harder than the byssal collagens that it covers.[42] Analogous proteins with slight differences in the amino acid composition have been reported to be present in *M. galloprovincialis* (Mgfp-1), *M. coruscus* (Mcfp-1) and *M. californianus* (Mcfp-1).[70–72]

Mefp-2 and Mefp-4 are located in the bulk of the mussel's adhesion plaque. Unlike Mefp-1, Mefp-2 is a smaller adhesive protein (40–47 kDa) with approximately 2–3 mol% of DOPA.[64,73] Mefp-2 contains abundant secondary structure and is relatively resistant to a variety of proteases, which is an important characteristic for the integrity of the byssal plaque. In addition, Mefp-2 is an adhesive protein that plays a stabilization role due to its high cysteine content (6–7 mol%).[30]

Mefp-4 has a molecular mass of ∼80 kDa and a DOPA content of 4 mol%.[64,74,75] Mefp-4 also contains high levels of arginine, histidine and glycine.[64,73] This protein likely serves as a coupling agent in the thread plaque junction designated by the pre-collagens and the byssal plaque protein Mefp-2.[75]

The interface between the substratum and the adhesion plaque of mussels is formed by Mefp-3 and Mefp-5 proteins. Both proteins contain higher levels of DOPA. Mefp-3 contains 20–25 mol% DOPA, whereas Mefp-5 contain about 30 mol% DOPA.[73]

Mefp-3 is the smallest adhesive protein (∼5–7 kDa), with a large number of arginine residues.[31] Other mussels contain analogous proteins. Mgfp-3A and

Mgfp-3B have been identified in the adhesive plaque of *M. galloprovincialis*, whereas Mcfp-3 (12 variants) have been found in *M. californianus*.[76,77]

Mefp-5 is a relatively small protein (9.5 kDa). It has many serine residues that can partly be modified to phosphoserines. A protein analogous to Mefp-5 has been reported to be present in *M. galloprovincialis* (Mgfp-5)[65] and in *M. californianus* (Mcfp-5: 2 variants).[63]

Mcfp-6 was identified along with Mcfp-5 by Zhao and Waite.[63] This small protein (11.6 kDa) contains a relatively large amount of tyrosine (~ 20 mol%) and a small amount of DOPA (< 5 mol%). It is suggested that its function is to provide a cohesive link between the surface-coupling DOPA-rich proteins and the bulk of the plaque proteins.

14.5   Potential Applications and Biomimetics

14.5.1   Biomimetics

Some characteristics of mussels have attracted the interest of researchers, especially in the field of biomimetics. Mussels are able to produce their byssus fibres and plaques in a few minutes, and these fibres spontaneously adhere to rocks. The adhesive process can be described as follows. The concentrated adhesive proteins, which are stored in a Ca^{2+}/Mg^{2+} rich medium at pH 5, are found in the secretory granules located in the cement gland. The secretory granules await a signal transduction to release their contents. When mussels want to attach themselves to hard surfaces, proteins are released onto the surface, flowing rapidly. It has been suggested that the condensed state of the proteins helps to dehydrate the hard surface.[43] Finally, the DOPA residues will be oxidized to quinones, producing a cross-linking process that leads to solidification.

One of the most interesting techniques that mussels use is the fast bonding process to metal and mineral surfaces in a wet saline environment at ambient temperature. This behaviour can be observed in other marine organisms, such as barnacles, tubeworms, sea cucumbers and limpets. This natural adhesive process has not been imitated by synthetic polymers since they cannot displace surface water. The effect of moisture affects the stability of interfacial adhesion between metal and polymer composites.

Waite *et al.* indicated that at least 10 proteins are present in the adhesive plaques, but only two of these, Mefp-3 and Mefp-5, are implicated as interfacial.[43] Both are among the most modified of all byssus proteins, with 42% and 37% of modified amino acids, respectively. The main modification in both proteins is tyrosine hydroxylation to DOPA. It has been shown that the *o*-dihydroxyphenolic moiety of DOPA is implicated in the strong H-bonding to hydroxyapatite and coordination complexes with the oxides of iron,[78] zinc,[79] titanium,[80] aluminum[81] and gold surfaces.[82]

The different DOPA-containing precursors are stockpiled in specific places in the byssus fibres. In the cuticle, DOPA mediates formation of covalent diDOPA cross-links of bis- and triscatecholato-Fe^{III} complexes with Mefp-1,[83–85] and even triscatecholato-Fe^{III} complexes that serve as vehicles for diDOPA cross-link formation by a one-electron redox exchange.[86] Mefp-3 and Mefp-5 are deposited at the interface with a hard surface and remain with their DOPA intact (surface metal chelates).[43]

Mussel byssus is a source of inspiration for the design of novel biomimetic polymers and composites. Waite suggests that mussel adhesive is essentially analogous to composite thermosets. It consists of fibres that are dispersed in a cross-linked resin, rendering it resistant to heat and solvents. The fibres are formed by collagen and the resin is proteins with basic isoelectric points, high levels of the amino acid DOPA and an extended, flexible conformation.[25]

One of the attributes why mussel adhesion has attracted interest for potential applications is that mussels can adhere themselves to a variety of substrates.[43] The adhesion of these glue proteins to various solid materials, such as plastic, glass, metals, Teflon and even living body substances such as porcine skin and diverse types of mammalian cells, has been verified.[62,64,65,87–93] Another advantage of mussel byssal adhesion is that it is fully functional under water or in humid conditions.[4] All stages of the bonding process occur rapidly under ambient and wet conditions. Finally, byssal adhesion appears to be a smart process in the sense that it exhibits compliance or modulus matching.[40] In addition, byssal threads are environmentally friendly and biodegradable materials. Therefore, mussel adhesion has a wide range of potential uses in diverse biotechnological applications.

14.5.2   Surgical Adhesives

Ideal surgical adhesives should adhere to the tissue substrate, providing adequate strength in the presence of physiological fluids. They should also be biodegradable and should not elicit an immune response.[93] Ninan *et al.* examined the adhesive characteristics of glue proteins extracted from mussels without the use of a curing agent.[93] These glue proteins were used to bond porcine skin in an end-to-end joint cured in controlled environments similar to common surgical environments: dry and humid environments such as those used for external and internal incision. When comparing the results with those obtained from the commercially adhesive fibrin Tisseelt, they concluded that joints bonded with mussel extract showed adhesive strengths similar to those bonded with fibrin for cures times between 12 and 24 hours. This extract can be used as a potential biomaterial for surgical applications if the rate of cure can be accelerated and show biocompatibility properties as well.

14.5.3   Dental Adhesives

Mussel adhesives can be used as medical and dental adhesives as they are harmless to the human body and do not impose immunogenicity.[94,95] Tay and Pashley suggested that future medical and dental adhesives may contain domains derived from protein-based, underwater bioadhesives secreted by aquatic animals such as mussels, making them less dependent on the surface energy of the bonding substrate and less susceptible to hydrolytic degradation.[93,96-99]

Because DOPA residues are known for their ability to impart adhesive and curing properties to mussel adhesive proteins, hybrid systems containing Mefp-1 decapeptides and synthetic polymers have been used for the production of hydrogels for medical applications. Lee *et al.* prepared hybrid systems containing linear and branched DOPA-modified poly(ethylene glycol)s (PEG DOPAs), resulting in improved adhesive and cohesive properties.[100] This gel can potentially function as a bioadhesive material for medical purposes. Other hybrid systems that are made of Mefp-based peptides and synthetic polymers such as different acrylates have been developed at the Fraunhofer Institute for Manufacturing Technology and Advanced Materials by solid-phase peptide synthesis for dental applications. These hybrid systems showed improved tensile shear strength compared with pure acrylate systems.

14.5.4 Biosensors and Immunosensors

Biosensors combine the molecular recognition capabilities of biology with the signal processing capabilities of electronic devices. In order to fabricate biosensors, adhesives are needed to integrate biological components into microfabricated devices. The immobilization of mussel adhesive proteins such as the Mefp class and collagens on solid supports may be exploited for uses in the design of biosensors, immunosensors or artificial tissue scaffolding and constructs.[101,102] Saby and Luong constructed a glucose oxidase (GOD) based electrode using an adhesive protein isolated from *M. edulis*.[101] Then the mussel adhesive protein was oxidized to form a stable film on platinum, gold or glassy carbon (GC) electrodes. They were covalently linked to the GOD. The modified electrodes were able to detect tetrachlorobenzo-1,4-quinone (TCBQ) under specific conditions. In the case of GC/mussel adhesive protein/GOD electrodes, they exhibited a linear response for TCBQ, ranging from 10 nM to 1 µM (3.96 nA nM^{-1}) with a detection limit of 10 nM, retaining about 90% and 87% of its activity after 150 and 250 repeated injections, respectively. Yamada *et al.* suggest that the identification and subsequent use of the polyphenol oxidase derived from *M. edulis* byssal structures may also be exploited as a thickening agent for numerous industries and applications.[103] Hwang *et al.* prepared modified recombinant Mgfp-5 that can be used as a novel immobilization agent for biochip systems or as a mediator for drug delivery.[88]

14.5.5 Cell Culture and Tissue Engineering

Several cell-adhesion materials have been developed in order to coat artificial surfaces and promote cell adhesion. Among them are extracellular matrix proteins, poly-L-lysine (PLL), Cell-Tak and mussel adhesive protein (MAP).

Extracellular matrix proteins have been widely used in cell culture and tissue engineering because they have various cell-adhesion motifs. However, they are expensive and are associated with infection risks since they are extracted from foreign animals. By contrast, PLL is cheaper than extracellular matrix proteins but it is not always an effective cell-adhesion material due to cytotoxicity and abnormal cell spreading.[104]

A commercially available cell-adhesive agent based on mussel adhesive proteins is Cell-Tak (BD Bioscience Clontech). It is an extracted mixture derived from Mefp-1 and Mefp-2. Cell-Tak has less cytotoxicity and promotes improved cell-adhesion ability among various cell types, including mammalian and human cell lines.[105] Although Mefp-1 and Mefp-2 have no higher DOPA levels than Mefp-3 and Mefp-5, these products have been used to improve the attachment of cultured cells and tissues. It is now quite evident that the role of Mefp-1 is related mainly to its coating function rather than with its adhesion properties, owing to its lower DOPA levels.[42] Even though mussel adhesive proteins have many remarkable properties, their practical applications are restricted by an inefficient generation process and an extremely low production rate.[106]

In order to overcome these disadvantages, Hwang *et al.* designed and constructed a hybrid of the MAP fp-151, which is a fusion protein with six type 1 (fp-1) decapeptide repeats at each type 5 (fp-5) terminus.[89] It was found that fp-151 has advantages such as high production yield in *Escherichia coli* and simple purification. Several cell-adhesion experiments have demonstrated the potential use of fp-151 for cell or tissue adhesion.

Hwang *et al.* have designed a new cell-adhesive protein called fp-151-RGD that is a fusion of the Arg-Gly-Asp motif to hybrid MAP fp-151.[91] The Arg-Gly-Asp (RGD) sequence is the most effective cell-adhesion recognition motif that can be found in materials, such as collagen, fibronectin and tenascin C, and has been used to stimulate cell adhesion on artificial surfaces.[107–110] The results showed that fp-151-RGD had the advantages of fp-151, but also showed superior cell adhesion and spreading abilities under serum-free conditions regardless of mammalian cell type compared with other commercially produced cell-adhesion materials such as poly-L-lysine (PLL) and the naturally extracted MAP mixture Cell-Tak. These properties might be explained due to three cell-binding mechanisms: cationic binding, DOPA adhesion and the cell-adhesion motif. The results demonstrated the suitability of fp-151-RGD as a cell-adhesion material in cell culture and tissue engineering. In addition, fp-151 has the potential to be fused with other specific motifs instead of RGD motifs, so that specific target cells can be tightly bonded to an artificial extracellular matrix or biomaterial; thus fp-151-based fusion proteins with specific cell-recognition motifs can be used in tissue engineering as specific cell-adhesion biomaterials.

Proteins with a high DOPA content have been studied as cell-adhesion biomaterials. Hwang *et al.* reported the production of recombinant *M. gallo-provincialis* foot protein type-5 (Mgfp-5) in *E. coli*,[65] showing that the recombinant protein had better adhesive properties than Cell-Tak in terms of surface adhesion and cell immobilization. They also used purified and tyrosinase-modified recombinant Mgfp-5 to adhere living anchorage-independent cells such as insect drosophila S2 cells and human MOLT-4 cells onto glass slides, finding that the modified recombinant Mgfp-5 can be used as an adhesive biomaterial for cells.

14.5.6 Biotechnological Applications

Efficient target gene delivery into eukaryotic cells is a key issue to gene therapy and biotechnological research. In this field, various histone proteins have been analyzed as potential gene delivery materials because they display higher transfection efficiency in mammalian cells compared to the widely used transfection agent Lipofectamine 2000. In order to find new alternatives, a recombinant mussel adhesive protein has been assessed as a potential gene delivery material. Hwang *et al.* investigated the use of the recombinant mussel adhesive protein fp-151 as a gene delivery material,[90] because hybrid fp-151 displays a similar basic amino acid content and high theoretical p*I* values as histone HI proteins, and exhibits efficient DNA binding ability. Hwang *et al.* transfected mammalian cells (human 293T and mouse NIH/3T3) with foreign genes using the hybrid fp-151 as the gene delivery carrier. The results showed that this hybrid protein displays a transfection efficiency comparable with the efficiency of the widely used Lipofectamine 2000. Therefore, mussel adhesive protein may also be used as a potential protein-based mediator for efficient gene delivery.

14.6 Conclusions

In this chapter we have reviewed some important characteristics and properties of the mussel byssus. The mussel byssus is a tough biopolymer fibre that occurs in nature and shows remarkable mechanical as well as adhesive properties. The composition of the proteins that allow mussel byssus fibres to adhere themselves to many different substrates in dry and humid environments has been reviewed and an up-to-date summary of the main applications that are inspired by or take advantage of these adhesive proteins have been reported. These applications include surgical and dental adhesives, biosensors and adhesives for tissue engineering.

The mechanical properties of byssal fibres have also been reviewed, since their toughness is six times greater than that of the human tendon collagen and is comparable with that of Kevlar. It has been shown that the mechanical behaviour of such fibres is influenced by environmental conditions such as temperature, humidity and salinity, among others. For instance, dry byssal fibres show an elastoplastic stress–strain curve, while humid byssus display an elastomeric-like behaviour. As in the case of other protein-based fibres such as spider silk and silkworm silk, the understanding of the underlying principles that allow such remarkable mechanical behaviour is important for the development of new applications for mussel byssus.

References

1. S. L. Brazee and E. Carrington, *Biol. Bull.*, 2006, **1**, 263–274.
2. J. M. Gosline, M. Lillie, E. Carrington, P. A. Guerette, C. S. Ortlepp and K. N. Savage, *Philos. Trans. R. Soc. London, Ser. B*, 2002, **357**, 121–132.
3. E. Carrington and J. M. Gosline, *Am. Malacol. Bull.*, 2004, **18**, 135–142.
4. J. H. Waite, *Results Probl. Cell Differ.*, 1992, **19**, 27–54.
5. G. M. Moeser and E. Carrington, *J. Exp. Biol.*, 2006, **209**, 1996–2003.
6. N. Aldred, L. K. Ista, M. E. Callow, J. A. Callow, G. P. Lopez and A. S. Clare, *J. R. Soc., Interface*, 2006, **3**, 37–43.
7. T. Pearce and M. LaBarbera, *J. Exp. Biol.*, 2009, **212**, 1442–1448.
8. O. P. Troncoso, F. G. Torres and C. J. Grande, *Acta Biomater.*, 2008, **4**, 1114–1117.
9. J. A. Allen, M. Cook, D. J. Jackson, S. Preston and E. M. Worth, *J. Moll. Stud.*, 1976, **42**, 279–289.
10. J. E. Smeathers and J. F. V. Vincent, *J. Moll. Stud.*, 1979, **45**, 219–230.
11. A. Martínez-Lage, A. González-Tizón and J. Méndez, *Heredity*, 1995, **74**, 369–375.
12. C. Sun, J. M. Lucas and J. H. Waite, *Biomacromolecules*, 2002, **3**, 1240–1248.
13. E. M. Gosling, in *The Mussel Mytilus: Ecology, Physiology, Genetics and Culture*, ed. E. M. Gosling, Elsevier, Amsterdam, 1992, pp. 1–20.
14. E. C. Bell and J. M. Gosline, *J. Exp. Biol.*, 1996, **199**, 1005–1017.
15. M. J. Harrington and J. H. Waite, *J. Exp. Biol.*, 2007, **210**, 4307–4318.
16. C. Y. Lee, S. S. Lim and M. D. Owen, *Can. J. Zool.*, 1990, **68**, 2005–2009.
17. R. Seed and T. H. Suchanek, in *The Mussel Mytilus: Ecology, Physiology, Genetics and Culture*, ed. E. M. Gosling, Elsevier, Amsterdam, 1992, pp. 87–169.
18. P. Dolmer and I. Svane, *Ophelia*, 1994, **40**, 63–74.
19. J. R. E. Harger, *Veliger*, 1970, **12**, 401–414.
20. J. D. Witman and T. H. Suchanek, *Mar. Ecol. Prog. Ser.*, 1984, **16**, 259–268.
21. J. P. A. Gardner and D. O. F. Skibinski, *Mar. Ecol. Prog. Ser.*, 1991, **71**, 235–243.
22. G. L. Willis and D. O. Skibinski, *Mar. Biol.*, 1992, **112**, 403–408.
23. H. A. Price, *J. Mar. Biol. Assoc. U. K.*, 1980, **60**, 1035–1037.
24. H. A. Price, *J. Mar. Biol. Assoc .U. K.*, 1982, **62**, 147–155.
25. J. H. Waite, *Integr. Comp. Biol.*, 2002, **42**, 1172–1180.
26. T. L. Coombs and P. J. Keller, *Aquat. Toxicol.*, 1981, **1**, 291–300.
27. X. X. Qin, K. J. Coyne and J. H. Waite, *J. Biol. Chem.*, 1997, **272**, 32623–32627.
28. X. X. Qin and J. H. Waite, *Proc. Natl. Acad. Sci. U. S. A.*, 1998, **95**, 10517–10522.
29. K. J. Coyne, X. X. Qin and J. H. Waite, *Science*, 1997, **277**, 1830–1832.
30. K. Inoue, Y. Takeuchi, D. Miki and S. Odo, *J. Biol. Chem.*, 1995, **270**, 6698–6701.

31. V. V. Papov, T. V. Diamond, K. Biemann and J. H. Waite, *J. Biol. Chem.*, 1995, **270**, 20183–20192.
32. L. M. Rzepecki, K. M. Hansen and J. H. Waite, *Biol. Bull.*, 1992, **183**, 123–137.
33. J. H. Waite, *J. Biol. Chem.*, 1983, **258**, 2911–2915.
34. J. H. Waite and X. X. Qin, *Biochemistry*, 2001, **40**, 2887–2893.
35. L. Eckroat and L. M. Steele, *Am. Malacol. Bull.*, 1993, **10**, 103–108.
36. N. Holten-Andersen, G. Fantner, S. Hohlbauch, J. H. Waite and F. W. Zok, *Nat. Mater.*, 2007, **6**, 669–672.
37. C. Sun and J. H. Waite, *J. Biol. Chem.*, 2005, **280**, 39332–39336.
38. H. Zhao and J. H. Waite, *Biochemistry*, 2005, **44**, 15915–15923.
39. H. Lee, F. Scherer and P. B. Messersmith, *Proc. Natl. Acad. Sci. U. S. A.*, 2006, **103**, 12999–13003.
40. J. H. Waite, H. C. Lichtenegger, G. D. Stucky and P. Hansma, *Biochemistry*, 2004, **43**, 7653–7662.
41. N. Holten-Andersen, T. E. Mates, M. S. Toprak, G. D. Stucky, F. W. Zok and J. H. Waite, *Langmuir*, 2009, **25**, 3323–3326.
42. N. Holten-Andersen and J. H. Waite, *J. Dent. Res.*, 2008, **87**, 701–709.
43. J. H. Waite, N. Holten-Andersen, S. Jewhurst and C. Sun, *J. Adhes.*, 2005, **81**, 297–317.
44. S. Suresh, *Science*, 2001, **292**, 2447–2451.
45. M. Wiegemann, *Aquat. Sci.*, 2005, **67**, 166–176.
46. J. H. Waite, E. Vaccaro, C. Sun and J. M. Lucas, *Phil. Trans. R. Soc. London, Ser. B*, 2002, **357**, 143–153.
47. M. W. Denny, *Limnol. Oceanogr.*, 1985, **30**, 171–1187.
48. E. C. Bell and M. W. Denny, *J. Exp. Mar. Biol. Ecol.*, 1994, **181**, 9–29.
49. E. Carrington, *Limnol. Oceanogr.*, 2002, **47**, 1723–1733.
50. L. E. DeFrate and G. Li, *Biomech. Model. Mechanobiol.*, 2007, **6**, 245–251.
51. G. A. Holzapfel, *Nonlinear Solid Mechanics: A Continuum Approach for Engineering*, Wiley, New York, 2000, p. 455.
52. G. A. Johnson, G. A. Livesay, S. L. Woo and K. R. Rajagopal, *J. Biomech. Eng.*, 1996, **118**, 221–226.
53. J. A. Weiss and J. C. Gardiner, *Crit. Rev. Biomed. Eng.*, 2001, **29**, 303–371.
54. M. Tsukada, Y. Gotoh, H. Yasui, G. Freddi and H. Usuki, *J. Seric. Sci. Jpn.*, 1995, **64**, 435–445.
55. N. Aldred, T. Wills, D. N. Williams and A. S. Clare, *J. R. Soc., Interface*, 2007, **4**, 1159–1167.
56. J. Kong and S. Yu, *Acta Biochim. Biophys. Sin.*, 2007, **39**, 549–559.
57. A. Hagenau, H. A. Scheidt, L. Serpell, D. Huster and T. Scheibel, *Macromol. Biosci.*, 2009, **9**, 162–168.
58. A. Barth and C. Zscherp, *Q. Rev. Biophys.*, 2002, **35**, 369–430.
59. J. H. Waite, *J. Comp. Physiol. B*, 1986, **156**, 491–496.
60. C. V. Benedict and J. H. Waite, *J. Morphol.*, 1986, **189**, 261–270.
61. A. Tamarin, P. Lewis and J. J. Askey, *J. Morphol.*, 1979, **149**, 199–221.

62. D. R. Filpula, S. M. Lee, R. P. Link and S. L. Strausberg, *Biotechnol. Prog.*, 1990, **6**, 171–177.
63. H. Zhao and J. H. Waite, *J. Biol. Chem.*, 2006, **281**, 26150–26158.
64. H. J. Cha, D. S. Hwang and S. Lim, *Biotechnol. J.*, 2008, **3**, 631–638.
65. D. S. Hwang, H. J. Yoo, J. J. Jun, W. K. Moon and H. J. Cha, *Appl. Environ. Microbiol.*, 2004, **70**, 3352–3359.
66. T. J. Deming, *Curr. Opin. Chem. Biol.*, 1999, **3**, 100–105.
67. M. Yu and T. J. Deming, *Macromolecules*, 1998, **31**, 4739–4745.
68. M. Yu, J. Hwang and T. J. Deming, *J. Am. Chem. Soc.*, 1999, **121**, 5825–5826.
69. J. H. Waite and M. L. Tanzer, *Science*, 1981, **212**, 1038–1040.
70. K. Inoue and S. Odo, *Biol. Bull.*, 1994, **186**, 349–355.
71. K. Inoue, Y. Takeuchi, S. Takeyama, E. Yamaha, F. Yamazaki, S. Odo and S. Harayama, *J. Mol. Evol.*, 1996, **43**, 348–356.
72. J. H. Waite, *J. Comp. Physiol. B*, 1986, **156**, 491–496.
73. H. G. Silverman and F. F. Roberto, *Mar. Biotechnol.*, 2007, **9**, 661–681.
74. V. Vreeland, J. H. Waite and L. Epstein, *J. Phycol.*, 1998, **34**, 1–8.
75. S. C. Warner and J. H. Waite, *Mar. Biol.*, 1999, **134**, 729–734.
76. K. Inoue, Y. Takeuchi, D. Miki, S. Odo, S. Harayama and J. H. Waite, *Eur. J. Biochem.*, 1996, **239**, 172–176.
77. H. Zhao, N. B. Robertson, S. A. Jewhurst and J. H. Waite, *J. Biol. Chem.*, 2006, **281**, 11090–11096.
78. M. J. McWirter, P. J. Bremer, I. L. Lamont and A. J. McQuillan, *Langmuir*, 2003, **19**, 3575–3577.
79. G. Ramakrishna and H. N. Ghosh, *Langmuir*, 2003, **19**, 3006–3012.
80. Y. Liu, J. I. Dadap, D. Zimdars and K. B. Eisenthal, *J. Phys. Chem. B*, 1999, **103**, 2480–2486.
81. S. L. Simpson, K. J. Powell and S. J. Sjoberg, *J. Colloid Interface Sci.*, 2000, **229**, 568–574.
82. J. L. Dalsin, B. H. Hu, B. P. Lee and P. B. Messersmith, *J. Am. Chem. Soc.*, 2003, **125**, 4253–4258.
83. L. A. Burzio and J. H. Waite, *Biochemistry*, 2000, **39**, 11147–11153.
84. L. M. McDowell, L. A. Burzio, J. H. Waite and J. Schaefer, *J. Biol. Chem.*, 1999, **274**, 20293–20295.
85. S. W. Taylor, D. B. Chase, M. H. Emptage, M. J. Nelson and J. H. Waite, *Inorg. Chem.*, 1996, **35**, 7572–7577.
86. M. J. Sever, J. T. Weisser, J. Monahan, S. Srinivasan and J. J. Wilker, *Angew. Chem. Int. Ed.*, 2004, **43**, 448–450.
87. M. Kitamura, K. Kawakami, N. Nakamura, K. Tsumoto, H. Uchiyama, Y. Ueda, I. Kumagai and T. Nakaya, *J. Polym. Sci.*, 1999, **37**, 729–736.
88. D. S. Hwang and H. J. Cha, *J. Biotechnol.*, 2007, **127**, 727–735.
89. D. S. Hwang, Y. Gim, H. J. Yoo and H. J. Cha, *Biomaterials*, 2007, **28**, 3560–3568.
90. D. S. Hwang, K. R. Kim, S. Lim, Y. S. Choi and H. J. Cha, *Biotechnol. Bioeng.*, 2009, **102**, 616–623.
91. D. S. Hwang, S. B. Sim and H. J. Cha, *Biomaterials*, 2007, **28**, 4039–4046.

92. B. P. Frank and G. Belfort, *Langmuir*, 2001, **17**, 1905–1912.
93. L. Ninan, J. Monahan, R. L. Stroshine, J. J. Wilker and R. Y. Shi, *Biomaterials*, 2003, **24**, 4091–4099.
94. J. Dove and P. Sheridan, *J. Am. Dent. Assoc.*, 1986, **112**, 879.
95. D. A. Grande and M. I. Pitman, *Bull. Hosp. Jt. Dis. Orthop. Inst.*, 1988, **48**, 140–148.
96. F. R. Tay and D. H. Pashley, *J. Adhes. Dent.*, 2002, **4**, 91–103.
97. J. P. Fulkerson, L. A. Norton, G. Gronowicz, P. Picciaino, J. M. Massicotte and C. W. Nissen, *J. Orthopaed. Res.*, 1990, **8**, 793–798.
98. J. B. Robin, P. Picciano, R. S. Kusleika, J. Salazar and C. Benedict, *Arch. Ophthalmol.*, 1988, **106**, 973–977.
99. S. P. Schmidt, J. R. Resser, R. L. Sims, D. L. Mullins and D. J. Smith, *Wounds*, 1994, **6**, 62–67.
100. B. P. Lee, J. L. Dalsin and P. B. Messersmith, *Biomacromolecules*, 2002, **3**, 1038–1047.
101. C. Saby and J. H. T. Luong, *Electroanalysis*, 1998, **10**, 1193–1199.
102. J. D. Newman and S. J. Setford, *Mol. Biotechnol.*, 2006, **32**, 249–268.
103. K. Yamada, T. H. Chen, G. Kumar, O. Vesnovsky, L. D. T. Topoleski and G. F. Payne, *Biomacromolecules*, 2000, **1**, 252–258.
104. A. Bershadsky, A. Chausovsky, E. Becker, A. Lyubimova and B. Geiger, *Curr. Biol.*, 1996, **6**, 1279–1289.
105. C. V. Benedict and P. T. Picciano, in *Adhesives from Renewable Resources*, ed. R. W. Hemingway, A. H. Conner and S. J. Branham, American Chemical Society, Washington, 1989, pp. 465–483.
106. R. L. Strausberg, D. M. Andersen, D. R. Filpula, M. Finkelman, R. P. Link and R. McCandliss, in *Adhesives from Renewable Resources*, ed. R. W. Hemingway, A. H. Conner and S. J. Branham, American Chemical Society, Washington, 1989, pp. 452–464.
107. U. Hersel, C. Dahmen and H. Kessler, *Biomaterials*, 2003, **24**, 4385–4415.
108. M. D. Pierschbacher and E. Ruoslahti, *Nature*, 1984, **309**, 30–33.
109. S. M. Cutler and A. J. Garcia, *Biomaterials*, 2003, **24**, 1759–1770.
110. J. H. Jang, J. H. Hwang and C. P. Chung, *Biotechnol. Lett.*, 2004, **26**, 1831–1835.

Subject Index

The index covers both volumes. The volume number is given in **bold** before the page numbers. Page numbers in *italics* refer to figures or tables.